Climate Change, World Consequences, and the Sustainable Development Goals for 2030

Ana Pego
Nova University of Lisbon, Portugal

A volume in the Practice, Progress, and
Proficiency in Sustainability (PPPS) Book Series

Published in the United States of America by
IGI Global
Engineering Science Reference (an imprint of IGI Global)
701 E. Chocolate Avenue
Hershey PA, USA 17033
Tel: 717-533-8845
Fax: 717-533-8661
E-mail: cust@igi-global.com
Web site: http://www.igi-global.com

Copyright © 2023 by IGI Global. All rights reserved. No part of this publication may be reproduced, stored or distributed in any form or by any means, electronic or mechanical, including photocopying, without written permission from the publisher. Product or company names used in this set are for identification purposes only. Inclusion of the names of the products or companies does not indicate a claim of ownership by IGI Global of the trademark or registered trademark.
               Library of Congress Cataloging-in-Publication Data

Names: Pego, Ana, 1969- editor.
Title: Climate change, world consequences, and the Sustainable Development
   Goals for 2030 / Ana Pego, editor.
Description: Hershey PA : Engineering Science Technology, [2022] | Includes
   bibliographical references and index. | Summary: "This book discusses
   the impact of climate change on the environment and the prospects for
   citizens, cities and industry, drawing on the significance of climate
   change for ecosystems and the well-being of citizens, as well as case
   studies from many countries, for an in-depth analysis of environmental
   policy"-- Provided by publisher.
Identifiers: LCCN 2022007807 (print) | LCCN 2022007808 (ebook) | ISBN
   9781668448298 (hardcover) | ISBN 9781668448304 (paperback) | ISBN
   9781668448311 (ebook)
Subjects: LCSH: Sustainable Development Goals. | Sustainable development. |
   Climatic changes--Economic aspects.
Classification: LCC HC79.E5 C5944 2022 (print) | LCC HC79.E5 (ebook) |
   DDC 338.9/27--dc23/eng/20220503
LC record available at https://lccn.loc.gov/2022007807
LC ebook record available at https://lccn.loc.gov/2022007808

This book is published in the IGI Global book series Practice, Progress, and Proficiency in Sustainability (PPPS) (ISSN: 2330-3271; eISSN: 2330-328X)

British Cataloguing in Publication Data
A Cataloguing in Publication record for this book is available from the British Library.

All work contributed to this book is new, previously-unpublished material. The views expressed in this book are those of the authors, but not necessarily of the publisher.

For electronic access to this publication, please contact: eresources@igi-global.com.

Practice, Progress, and Proficiency in Sustainability (PPPS) Book Series

Ayman Batisha
International Sustainability Institute, Egypt

ISSN:2330-3271
EISSN:2330-328X

Mission

In a world where traditional business practices are reconsidered and economic activity is performed in a global context, new areas of economic developments are recognized as the key enablers of wealth and income production. This knowledge of information technologies provides infrastructures, systems, and services towards sustainable development.

The **Practices, Progress, and Proficiency in Sustainability (PPPS) Book Series** focuses on the local and global challenges, business opportunities, and societal needs surrounding international collaboration and sustainable development of technology. This series brings together academics, researchers, entrepreneurs, policy makers and government officers aiming to contribute to the progress and proficiency in sustainability.

Coverage

- Knowledge clusters
- Environmental informatics
- Innovation Networks
- Strategic Management of IT
- Eco-Innovation
- Sustainable Development
- E-Development
- Technological learning
- Outsourcing
- Global Content and Knowledge Repositories

IGI Global is currently accepting manuscripts for publication within this series. To submit a proposal for a volume in this series, please contact our Acquisition Editors at Acquisitions@igi-global.com or visit: http://www.igi-global.com/publish/.

The Practice, Progress, and Proficiency in Sustainability (PPPS) Book Series (ISSN 2330-3271) is published by IGI Global, 701 E. Chocolate Avenue, Hershey, PA 17033-1240, USA, www.igi-global.com. This series is composed of titles available for purchase individually; each title is edited to be contextually exclusive from any other title within the series. For pricing and ordering information please visit http://www.igi-global.com/book-series/practice-progress-proficiency-sustainability/73810. Postmaster: Send all address changes to above address. Copyright © 2023 IGI Global. All rights, including translation in other languages reserved by the publisher. No part of this series may be reproduced or used in any form or by any means – graphics, electronic, or mechanical, including photocopying, recording, taping, or information and retrieval systems – without written permission from the publisher, except for non commercial, educational use, including classroom teaching purposes. The views expressed in this series are those of the authors, but not necessarily of IGI Global.

Titles in this Series

For a list of additional titles in this series, please visit: www.igi-global.com/book-series/practice-progress-proficiency-sustainability/73810

Leadership Approaches to the Science of Water and Sustainability
Kristin Joyce Tardif (University of Arkansas, Fort Smith, USA)
Engineering Science Reference • © 2022 • 219pp • H/C (ISBN: 9781799896913) • US $215.00

Eco-Friendly and Agile Energy Strategies and Policy Development
Mir Sayed Shah Danish (University of the Ryukyus, Japan) and Tomonobu Senjyu (University of the Ryukyus, Japan)
Engineering Science Reference • © 2022 • 272pp • H/C (ISBN: 9781799895022) • US $240.00

Handbook of Research on Global Institutional Roles for Inclusive Development
Neeta Baporikar (Namibia University of Science and Technology, Namibia & University of Pune, India)
Information Science Reference • © 2022 • 398pp • H/C (ISBN: 9781668424483) • US $270.00

Analyzing Sustainability in Peripheral, Ultra-Peripheral, and Low-Density Regions
Rui Alexandre Castanho (WSB University, Poland)
Engineering Science Reference • © 2022 • 334pp • H/C (ISBN: 9781668445488) • US $240.00

Frameworks for Sustainable Development Goals to Manage Economic, Social, and Environmental Shocks and Disasters
Cristina Raluca Gh. Popescu (University of Bucharest, Romania & The Bucharest University of Economic Studies, Romania)
Engineering Science Reference • © 2022 • 296pp • H/C (ISBN: 9781668467503) • US $240.00

Handbook of Research on Building Inclusive Global Knowledge Societies for Sustainable Development
Cristina Raluca Gh. Popescu (University of Bucharest, Romania & The Bucharest University of Economic Studies, Romania)
Information Science Reference • © 2022 • 405pp • H/C (ISBN: 9781668451090) • US $295.00

Handbook of Research on Principles and Practices for Orchards Management
Waleed Fouad Abobatta (Horticulture Research Institute, Agriculture Research Center, Egypt) Ahmed Farag (Central Laboratory for Agricultural Climate, Agriculture Research Center, Egypt) and Mohamed Abdel-Raheem (National Research Center, Egypt)
Engineering Science Reference • © 2022 • 382pp • H/C (ISBN: 9781668424230) • US $295.00

701 East Chocolate Avenue, Hershey, PA 17033, USA
Tel: 717-533-8845 x100 • Fax: 717-533-8661
E-Mail: cust@igi-global.com • www.igi-global.com

Table of Contents

Preface xiv

Chapter 1
A Study on the Municipal Waste Management Practices of European Union Countries 1
    Fatma Gül Altın, Mehmet Akif Ersoy University, Turkey

Chapter 2
An Economic Assessment of Implementing Dominica's Nationally Determined Contributions 21
    Don Charles, University of the West Indies, Trinidad and Tobago

Chapter 3
Climate Change: The New Normal Management 46
    Ezatollah Karami, Shiraz University, Iran
    Marzieh Keshavarz, Payame Noor University, Iran

Chapter 4
Climate Change Information Sources: Fact-Checking and Attitude 73
    Adebowale Jeremy Adetayo, Adeleke University, Nigeria
    Blessing Damilola Abata-Ebire, Federal Polytechnic, Nigeria
    Yetunde Omodele Oladipo, University of Salford, UK

Chapter 5
Deployment and Optimization of Virtual Power Plants and Microgrids: An Opportunity for the
Energetic Transition in Algeria 89
    Abdelmadjid Recioui, University of Boumerdes, Algeria

Chapter 6
Does the Development of New Energy Vehicles Promote Carbon Neutralization? Case Studies in
China 109
    Poshan Yu, Soochow University, China & Australian Studies Centre, Shanghai University,
        China
    Shucai Xu, School of Vehicle and Mobility, Tsinghua University, China
    Ziling Cheng, Independent Researcher, China
    Michael Sampat, Independent Researcher, Canada

Chapter 7
Global Citizen Science Programs and Their Contribution to the Sustainable Development Goals ... 132
Alexandros Skondras, Aristotle University of Thessaloniki, Greece
Eleni Karachaliou, Aristotle University of Thessaloniki, Greece
Ioannis Tavantzis, Aristotle University of Thessaloniki, Greece
Nikolaos Tokas, Aristotle University of Thessaloniki, Greece
Margarita Angelidou, Aristotle University of Thessaloniki, Greece
Efstratios Stylianidis, Aristotle University of Thessaloniki, Greece

Chapter 8
Global Warming on Business Planning in Mexico and the Impact of Best Practices on Quality of Life ... 152
José G. Vargas-Hernández, Posgraduate and Research Department, Tecnológico Mario Molina Unidad Zapopan, Mexico
Pedro Antonio López de Haro, Universidad Autónoma Indígena de México, Mexico

Chapter 9
Green Intellectual Capital as a Catalyst for the Sustainable Development Goals: Evidence From the Spanish Wine Industry ... 163
Javier Martínez Falcó, University of Alicante, Spain
Bartolomé M. Marco-Lajara, University of Alicante, Spain
Patrocinio Zaragoza-Sáez, University of Alicante, Spain
Lorena Ruiz-Fernández, University of Alicante, Spain

Chapter 10
Human Being and the Homeostasis Restoration of the Biosphere of the Earth and of Anthropocenosis: Rethinking SDGs as for Dangerous Planetary Changes ... 183
Andrey I. Pilipenko, The Russian Presidential Academy of National Economy and Public Administration, Russia
Olga I. Pilipenko, Independent Researcher, Russia
Zoya A. Pilipenko, Bank of Russia, Russia

Chapter 11
Return to the Basics: Vernacular Architecture as a Tool to Address Climate Change ... 215
Seyda Emekci, Ankara Yildirim Beyazit University, Turkey

Chapter 12
Turkey's Financial Alignment With the European Green Deal ... 229
Meryem Filiz Baştürk, Uludag University, Turkey

Chapter 13
Urban Transition via Municipal Transformation ... 243
Konstantinos Asikis, Urban Planning Laboratory, Archtecture School, National Technical University of Athens, Greece
Alexia Spyridonidou, European Public Law Organisation, Greece
Sofia Aivalioti, Bax & Company, Spain

Compilation of References	258
About the Contributors	297
Index	304

Detailed Table of Contents

Preface xiv

Chapter 1
A Study on the Municipal Waste Management Practices of European Union Countries 1
    Fatma Gül Altın, Mehmet Akif Ersoy University, Turkey

Industrialization, increasingly crowded cities, and the change in consumption habits have caused municipal waste to be an important part of sustainable development. In this study, the practices of EU countries regarding municipal waste management were evaluated using clustering and data envelopment analysis (DEA). The data set for 2019 was collected from Eurostat and eight variables were identified for the analyses. In the first stage, 28 EU countries were divided into four groups using eight variables and EM algorithm. In the second stage, the efficiency scores of the 28 EU countries' recovery and recycling practices were calculated using DEA. A single input and three outputs were determined for the DEA, and the overall, technical, and scale efficiency values were determined using the output-oriented DEA models. The findings show that countries that are efficient in terms of recovery and recycling practices are in the first and second clusters.

Chapter 2
An Economic Assessment of Implementing Dominica's Nationally Determined Contributions 21
    Don Charles, University of the West Indies, Trinidad and Tobago

The Government of the Commonwealth of Dominica has joined with the international community to pursue the reduction in its country's GHG emissions. Its Nationally Determined Contributions (NDCs) highlighted its intention to reduce emissions in the energy sector by 98.6%; the transport sector by 16.9%; the commercial, residential, agriculture, forestry, and fishing sector by 8.1%; and the solid waste sector by 78.6%. The achievement of the NDCs is conditional upon the country receiving timely access to international climate change financing, technology development, and capacity building support. However, climate finance is shrouded with bureaucracy and is very difficult for developing countries to access. This study assesses the economic impacts of implementing the NDCs in Dominica. It quantifies the cost of implementing the NDCs in the identified sectors, identifies the risks and critical factors to be considered in the implementation of the NDCs, and provides recommendations for the mobilization of finance to facilitate the NDCs implementation.

Chapter 3
Climate Change: The New Normal Management 46
    Ezatollah Karami, Shiraz University, Iran
    Marzieh Keshavarz, Payame Noor University, Iran

Climatic and anthropogenic changes have affected water availability, food security, poverty, and migration in many parts of the world. While there will be no return to the normal climate and living, in arid and semi-arid regions, climate change has reached a point that can be best described as a new normal. Since the past cannot provide adequate guidance for the future, to manage the new normal, past knowledge should be unlearned. New normal management is a non-linear, complex, and non-deterministic behavior that considers the non-routine and uncertain features of climate change. The potential application of new normal management was assessed through in-depth review of the literature and three cases in Iran. The findings revealed significant differences between the crisis and new normal management practices and consequences. Accordingly, new normal management is a promising approach in facing climate change. To integrate new normal management into practice, political will, mobilization of resources, unlearning and relearning, and multilateral coordination are required.

Chapter 4
Climate Change Information Sources: Fact-Checking and Attitude..73
    Adebowale Jeremy Adetayo, Adeleke University, Nigeria
    Blessing Damilola Abata-Ebire, Federal Polytechnic, Nigeria
    Yetunde Omodele Oladipo, University of Salford, UK

This study sought to investigate the relationship between climate change information sources, fact-checking, and attitude among students at Adeleke University. A descriptive survey design was adopted for the study using a questionnaire. The data were analysed using descriptive and inferential statistics. The majority of the 688 survey respondents had an unworried attitude about climate change. Students were discovered to obtain climate change information through Google, television, friends, family, Facebook, radio, YouTube, and Instagram. Students were discovered to often fact-check climate change information. Facebook, Twitter, Instagram, YouTube, radio, church/mosque, friends, religious leaders, and fact-checking have a significant relationship with climate change attitudes. The study concluded that using social media and religious aspects as a source of climate change information may associate to an unworried attitude about climate change. As a result, it suggests addressing religious concerns about climate change.

Chapter 5
Deployment and Optimization of Virtual Power Plants and Microgrids: An Opportunity for the
Energetic Transition in Algeria..89
    Abdelmadjid Recioui, University of Boumerdes, Algeria

In Algeria, the government trend is towards the adoption of clean sources in electrical energy production due to the imoprtant solar and wind potential that is available in the desert. Virtual power plants (VPP) are modular designed entities based on software communication technologies which efficiently integrate, organize, and manage decentralized generation, storage, and consumption through a smart energy management system (EMS). VPP can only be created if there is a market to sell its power and services whereas microgrids (MG) can be created anywhere and are not market dependent. In this chapter, a description of VPP and MG is presented. The key components are described, and a comparative study is done to assess which option to adopt for the Algerian context. A review about the deployment experiences, trends, and operation optimization is presented. The aim is to assess how to deploy them in a collaborative way to fit the Algerian future energy sector perspective.

Chapter 6
Does the Development of New Energy Vehicles Promote Carbon Neutralization? Case Studies in China... 109

    Poshan Yu, Soochow University, China & Australian Studies Centre, Shanghai University, China
    Shucai Xu, School of Vehicle and Mobility, Tsinghua University, China
    Ziling Cheng, Independent Researcher, China
    Michael Sampat, Independent Researcher, Canada

This chapter aims to study whether and how current practices based on the development of new energy vehicles can help promote carbon neutrality in China and in turn contribute to the improvement of global issues related to climate change. Meanwhile, this chapter will explore the role of the government in promoting the development of new energy vehicles in the aspect of sustainable development through policies. How do these institutions promote the development of new energy vehicles in various companies and different provinces in China? How will these developments in turn affect stakeholders in various sectors? This chapter will investigate the impact of China's devotion to the field of new energy vehicles on the sustainable development and low-carbon economy issue. Cases from China will be used to illustrate the improvement the new energy vehicle will make to the low-carbon economy. This chapter will provide suggestions for the government to deal with the problems that occur in the field of new energy vehicles and solve the confronted problems in the aspect of climate change.

Chapter 7
Global Citizen Science Programs and Their Contribution to the Sustainable Development Goals ... 132

    Alexandros Skondras, Aristotle University of Thessaloniki, Greece
    Eleni Karachaliou, Aristotle University of Thessaloniki, Greece
    Ioannis Tavantzis, Aristotle University of Thessaloniki, Greece
    Nikolaos Tokas, Aristotle University of Thessaloniki, Greece
    Margarita Angelidou, Aristotle University of Thessaloniki, Greece
    Efstratios Stylianidis, Aristotle University of Thessaloniki, Greece

The aim of this chapter is to review how citizen science programs can contribute to each SDG, supported also by successful cases. Prerequisites have been the inclusion of at least one case from all five global areas and the field of interest is well spread through various study areas while the cases are only RFO or RPO projects. The design and implementation of this research is structured as follows. Each SDG is analyzed in a separate chapter. To begin with, the main goals of the relevant SDG are mentioned. The chapter continues with the relevance of citizen science projects to the respective goal. Ultimately, an example on how a citizen science project can make the goal effective is chosen. The case studies are selected based on their location, the timeframe of the project, and are structured beginning with the people or institutions involved, continuing with the project description and concluding with the project's contribution to the relevant SDG. The chapter then proceeds to a comparative case study analysis, with results and discussion in order to extract conclusions.

Chapter 8
Global Warming on Business Planning in Mexico and the Impact of Best Practices on Quality of Life 152

    José G. Vargas-Hernández, Posgraduate and Research Department, Tecnológico Mario Molina Unidad Zapopan, Mexico
    Pedro Antonio López de Haro, Universidad Autónoma Indígena de México, Mexico

The aim of this chapter is to identify and analyze some of the best practices in relation to the impact of global warming on business planning in Mexico. The phenomenon of global warming as changes in the business environment has impacted on the strategic planning processes of Mexican companies in the creation of objectives and methodologies in various fields and sectors, highlighting the importance of adapting to the needs and the limitations that may arise with climate change in the various markets we are focused.

Chapter 9
Green Intellectual Capital as a Catalyst for the Sustainable Development Goals: Evidence From the Spanish Wine Industry 163
    Javier Martínez Falcó, University of Alicante, Spain
    Bartolomé M. Marco-Lajara, University of Alicante, Spain
    Patrocinio Zaragoza-Sáez, University of Alicante, Spain
    Lorena Ruiz-Fernández, University of Alicante, Spain

The purpose of this research is to analyze the different effects of the green intellectual capital (GIC) of wine companies on the fulfillment of the Sustainable Development Goals (SDGs), contributing to academic literature in a remarkable way, since, to the authors' knowledge, there is no previous research that has addressed this relationship. In order to achieve the proposed objective, the research follows a qualitative approach, since the single case study was used. The research results demonstrate how the three dimensions of the GIC (green human capital, green structural capital, green relational capital) act as catalysts for the fulfillment of SDGs 3, 5, 6, 7, 8, 9, 11, 12, 13, 15, and 17.

Chapter 10
Human Being and the Homeostasis Restoration of the Biosphere of the Earth and of Anthropocenosis: Rethinking SDGs as for Dangerous Planetary Changes 183
    Andrey I. Pilipenko, The Russian Presidential Academy of National Economy and Public Administration, Russia
    Olga I. Pilipenko, Independent Researcher, Russia
    Zoya A. Pilipenko, Bank of Russia, Russia

The Earth's biosphere could be interpreted as a meta-system that integrates anthropocenosis as its component. The anthropocenosis integrates numerous human-created systems in the economy, society, technologies that play the role of system elements of the biosphere. Nature's destruction is unequivocally connected with the activity of these elements. This is because the human-created systems have progressed and become more complicated due to the destruction of the Earth's biosphere. It means a violation of the dialectics of the interaction of the system integrity and its constituent elements. Homeostasis restoration of the Earth's biosphere and anthropocenosis makes the authors rethink the role of human being in the processes of system formation. Only an intellectually autonomous person with high social responsibility is able to integrate the goals of restoring the Earth's biosphere into his ethical values and realize them. So, the UN SDGs will be achieved only if individuals are formed with a conscious mission to restore the Earth's biosphere and its interaction with the anthropocenosis.

Chapter 11
Return to the Basics: Vernacular Architecture as a Tool to Address Climate Change 215
    Seyda Emekci, Ankara Yildirim Beyazit University, Turkey

Buildings use a substantial amount of primary energy, and as a consequence, they are a significant source of greenhouse gas emissions, which contribute to global warming and climate change. A considerable change in building design ideas, methods, technologies, and design and construction systems is necessary in light of the oncoming worldwide energy and environmental concerns. Vernacular architecture is a valuable resource that has the ability to contribute significantly to the definition of sustainable design principles. It is distinguished by a particular style of formal expression that has evolved in reaction to a variety of elements—geographical, climatic, social, and economic—that describe the local location or region in question. The purpose of this chapter is to examine instances of vernacular architecture that have been used to combat climate change and to qualitatively appraise the bioclimatic design solutions that have been applied.

Chapter 12
Turkey's Financial Alignment With the European Green Deal 229
Meryem Filiz Baştürk, Uludag University, Turkey

Signatory countries of the Paris Climate Change Agreement have committed to decreasing carbon emissions and reducing global warming by at least two degrees Celcius for fighting against climate change. Moreover, the European Union has declared to European Green Deal, and it has been taken one step further from the Paris Climate Change Agreement. European Green Deal aims to transform Europe into the first carbon-neutral continent in the world in 2050. EU member countries have prioritized achieving the reduced emission level and then reaching carbon-neutral societies in the next phase. European Green Deal has not only been related to EU member countries but also has affected Turkey as a prominent commercial partner and candidate member country. Ministry of Trade – Republic of Turkiye has released the Green Deal Action Plan to comply with the changes that will emerge with European Green Deal and related transformation policies. This study has been aimed to evaluate the third section of the Green Deal Action plan, which involves green finance.

Chapter 13
Urban Transition via Municipal Transformation 243
Konstantinos Asikis, Urban Planning Laboratory, Archtecture School, National Technical University of Athens, Greece
Alexia Spyridonidou, European Public Law Organisation, Greece
Sofia Aivalioti, Bax & Company, Spain

Municipalities have to play a substantial role for urban resilience and sustainable development not only as local administrators, but also as strategists for integrated territorial planning, enablers by generating the urban activity and growth, servants by providing public quality-of-life amenities, and investors (public investments, infrastructures, and soft projects). Cities need further specialization to deal with complexities of the modern challenges, including pandemics, climate change, resources depletion, and socio-economic issues. They need to become more responsive and effective to emergencies as well as the long-term sustainability. They have to keep up with social and technological innovation and transit to a new model of operation. In order to achieve these, municipalities have to be organized like the contemporary companies; plan intelligent policies regarding resources, social capital, and economic growth, with respect to SDGs; apply smart and healthy city projects; involve cities to the international programmes, initiatives, and networks; and be familiar with the new era funding tools.

Compilation of References 258

About the Contributors 297

Index 304

Preface

This book aims to address the key issues related to climate change, its implications for the world, and the 2030 Sustainable Development Goals. The IGI book offers innovative insights on climate change in relation to the SDGs for 2030, based on a balance within sectors and offers an innovative discussion for 2030. How will business achieve these goals? The content of this publication represents research that spans corporate social responsibility, economic policy, and environmental protection. This book serves as an important reference source for planners, policy makers, managers, entrepreneurs, students, researchers, and academics dealing with conceptual, technological, and design issues related to climate change and economic transformation around the world. The book also conducts an analysis of climate change to understand how society is dealing with it and its impact on economic sectors. It also examines current strategies for achieving the Sustainable Development Goals and mitigating negative impacts on the environment.

Chapter 1. A Study on Municipal Waste Management Practices of European Union Countries

Fatma Gül Altın

Industrialization, increasingly crowded cities, and the change in consumption habits have caused municipal waste to be an important part of sustainable development. In this study, the practices of EU countries regarding municipal waste management were evaluated using clustering and Data Envelopment Analysis (DEA). The data set for 2019 was collected from Eurostat and eight variables were identified for the analyses. In the first stage, 28 EU countries were divided into four groups using eight variables and EM algorithm. In the second stage, the efficiency scores of the 28 EU countries' recovery and recycling practices were calculated using DEA. A single input and three outputs were determined for the DEA, and the overall, technical and scale efficiency values were determined using the output-oriented DEA models. The findings show that countries that are efficient in terms of recovery and recycling practices are in the first and second clusters.

Preface

Chapter 2. An Economic Assessment of Implementing Dominica's Nationally Determined Contributions

Don Charles

The Government of the Commonwealth of Dominica has joined with the international community to pursue the reduction in its country's GHG emissions. Its Nationally Determined Contributions (NDCs) highlighted its intention to reduce emissions in the energy sector by 98.6%, the transport sector by 16.9%, the commercial, residential, agriculture, forestry and fishing sector by 8.1%, and the solid waste sector by 78.6%. The achievement the NDCs is conditional upon the country receiving timely access to international climate change financing, technology development, and capacity building support. However, climate finance is shrouded with bureaucracy, and is very difficult for developing countries to access. This study assesses the economic impacts of implementing the NDCs in Dominica. It quantifies the cost of implementing the NDCs in the identified sectors, identifies the risks and critical factors to be considered in the implementation of the NDCs, and provides recommendations for the mobilization of finance to facilitate the NDCs implementation.

Chapter 3. Climate Change: The New Normal Management

Ezatollah Karami, Marzieh Keshavarz

Climatic and anthropogenic changes have affected water availability, food security, poverty and migration in many parts of the world. While there will be no return to the normal climate and living, in arid and semi-arid regions, climate change has reached a point that can be best described as a new normal. Since the past cannot provide adequate guidance for the future, to manage the new normal, past knowledge should be unlearned. New normal management is a non-linear, complex and non-deterministic behaviour that considers the non-routine and uncertain features of climate change. The potential application of new normal management was assessed through in-depth review of the literature and three cases in Iran. The findings revealed significant differences between the crisis and new normal management practices and consequences. Accordingly, new normal management is a promising approach in facing climate change. To integrate new normal management into practice, political will, mobilization of resources, unlearning and relearning, and multilateral coordination are required.

Chapter 4. Climate Change Information Sources: Fact-Checking and Attitude

Adebowale Jeremy Adetayo, Blessing Damilola Abata-Ebire, Yetunde Omodele Oladipo

This study sought to investigate the relationship between climate change information sources, fact-checking and attitude among students at Adeleke University. A descriptive survey design was adopted for the study using a questionnaire. The data were analysed using descriptive and inferential statistics. The majority of the 688 survey respondents had an unworried attitude about climate change. Students were discovered to obtain climate change information through Google, television, friends, family, Facebook, radio, YouTube, and Instagram. Students were discovered to often fact-check climate change information. Facebook, Twitter, Instagram, YouTube, Radio, Church/Mosque, Friends, Religious leaders, and fact-checking have a significant relationship with climate change attitudes. The study concluded that using social media and religious aspects as a source of climate change information may associate to an unworried attitude about climate change. As a result, it suggests addressing religious concerns about climate change.

Chapter 5. Deployment and Optimization of Virtual Power Plants and Microgrids: An Opportunity for the Energetic Transition in Algeria

Abdelmadjid Recioui

In Algeria, the government trend is towards the adoption of these clean sources in electrical energy production due to the important solar and wind potential that is available in the desert. Virtual Power plants (VPP) are modular designed entities based on software communication technologies which efficiently integrate, organize and manages decentralized generation, storage and consumption through a smart energy management system (EMS). VPP can only be created if there is a market to sell its power and services whereas Microgrids (MG) can be created anywhere and are not market dependent. In this chapter, a description of VPP and MG is presented. The key components are described and a comparative study is done to assess which option to adopt for the Algerian context. A review about the deployment experiences, trends and operation optimization is presented. The aim is to assess how to deploy them in a collaborative way to fit the Algerian future energy sector perspectives.

Chapter 6. Does the Development of New Energy Vehicle Promote Carbon Neutralization? Case Studies in China

Poshan Yu, Shucai Xu, Ziling Cheng, Michael Sampat

This chapter aims to study whether and how current practices based on the development of new energy vehicles can help promote carbon neutrality in China in turn contribute to the improvement of global issues related to climate change. Meanwhile, this chapter will explore the role of the government in promoting the development of new energy vehicles in the aspect of sustainable development through policies. How do these institutions promote the development of new energy vehicles in various companies and different provinces in China? How will these developments in turn affect stakeholders in various sectors? This chapter will investigate the impact of China's devotion to the field of new energy vehicles on the sustainable development & low-carbon economy issue. Cases from China will be used to illustrate the improvement the new energy vehicle will make to the low-carbon economy. This chapter will provide suggestions for the government to deal with the problems that occur in the field of new energy vehicles and solve the confronted problems in the aspect of climate change.

Preface

Chapter 8. Global Warming on Business Planning in México and the Impact of Best Practices on Quality of Life

José Vargas-Hernández, Pedro López de Haro

Abstract: The aim of this paper is to identify and analyze some of the best practices in relation to the impact of global warming on business planning in Mexico. The phenomenon of global warming as changes in the business environment has impacted on the strategic planning processes of Mexican companies in the creation of objectives and methodologies in various fields and sectors, highlighting the importance of adapting to the needs and the limitations that may arise with climate change in the various markets we are focused. Keywords: Global warming, Strategic planning, Mexican companies.

Chapter 9. Green Intellectual Capital as a Catalyst for the Sustainable Development Goals: Evidence From the Spanish Wine Industry

Bartolomé Marco-Lajara, Patrocinio Zaragoza-Sáez,
Javier Martínez Falcó, Lorena Ruiz-Fernández

The purpose of this research is to analyze the different effects of the Green Intellectual Capital (GIC) of wine companies on the fulfilment of the Sustainable Development Goals (SDGs), contributing to academic literature in a remarkable way, since, to our knowledge, there is no previous research that has addressed this relationship. In order to achieve the proposed objective, the research follows a qualitative approach, since the single case study was used. The research results demonstrate how the three dimensions of the GIC (Green Human Capital, Green Structural Capital, Green Relational Capital) act as catalysts for the fulfilment of SDGs 3, 5, 6, 7, 8, 9, 11, 12, 13, 15, and 17.

Chapter 10. Human Being and the Homeostasis Restoration of the Biosphere of the Earth and of Anthropocenosis: Rethinking SDGs as for Dangerous Planetary Changes

Andrey I. Pilipenko, Olga I. Pilipenko, Zoya A. Pilipenko

The Earth's biosphere could be interpreted a meta-system that integrates anthropocenosis as its component. The anthropogenesis integrates numerous human-created systems in the economy, society, technologies that play the role of system elements of the biosphere. The nature's destruction is unequivocally connected with the activity of these elements. This is because the human-created systems have progressed, become more complicated due to the destruction of the Earth's biosphere. It means a violation of the dialectics of the interaction of the system integrity and its constituent elements. Homeostasis restoration of the Earth's biosphere and anthropocenosis makes to rethink the role of human being in the processes of system formation. Only an intellectually autonomous person with high social responsibility is able to integrate the goals of restoring the Earth's biosphere into his ethical values and realize them. So, the UN SDGs will be achieved only if individuals are formed with a conscious mission to restore the Earth's biosphere and its interaction with the anthropocenosis

Chapter 11. Return Them to Basics: Vernacular Architecture as a Tool to Address Climate Change

Seyda Emekci

Buildings use a substantial amount of primary energy, and as a consequence, they are a significant source of greenhouse gas emissions, which contribute to global warming and climate change. A considerable change in building design ideas, methods, technologies, and design and construction systems is necessary in light of the oncoming worldwide energy and environmental concerns. Vernacular architecture is a valuable resource that has the ability to contribute significantly to the definition of sustainable design principles. It is distinguished by a particular style of formal expression that has evolved in reaction to a variety of elements – geographical, climatic, social, and economic – that describe the local location or region in question. The purpose of this chapter is to examine instances of vernacular architecture that has been used to combat climate change and to qualitatively appraise the bioclimatic design solutions that have been applied.

Chapter 12. Turkey's Financial Alignment With the European Green Deal

Meryem Filiz Baştürk

Signatory countries of the Paris Climate Change Agreement have committed to decreasing carbon emissions and reducing global warming by at least 2 degrees Celsius for fighting against climate change. Moreover, European Union has declared to European Green Deal, and it has been taken one step further from The Paris Climate Change Agreement. European Green Deal aims to transform Europe into the first carbon-neutral continent in the World in 2050. EU member countries have prioritized achieving the reduced emission level and then reaching carbon-neutral societies in the next phase. European Green Deal has not only been related to EU member countries but also has been affected Turkey as a prominent commercial partner and candidate member country. Ministry of Trade – Republic of Turkey has released the Green Deal Action Plan to comply with the changes that will emerge with European Green Deal and related transformation policies. This study has been aimed to evaluate the third section of the Green Deal Action plan, which involves green finance.

Chapter 13. Urban Transition via Municipal Transformation

Konstantinos Asikis, Alexia Spyridonidou, Sofia Aivalioti

Municipalities have to play a substantial role for urban resilience and sustainable development not only as local administrators, but also as strategists for integrated territorial planning, enablers by generating the urban activity and growth, servants by providing public quality-of-life amenities, investors (public investments, infrastructures and soft projects). Cities need further specialization to deal with complexities of the modern challenges, including pandemics, climate change, resources depletion and socio-economic issues. They need to become more responsive and effective to emergencies as well as the long-term sustainability. They have to keep up with social and technological innovation and transit to a new model of operation. In order to achieve these, Municipalities have to be organized like the contemporary com-

Preface

panies, plan intelligent policies regarding resources, social capital and economic growth, with respect to SDGs, apply smart & healthy city projects involve cities to the international programmes, initiatives and networks, be familiar with the new era funding tools.

Ana Pego
Nova University of Lisbon, Portugal

Chapter 1
A Study on the Municipal Waste Management Practices of European Union Countries

Fatma Gül Altın
https://orcid.org/0000-0001-9236-0502
Mehmet Akif Ersoy University, Turkey

ABSTRACT

Industrialization, increasingly crowded cities, and the change in consumption habits have caused municipal waste to be an important part of sustainable development. In this study, the practices of EU countries regarding municipal waste management were evaluated using clustering and data envelopment analysis (DEA). The data set for 2019 was collected from Eurostat and eight variables were identified for the analyses. In the first stage, 28 EU countries were divided into four groups using eight variables and EM algorithm. In the second stage, the efficiency scores of the 28 EU countries' recovery and recycling practices were calculated using DEA. A single input and three outputs were determined for the DEA, and the overall, technical, and scale efficiency values were determined using the output-oriented DEA models. The findings show that countries that are efficient in terms of recovery and recycling practices are in the first and second clusters.

INTRODUCTION

Sustainable development is a concept that is at the center of our age, both to understand the environment and to solve global problems. The focus of the concept of sustainable development is the earth, which is getting more crowded day by day. In 2021, the world population is around 7.7 billion. That is 9.5 times more than the estimated 800 million people who lived in 1750 at the start of the Industrial Revolution. (Sachs, 2015, p. 1, Worldometer, 2021). The world population is expected to reach 10.9 billion in 2100 (United Nations, 2021).

The United Nations (UN) Conference on the Human Environment held in Stockholm in 1972 emphasized the importance of environmental protection for sustainable development (Rogers et al., 2008:9).

DOI: 10.4018/978-1-6684-4829-8.ch001

In 1983, a major study was started by the World Commission on Environment and Development, which made sustainable development recognized as the most important concept and practice of our time (Blewitt, 2018, p. 11). In 1987, the result of this study was published as Our Common Future, also known as the Brundtland Report (Elliott, 2013, p. 8). In the report, the concept of sustainable development ensured a perspective for the integration of environmental policies and development strategies (WCED, 1987).

The next important evolution for the concept of sustainable development was the UN Conference on Environment and Development in Rio de Janeiro in 1992, also known as the Earth Summit (Jansen, 2003, p. 231). The other steps Millennium Summit (2000), World Summit on Sustainable Development (2002) and United Nations Conference on Sustainable Development (2012) taken in the following years and the 2030 Agenda for Sustainable Development and its 17 Sustainable Development Goals were adopted by all United Nations Member States in 2015. This was also an urgent call to action by all developed and developing countries for a global partnership (United Nations, 2021).

12th of the Vision 2030 Goals is "Responsible Consumption and Production". The main target of this goal is to reduce waste generation through prevention, reduction, recycling and reuse by 2030 (United Nations, 2021). In parallel with this aim, in the report published by the European Environment Agency in 2013, the main objectives and goals of the EU environmental policy for the period 2010-2050 were explained. In the report, it is stated that the aim of the EU waste policy is to contribute to a sustainable economy by providing as much resources from waste as possible. While only 38% of waste is recycled in the EU, five tons of waste is generated each year by an average European. In some EU countries, more than 60% of household waste still goes to landfill (European Commission, 2021).

Municipal waste management constitutes the environmental dimension of sustainable development areas. Due to crowded cities and increased consumption, urbanization has harmful effects on the environment. This highlights the importance of sustainable development in the functioning of cities in order to protect both the city dwellers and the ecosystem (Mesjasz-Lech, 2014, pp. 244-245). Municipal waste is expressed as waste collected and processed by municipalities. Municipal waste includes household wastes and wastes whose nature and composition are similar to household wastes (European Commission, 2016). In 2015, the European Commission set new goals for municipal waste of 60% recycling and preparing for reuse by 2025 and 65% by 2030 (European Environment Agency, 2021). The EU waste management is based on the five-stage "waste hierarchy" specified in the Waste Framework Directive. The waste hierarchy was created to show a preference for waste management and disposal. According to Figure 1, preventing waste is the most preferred alternative. On the other hand, the storage of waste should be seen as a last solution.

In this study, it is aimed to investigate whether there are differences in the municipal waste management practices of EU countries. In this direction, while EU countries are clustered using the EM algorithm in terms of municipal waste management practices, the municipal waste management efficiency scores of the countries are calculated using DEA. The rest of this study is created as follows: In the subsequent section a brief literature review is given. In section 3, detailed explanations are given about the Expectation Maximization clustering algorithm, Dunn Index and Data Envelopment Analysis approaches. In section 4, after the data are described, the clustering and efficiency analyses are made. Finally, the results are evaluated and suggestions for further studies are presented.

Figure 1. Waste Hierarchy
Source: European Commission (2022)

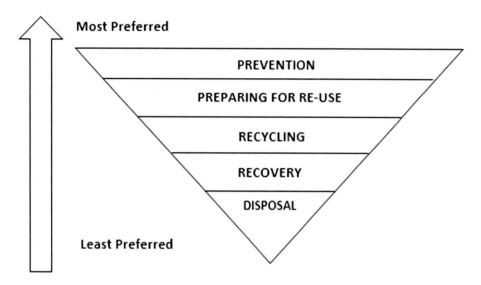

BACKGROUND

When the literature is searched, it is seen that there are many studies examining the waste management. The most recent studies in which the waste management practices of the EU countries evaluated using clustering and the DEA method are summarized below.

Fan et al. (2022) developed an artificial neural network (ANN) model supported by SHAP to predict the plastic waste generation of 27-EU countries in 2030. In the study, 12 variables belonging to the years 2010-2018 were used for the analysis. Forecasting analyses were made using 5 different scenarios regarding the plastic waste generation of EU countries. The results were evaluated in terms of clusters obtained using the K-Means algorithm. The findings argue that the amount of plastic waste determined in 2030 using the ANN model and a targeted recycling rate of 55% for EU countries are insufficient to reduce the environmental impact of plastic waste.

Hu et al. (2022) measured the efficiency of the recycling production system in EU countries using the network DEA method with imprecise data. In the study, data on the uncertain input-output variables for 2017 were collected from the world bank database. In the analysis, the efficiency scores of the countries at the lower and upper bounds were calculated based on the triangular fuzzy numbers and the α-cut of the membership function. The findings show that there is a large difference in fuzzy efficiencies between the production efficiency and recycling efficiency stages of most countries. Therefore, this situation leads to the conclusion that the circular economy system has low efficiency.

Gomonov et al. (2021) aimed to identify dynamic clustering approaches to ensure the management of cyclical production and consumption processes of EU countries. Considering the changes in the circular economy between 2000 and 2019, two indicators were chosen. The clustering quality of nine different clustering algorithms for the two determined indicators was evaluated using three different indexes. Then, K-Means clustering algorithm was chosen in terms of clustering quality and EU countries were grouped separately for the two indicators. In the study, the optimal number of clusters was determined using the Elbow method.

Rios and Picazo-Tadeo (2021) evaluated the performance of EU (28) countries in the treatment of municipal solid waste using DEA. The 2017 data of the countries (recycling, recovery, landfill and incineration) were used to calculate the efficiency scores. In terms of efficiency scores, the performance of Scandinavian countries was higher than the performances of Eastern European and Southern European member countries. In addition, it was stated that there was a positive relationship between the performance of the countries in municipal solid waste treatment and their economic development levels.

Söküt Açar and Ayman Öz (2020) divided EU member states and candidate Turkey into groups in terms of environmental waste indicators using hierarchical and non-hierarchical clustering algorithms. In the study, the optimal number of clusters was calculated using the Silhouette index. Eight environmental waste indicators were identified for clustering analyses. It was seen from the analyses that non-hierarchical clustering algorithms gave more successful results.

Giannakitsidou et al. (2020) evaluated the municipal solid waste management performances of 26 EU countries in two aspects: environmental performance and circular economy performance. In the study, input-output variables for the years 2014-2017 were used and efficiency scores were calculated using DEA. The results of the analysis showed that there were large differences between the performances of European countries. Former EU members such as Spain and France performed worse than newer members such as Slovenia or Poland, in terms of both environmental efficiency and circular economy efficiency. In addition, Belgium was the best performing country in both respects.

İçöz and Er (2020) divided EU and EU candidate countries into groups in terms of municipal waste treatment types using correspondence analysis and cluster analysis. The 2012 data used in the study were obtained from a Eurostat report. Four variables (recycled, composted, landfilled and incinerated) were determined for the analyses. Using the hierarchical clustering algorithm, six clusters were created for the European countries in terms of urban waste treatment types.

Castillo-Gimenez et al. (2019) examined the municipal waste treatment activities of EU (28) countries in two stages. In the first stage, the municipal waste processing performance of the countries was calculated in terms of landfill, incineration, recycling, and composting and digestion using a composite indicator of performance. This indicator was calculated using DEA and Multi-Criteria-Decision-Making (MCDM) methods. In the second stage, kernel densities and cluster analysis methods were applied. EU countries were divided into four clusters by using a hierarchical clustering algorithm in terms of municipal waste activities. According to the results of the analysis, the best performers were Northern and Central European countries, while the worst performers were Eastern European countries.

RESEARCH METHODOLOGY

In this section, information is given about the Expectation Maximization (EM) algorithm, Dunn Index and Data Envelopment Analysis used in the study. The methodological framework of the study is shown in Figure 2.

Figure 2. Methodological Framework

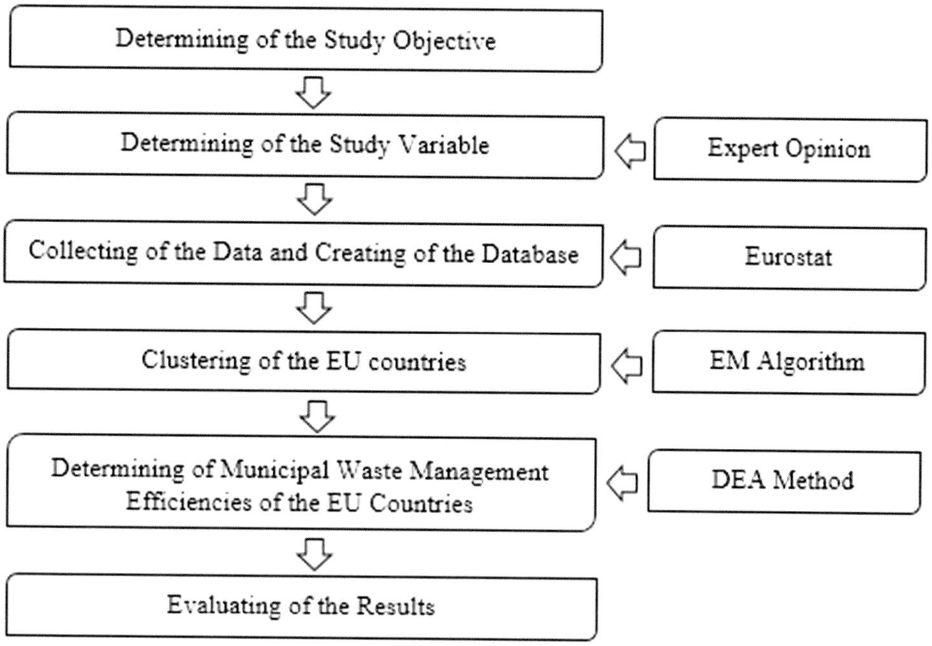

```
Determining of the Study Objective
            ⇩
Determining of the Study Variable  ⇐  Expert Opinion
            ⇩
Collecting of the Data and Creating of the Database  ⇐  Eurostat
            ⇩
Clustering of the EU countries  ⇐  EM Algorithm
            ⇩
Determining of Municipal Waste Management
Efficiencies of the EU Countries  ⇐  DEA Method
            ⇩
Evaluating of the Results
```

Expectation Maximization Clustering Algorithm

Data mining refers to the process of discovering valuable information from big data sets, involving methods at the intersection of artificial intelligence, machine learning, statistics, and database system (Lu, 2020, p. 240). The purpose of data mining is to find the hidden pattern in the existing data set and identify potentially useful and understandable correlations and patterns (Chung and Gray, 1999, p. 11). The reason for the widespread use of data mining in the information industry in recent years is the availability of large amounts of data and the need to transform this data into useful information (Ramasamy and Nirmala, 2020, p. 1). Clustering (unsupervised) and classifications (supervised) are two different types of learning methods in the data mining (Majumdar et al., 2017, p. 2). Clustering has a long history in data mining and is a common data mining tool used in many fields ranging from computer science, engineering, medical sciences to social sciences and economics (Dinh and Huynh, 2020, p. 2610).

Cluster analysis is a multivariate statistical method used to classify objects that are homogeneous within clusters or heterogeneous between clusters (Öner and Bulut, 2021, p. 4587). Clustering algorithms, which are a branch of unsupervised learning for pattern recognition, have many types in the literature (Yang et al., 2017, p. 101). In general, clustering algorithms can be divided into two groups as probability model-based approaches and non-parametric approaches. For probability model-based approaches, the Expectation Maximization (EM) algorithm is used. Non-parametric algorithms are divided into two: hierarchical and non-hierarchical algorithms. The k-means algorithm is the most widely used among these clustering methods (Maitra et al., 2012, p. 378).

EM is a clustering algorithm that finds the maximum likelihood parameters in cases where the equations cannot be solved straightly, giving fast and reliable results for large data sets (Bini and Mathew, 2016, p. 1251). Because of its simplicity and numerical stability, it is a method often used to calculate maximum likelihood or maximum a posteriori estimates (Lange, 1995, p. 425). However, the EM algorithm also has several disadvantages. For example, problem solving is often difficult to begin with, and the quality of the final solution depends on the quality of the initial solution (Xu and Jordan, 1996, p. 129).

Assume $\{X_1, X_2,...,X_n\}$ are the d independent n-dimensional observations (Yang et al., 2012, p. 3951):

$$f(x; \alpha,\theta) = \sum_{k=1}^{c} \alpha_k f(x; \theta_k) \tag{1}$$

where $\alpha_k > 0$ refers mixing proportions with the constraint $\sum_{k=1}^{c} \alpha_k = 1$ and $f(x; \theta_k)$ refers the density of x from kth class with corresponding parameters θ_k. Assume $Z = \{Z_1, Z_2,...,Z_n\}$ are the missing data in which $Z_i \in \{1,2,...,c\}$. If $Z_i = k$, it means that the ith data point belongs to the kth class. So, the joint pdf of the complete data $\{X_1, X_2,...,X_n, Z_1, Z_2,...,Z_n\}$ becomes.

$$f(x_1,...,x_n, z_1,...,z_n; \alpha,\theta) = \prod_{i=1}^{n}\prod_{k=1}^{c}[\alpha_k f(x_i; \theta_k)]^{z_{kd}} \tag{2}$$

where $z_{ki} = \begin{cases} 1, & \text{if } Z_i = k \\ 0, & \text{if } Z_i \neq k \end{cases}$. The log likelihood function is obtained as follows:

$$L(\alpha,\theta; x_1,...,x_n, z_1,...,z_n) = \sum_{i=1}^{n}\sum_{k=1}^{c} Z_{ki} ln[\alpha_k(x_i; \theta_k)] \tag{3}$$

EM is an iterative algorithm to maximize model parameters even when observations have missing data (Dempster et al., 1977:1). Each iteration has two steps: E-step (expectation), which requires marginalizing the missing data in the current model, M-step (maximization), in which the optimum parameters of a new model are determined (Duda et al., 2001, p. 53).

E-step: Since values z_{ki} are unknown, the conditional expected value $E(Z_{ki}| x_i; \alpha,\theta)$ is substituted for zki These conditional expected values are as given in (4) (Jacques and Preda, 2014, p. 98).

$$\hat{z}_{ki} = E(Z_{ki}|x_i; \alpha,\theta) = \frac{\alpha_k f(x_i; \theta_k)}{\sum_{s=1}^{c} \alpha_s f(x_i; \theta_s)} \tag{4}$$

M-step: In this step, the function in (5) is maximized such that $\sum_{k=1}^{c} \alpha_k = 1$.

$$\tilde{L}(\alpha,\theta; x_1,...,x_n) = \sum_{i=1}^{n}\sum_{k=1}^{c} \hat{z}_{ki} ln[\alpha_k f(x_i; \theta_k)] \tag{5}$$

The updated equation can be obtained by mixing proportions with (6) (Yang et al., 2012, p. 3951).

$$\alpha_k = \frac{\sum_{i=1}^{n} \hat{z}_{ki}}{n} \qquad (6)$$

In the next step, the d-variable Gaussian mixture model is considered.

$$f(x; \alpha, \theta) = \sum_{k=1}^{c} \alpha_k f(x; \theta_k)$$
$$= \sum_{k=1}^{c} \alpha_k (2\pi)^{-(\frac{d}{2})} |\Sigma_k|^{-(\frac{1}{2})} e^{-(\frac{1}{2})(x-\mu_k)' \Sigma_k^{-1}(x-\mu_k)} \qquad (7)$$

The parameter θ_k contains a mean vector μ_k and a covariance matrix $\Sigma_k x$. Then the update equations for those parameters are as follows:

$$\mu_k = \frac{\sum_{i=1}^{n} \hat{z}_{ki} x_i}{\sum_{i=1}^{n} \hat{z}_{ki}} \qquad (8)$$

$$\Sigma_k x = \frac{\sum_{i=1}^{n} \hat{z}_{ki} (x_i - \mu_k)((x_i - \mu_k))^T}{\sum_{i=1}^{n} \hat{z}_{ki}} \qquad (9)$$

Dunn Index

Dunn Index is one of the methods used to determine the optimal number of clusters. The method is based on the relationship between the degree of compactness and the degree of separation of clusters (Dunn, 1974, p. 95). A high value of the Dunn Index indicates a good clustering (Ben Ncir et al., 2021, p. 2). The main purpose of the Dunn's Index is to minimize the distance between clusters and to maximize the intra-cluster distance. If the clusters in the dataset are compact and well-separated, the distance between the clusters is high and the diameter of the clusters is small (Nawrin et al., 2017, p. 267).

Although the Dunn index is very simple, it has two main disadvantages. The first of these is very sensitive to the presence of noise in the data sets (Pal and Biswas, 1997, p. 850). Second, it is due to the minimum or maximum intermediate distance, which is not compatible with the actual structure of the underlying data set. This occurs especially for complex datasets such as the DNA microarray dataset. In complex datasets, clusters can be of any arbitrary shape and size, and therefore well-separated clusters are almost nonexistent (Shen et al., 2005, p. 1176). The Dunn index is formulated in equation 1 (Zhou and Xu, 2018, p. 80).

$$D_{nc} = \min_{i=i,\ldots,nc} \{\min_{j=i+1,\ldots,nc} (\frac{d(c_i,c_j)}{\max_{k=1,\ldots,nc} \text{diam}(c_k)})\}, \qquad (10)$$

where d (c_i, c_j) denotes the dissimilarity function between two clusters ci and cj

$$d(c_i, c_j) = \min_{x \in ci, y \in cj} d(x, y)$$

and diam (c) is defined as the diameter of cluster c:

$$\text{diam}(C) = \max d_{x,y \in C}(x,y).$$

Data Envelopment Analysis

Data Envelopment Analysis (DEA) is a non-parametric linear programming-based mathematical programming approach used to measure the relative efficiency of homogeneous and comparable Decision-Making Units (DMUs) (Tone et al., 2020, p. 560). The main advantage of the method is that it can be used to calculate relative efficiency for situations where multiple input and output variables make comparison difficult (Matsumoto et al., 2020, p. 3). DEA was first introduced by Charnes, Cooper, and Rhodes (1978) and has been enriched and modified over the years. DEA has been used in numerous studies in various fields, including sustainability research (Zhou et al., 2018, p. 2).

The first efficiency measurement studies using multiple inputs and single output were carried out by Farrell (1957). The basis of DEA is based on Farrell's production frontier approach and was developed to measure the efficiency of public organizations (Kneip et al., 1998, p. 783). There are two DEA models commonly used in the literature. The first is the CCR model based on the assumption of constant return to scale developed by Charnes, Cooper, and Rhodes (1978), the other is the BCC model based on the assumption of variable returns to scale developed by Banker, Charnes and Cooper (1984) (Charnes et al., 1994, p. 23). DEA models are linear programming-based models that aim to maximize or minimize the objective function under certain constraints (Angiz et al., 2010, p. 5153).

It is possible to develop models that provide input-oriented or output-oriented projections for both CCR and BCC models. In the output-oriented model, it is aimed to maximize the proportional increase in output variables while the amount of input variables is constant. (Dinc and Haynes, 1999, p. 475). Due to the use of output-oriented models in this study, information about the output oriented CCR and BCC models is given below.

The CCR Model

The production possibility set (P_C) of the CCR model is expressed as follows (Tone, 2001:32):

$$P_C = \{(x, y) \mid x \geq X\lambda, y \leq Y\lambda, \lambda \geq 0\}, \qquad (11)$$

Where λ is a semipositive vector in R^n.

The overall efficiencies of the DMUs are calculated using the CCR model. In order for DMU to be efficient according to the CCR model, it must have both technical and scale efficiencies (Charnes et al., 1994, p. 23). The output-oriented dual (envelopment) CCR model under a constant-returns to scale assumption is shown below (Luptacik, 2010, p. 150):

$$\max h_0 = \phi + \varepsilon \left(\sum_{i=1}^{m} s_i^- + \varepsilon \sum_{r=1}^{s} s_r^+ \right) \tag{12}$$

Subject to:

$$\sum_{j=1}^{n} x_{ij}\beta_j - x_{i0} + s_i^- = 0$$

$$-\sum_{j=1}^{n} y_{rj}\beta_j + \phi y_{r0} + s_i^+ = 0$$

$$\beta_j, s_i^-, s_r^+ \geq 0, \; j=1,2,\ldots,n; \; r=1,2,\ldots,s; \; i=1,2,\ldots,m$$

In the dual model, DMU is efficient when $\phi=1$ and $s_i^-, s_r^+ = 0$. The DMU, which is efficient according to the input-oriented CCR model, is also efficient according to the output oriented CCR model. There is a relationship $\phi=1/\theta$ between the two models. Therefore, while it is always $\theta \leq 1$ in the input oriented CCR model, it is always $\phi \geq 1$ in the output oriented CCR model.

The BCC Model

By the BCC model, both technical and scale efficiencies of the DMU are calculated. The production possibility set (P_B) of the BCC model is expressed as follows (Tone, 2001, p. 34):

$$P_B = \{(x,y)| \; x \geq X\lambda, \; y \leq Y\lambda, \; e\lambda=1, \; \lambda \geq 0\}. \tag{13}$$

The output-oriented dual (envelopment) BCC model under a variable-returns to scale assumption is shown below (Luptacik, 2010, p. 163):

$$\max h_0 = \varphi + \varepsilon \left(\sum_{i=1}^{m} s_i^- + \sum_{r=1}^{s} s_r^+ \right) \tag{14}$$

Subject to:

$$\sum_{j=1}^{n} x_{ij}\beta_j - x_{i0} + s_i^- = 0 \quad i=1,2,\ldots,m$$

$$\sum_{j=1}^{n} y_{rj}\beta_j - \varphi y_{r0} + s_i^+ = 0 \quad r=1,2,\&,s$$

$$\sum_{j=1}^{n} \beta_j = 1$$

$$\beta_j, s_i^-, s_r^+ \geq 0, j=1,2,\ldots,n; r=1,2,\ldots,s; i=1,2,\ldots,m$$

In the model DMU_0 is efficient for $\varphi=1$ and $s_i^-, s_r^+ = 0$, it is not efficient for other cases and is $\varphi >$ 1. The difference between CCR and BCC models is that the convexity constraint ($\sum_{j=1}^{n} \beta_j = 1$) is added to the BCC model (Thanassoulis, 2001, pp. 129-130).

ANALYSIS OF THE DATA

In this section, firstly, explanations are given about the data set used in the study. Then, EU countries were clustered in terms of the dataset obtained for municipal waste management practices using the EM algorithm. Finally, the efficiencies of municipal waste management practices of EU countries were evaluated using the DEA analysis.

The Data Set

In this study, the data set on municipal waste management practices of EU countries for 2020 was obtained from the Eurostat database. Eight variables (waste generated, waste treatment, disposal - incineration and recovery - energy recovery, disposal - landfill and other, disposal - incineration, recovery - energy recovery, recycling - material and recycling - composting and digestion) determined by taking expert opinions will be used in the analyses. In this study, kg per capita unit was used for the variables. Descriptions on the variables used in the study are given in Table 1.

Clustering of EU Countries using the EM Algorithm

In the study, firstly 28 EU countries were divided into groups using the EM clustering algorithm in terms of municipal waste management practices defined in Table 1. In the EM clustering algorithm, the number of clusters was determined by the researcher. In this study, Dunn's Index was used to determine the optimal number of clusters. Dunn Index values were calculated using the R-Studio program are shown in Table 2.

As seen in Table 2, the highest Dunn value is obtained when countries are divided into four, five or six groups.

Clustering is defined as the process of allocating entities so that similar entities are in the same classes. However, there are many problems with clustering techniques, such as determining the optimal

Table 1. Definitions of the variables

Code	Variable Name	Unit	Definition
v1	Waste Generated	kg per capita	Municipal waste is mainly collected from households and comprises similar wastes from sources such as commerce, offices and public institutions. The amount of municipal waste composes of waste collected by or on behalf of municipal authorities and disposed of through the waste management system.
v2	Waste Treatment	kg per capita	Waste treatment refers to the activities required to minimize the negative effects of waste on the environment. This indicator consists of municipal waste treatment for the treatment operations incineration (with and without energy recovery), recycling, composting and landfilling.
v3	Disposal – Incineration (D10) and Recovery - Energy Recovery (R1)	kg per capita	This indicator consists of the total of municipal waste incineration on land and its recovery to generate energy.
v4	Disposal - Landfill and Other (D1-D7, D12)	kg per capita	Landfill is expressed as deposit of waste into or onto land and it consists of specially engineered landfills and temporary storage of over one year on permanent sites. The indicator covers both landfills in internal sites where a generator of waste carries out its own waste disposal at the production site and in external sites.
v5	Disposal – Incineration (D10)	kg per capita	Disposal refers to any non-recovery operation, even if the reclamation of materials or energy is a secondary purpose of the operation. Some wastes still have to be disposed of, such as incineration without energy recovery.
v6	Recovery - Energy Recovery (R1)	kg per capita	Energy recovery is expressed as the incineration that fulfils the energy efficiency criteria set out in the Waste Framework Directive.
v7	Recycling - Material	kg per capita	Recycling refers to the recovery of waste materials by reprocessing them into original or other purpose products. However, this indicator does not include energy recovery and the reprocessing of materials for use as fuel.
v8	Recycling - Composting and Digestion	kg per capita	This indicator means compost/methanized municipal waste. The assumption of the indicator is that the only reasonable treatment of bio-waste is composting or anaerobic digestion.

Source: Eurostat, 2021

number of clusters (Everitt, 1980:75). Everitt (1974) developed the following equation (11) in order to determine the optimal number of clusters.

$$K = \sqrt{\frac{n}{2}} \tag{15}$$

Table 2. The Dunn Index Values with respect to different number of clusters

Number of Clusters	The Dunn Index Values
3	0.284
4	0.327*
5	0.327*
6	0.327*

Figure 3. Number of Countries in Each Cluster
Source: Created by the author

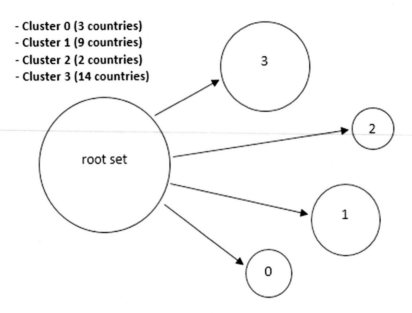

Using the equation (15), it was concluded that the optimal number of clusters was (3.74). Therefore, in the study, 28 EU countries were divided into four groups using the EM algorithm in terms of municipal waste management practices. After the optimal number of clusters was determined, cluster analysis was performed using the Rapid Miner program. In Figure 3, the number of countries in each cluster is shown.

According to the findings obtained from the EM Algorithm, there are 3 countries in Cluster 0, 9 countries in Cluster 1, 2 countries in Cluster 2 and 14 countries in Cluster 3. The countries in each cluster are shown in Table 3.

Cluster 0 consists of the Mediterranean countries Greece, Cyprus and Malta. The amounts of municipal waste generated per capita in Malta and Cyprus follow the countries in Cluster 0. On the other hand, the countries in Cluster 0 are the countries that use the most disposal practices in municipal waste management among EU countries. Disposal practices constitute the lowest step of the waste hierarchy shown in Figure 1. In addition, the recovery and recycling amounts of these countries are far behind Denmark and Luxembourg, which are in Cluster 2. Recovery and recycling practices constitute the second and third steps of the waste hierarchy. Although the countries in this cluster are not the EU's countries with

Table 3. Clusters

Clusters	Countries
Clusters 0	Greece, Cyprus, Malta
Clusters 1	Belgium, Germany, Ireland, France, Netherlands, Austria, Finland, Sweden, United Kingdom
Clusters 2	Denmark, Luxembourg
Clusters 3	Bulgaria, Czechia, Estonia, Spain, Croatia, Italy, Latvia, Lithuania, Hungary, Poland, Portugal, Romania, Slovenia, Slovakia

the lowest per capita income, it is possible to say that they are the EU's most unsuccessful countries in terms of municipal waste management.

It is seen that the per capita municipal waste amounts of the countries in Cluster 1 are close to the countries in Cluster 0. However, compared to the countries in Cluster 0, the countries in this cluster have much better results in terms of recovery and recycling practices. Germany ranks first among EU countries in terms of recycling-material method in municipal waste management. In addition, Austria is in the first place among EU countries in terms of recycling - composting and digestion method. In general, in terms of municipal waste management success, the countries in Cluster 1 rank second after the countries in Cluster 2. In addition, Ireland is the EU's second country with the highest per capita income in 2019. The per capita income of the other countries in this cluster (Belgium, Germany, France, Netherlands, Austria, Finland, Sweden and United Kingdom) comes after Denmark. Therefore, this situation shows the conclusion that the countries in Cluster 1 and Cluster 2 consist of the countries with the highest per capita income of the EU.

When Cluster 2 is examined, it has been determined that Denmark and Luxembourg are the countries that collected the most municipal waste per capita among EU countries in 2019 and that all the collected wastes are treated. These two countries are the most successful countries of the EU, especially in terms of recovery and recycling practices. Therefore, it is possible to say that Cluster 2 consists of the most successful countries in terms of municipal waste management. In addition, Luxembourg is the EU's country with the highest per capita income in 2019. Denmark is in third place.

It is seen that the amount of municipal waste collected per capita in the countries in Cluster 3 is close to the countries in Cluster 1. On the other hand, it has been determined that they are behind the countries in Cluster 1 in terms of recovery and recycling practices. In general, the disposal practices in municipal waste management of the countries in this cluster has a higher share than the countries in Cluster 1. Therefore, it is possible to say that the success of Cluster 3 in terms of municipal waste management practices lags behind Cluster 1. On the other hand, Cluster 3 consists of countries with the lowest per capita income among EU countries in 2019.

Calculation of Efficiency Scores of EU Countries using the DEA

In the second stage, the municipal waste management efficiencies of EU countries were determined using the Data Envelopment Analysis (DEA) method. In particular, the efficiency of the recovery and recycling practices, which constitute the third and fourth steps of the waste hierarchy of the countries shown in Figure 1, were evaluated. In this context, as the waste generated input variable; recovery - energy recovery, recycling - material and recycling - composting and digestion were used as output variables. In the analysis, the output-oriented model was preferred and the overall efficiency (CCR), technical efficiency (BCC) and scale efficiency (CCR/BCC) scores were calculated. EMS 1.3 package program was used to calculate efficiency scores. Efficiency scores of EU countries are shown in Table 4 below.

When Table 4 is examined, it is seen that Denmark, which is in Cluster 2, is efficient in terms of overall, technical and scale. It can be said that Luxembourg's overall (103.20%), technical (100.60%) and scale (102.58%) efficiency scores are almost efficient. In case Luxembourg increases the amount related to recovery or recycling practices by decreasing the amount related to disposal-landfill and other practices, it may become efficient in terms of overall, technical and scale efficiencies.

It is seen that Germany, Netherlands, Austria, Finland and Sweden from the 9 countries in Cluster 1 are efficient in terms of overall, technical and scale efficiencies. Belgium is efficient in terms of technical

Table 4. Efficiency scores of EU countries

Countries	Overall Efficiency Scores (CCR)	Technical Efficiency Scores (BCC)	Scale Efficiency (CCR / BCC)	Countries	Overall Efficiency Scores (CCR)	Technical Efficiency Scores (BCC)	Scale Efficiency (CCR / BCC)
Belgium	101.40%	100.00%	101.40%	Lithuania	123.10%	113.70%	108.27%
Bulgaria	161.30%	114.80%	140.51%	Luxembourg	103.20%	100.60%	102.58%
Czechia	195.00%	182.80%	106.67%	Hungary	180.20%	119.90%	150.29%
Denmark	100.00%	100.00%	100.00%	Malta	536.70%	471.00%	113.95%
Germany	100.00%	100.00%	100.00%	Netherlands	100.00%	100.00%	100.00%
Estonia	115.80%	100.00%	115.80%	Austria	100.00%	100.00%	100.00%
Ireland	117.40%	115.70%	101.47%	Poland	174.50%	100.00%	174.50%
Greece	299.10%	272.40%	109.80%	Portugal	191.70%	175.10%	109.48%
Spain	172.80%	158.90%	108.75%	Romania	555.70%	100.00%	555.70%
France	122.20%	122.10%	100.08%	Slovenia	112.70%	100.00%	112.70%
Croatia	179.30%	141.50%	126.71%	Slovakia	171.30%	131.70%	130.07%
Italy	121.40%	114.90%	105.66%	Finland	100.00%	100.00%	100.00%
Cyprus	353.80%	335.60%	105.42%	Sweden	100.00%	100.00%	100.00%
Latvia	133.20%	103.60%	128.57%	UK	120.30%	119.70%	100.50%

efficiency, while overall (101.40%) and scale (101.40%) efficiency scores are almost efficient. Ireland, France and United Kingdom are inefficient countries of this cluster in all three efficiency types. Compared to other countries in the cluster, it has been determined that these three countries use disposal-landfill and other practices more in municipal waste management. Therefore, these countries need to increase their recovery and recycling practices in order to improve their efficiency scores.

None of the countries in Cluster 0 (Malta, Cyprus and Greece) have efficient scores in terms of overall, technical and scale efficiencies. Since these three countries have the most disposal-landfill and other practices among EU countries, they should primarily reduce this amount significantly and give much more importance to recovery and recycling practices.

Of the 14 countries in Cluster 3, only Estonia, Poland, Romania and Slovenia have efficient scores in terms of technical efficiency. While Romania has efficient scores in terms of technical efficiency, it has the worst efficiency score in terms of overall efficiency. Similar to the countries in Cluster 0, the countries in this cluster need to reduce the disposal-landfill and other practices and improve the amounts of recovery and recycling practices.

CONCLUSION

Sustainable development has three basic elements: economic, social and environmental. Municipal waste management contributes both to the protection of the environment and to the circular economy. In this study, in the first stage, data on municipal waste management of 28 EU countries were divided into four groups using cluster analysis. Then, the municipal waste management performances of the countries were evaluated using DEA. In the study, data for 2019 were collected from the Eurostat database. While 8 variables were used in cluster analysis, one input and two output variables were determined for DEA analysis.

Cluster analysis and DEA results used in the study support each other. According to the results of the cluster analysis, it is seen that the countries that are successful in recovery and recycling practices are gathered in Cluster 1 and Cluster 2. It is seen that the countries in Cluster 1 (Denmark and Luxembourg) and Cluster 2 (Belgium, Germany, Ireland, France, Netherlands, Austria, Finland, Sweden, United Kingdom) consist of northern and western European countries. Greece, Cyprus and Malta, included in Cluster 0, are the southern European countries that are the most unsuccessful in their recovery and recycling practices. The level of success in the recovery and recycling practices of 14 eastern and southern European countries in Cluster 3 comes after Cluster 0.

In the second stage, the performances of EU countries in terms of recovery and recycling practices are measured by using DEA, taking into account the data obtained and Figure 1. It is seen that DEA results are similar to cluster analysis results. All of the countries that are efficient in terms of overall efficiency are included in Cluster 1 and Cluster 2. The efficiency scores of Greece, Cyprus and Malta, which are unsuccessful in recovery and recycling practices in Cluster 0, are also very low. The eastern and southern European countries efficiency scores in Cluster 3 are better than Cluster 0.

The European Commission has set a 65% goal for recycling and reuse of municipal waste by 2030. In this study, it is aimed to investigate whether there are differences in the municipal waste management practices of EU countries. In this direction, municipal waste management practices of the EU countries have been evaluated and it has been concluded that there are significant differences between the EU countries. In future studies, the EU's waste management issue can be expanded and examined from different perspectives (packaging waste, waste electrical and electronic equipment, end-of-life vehicles, etc.).

The results obtained in the study are valid within the scope of 2019, indicators determined for cluster analysis and input-output variables determined for DEA. However, it is possible to develop the municipal waste management of EU countries in next studies. In this context, the subject can be examined from different perspectives using more current data and different input-output combinations. In addition, new findings can be obtained by using different DEA models.

REFERENCES

Angiz, L. M. Z., Emrouznejad, A., & Mustafa, A. (2010). Fuzzy assessment of performance of a decision making-units using DEA: Non-Radial approach. *Expert Systems with Applications*, *37*(7), 5153–5157. doi:10.1016/j.eswa.2009.12.078

Ben Ncir, C.-E., Hamza, A., & Bouaguel, W. (2021). Parallel and scalable Dunn index for the validation of big data clusters. *Parallel Computing*, *102*, 1–10. doi:10.1016/j.parco.2021.102751

Bini, B. S., & Mathew, T. (2016). Clustering and regression techniques for stock prediction. *Procedia Technology*, *24*, 1248–1255. doi:10.1016/j.protcy.2016.05.104

Blewitt, J. (Ed.). (2015). Understanding sustainable development (3rd ed.). Routledge (Taylor & Francis Group).

Brundtland Commission. (1987). *WCED (Report of the world commission on environment and development: our common future)*. Oxford University Press.

Castillo-Gimenez, J., Montanes, A., & Picazo-Tadeo, A. J. (2019). Performance in the treatment of municipal waste: Are European Union Member States so different? *The Science of the Total Environment*, *687*, 1305–1314. doi:10.1016/j.scitotenv.2019.06.016 PMID:31412464

Charnes, A., Cooper, W. W., Lewin, A. Y., & Seiford, L. M. (Eds.). (1994). *Data envelopment analysis: theory, methodology and application*. Kluwer Academic Publisher. doi:10.1007/978-94-011-0637-5

Charnes, A., Cooper, W. W., & Rhodes, E. (1978). Measuring the efficiency of decision making units. *European Journal of Operational Research*, *2*(6), 429–444. doi:10.1016/0377-2217(78)90138-8

Chung, H. M., & Gray, P. (1999). Special section: Data mining. *Journal of Management Information Systems*, *16*(1), 11–16. doi:10.1080/07421222.1999.11518231

Dempster, A. P., Laird, N. M., & Rubin, D. B. (1977). Maximum likelihood from incomplete data via the EM algorithm. *Journal of the Royal Statistical Society. Series B. Methodological*, *39*(1), 1–38. doi:10.1111/j.2517-6161.1977.tb01600.x

Dinc, M., & Haynes, K. E. (1999). Sources of regional inefficiency an integrated shift-share data envelopment analysis and input-output approach. *The Annals of Regional Science*, *33*(4), 469–489. doi:10.1007001680050116

Dinh, D.-T., & Huynh, V.-N. (2020). k -PbC: An Improved Cluster Center Initialization for Categorical Data Clustering. *Applied Intelligence*, *50*(8), 2610–2632. doi:10.100710489-020-01677-5

Duda, R. O., Hart, P. E., & Stork, D. G. (Eds.). (2000). *Pattern classification* (2nd ed.). Wiley-Interscience.

Dunn, J. C. (1974). Well-separated clusters and optimal fuzzy partitions. *Journal of Cybernetics*, *4*(1), 95–104. doi:10.1080/01969727408546059

Elliott, J. A. (Ed.). (2013). *An introduction to sustainable development* (4th ed.). Routledge (Taylor & Francis Group).

European Commission. (2016). *Guidance on municipal waste data collection*. https://ec.europa.eu/eurostat/documents/342366/351758/Guidance+on+municipal+waste/3106067c-6ad6-4208-bbed-49c08f7c47f2

European Commission. (n.d.). https://ec.europa.eu/environment/topics/waste-and-recycling_en

European Environment Agency. (2021). *Municipal waste management across European Countries*. https://www.eea.europa.eu/publications/municipal-waste-management-across-european-countries

Eurostat. (n.d.). https://ec.europa.eu/eurostat/cache/metadata/en/cei_wm030_esmsip2.htm

Everitt, B. (Ed.). (1974). *Cluster analysis*. Heinmann Educational Books.

Everitt, B. (1980). Cluster analysis. *Quality & Quantity*, *14*(1), 75–100. doi:10.1007/BF00154794

Fan, Y.V., Jiang, P., Tan, R.R., Aviso, K.B., You, F., Zhao, X., Lee, C.T. & Klemes, J. J. (2022). Forecasting plastic waste generation and interventions for environmental hazard mitigation. *Journal of Hazardous Materials*, *424*(Part A), 1-14.

Farrell, M. J. (1957). The measurement of productivity efficiency. *Journal of the Royal Statistical Society. Series A (General), 120*(3), 253–281. doi:10.2307/2343100

Giannakitsidou, O., Giannikos, I., & Chondrou, A. (2020). Ranking European Countries on the basis of their environmental and circular economy performance: A DEA application in MSW. *Waste Management (New York, N.Y.), 109*, 181–191. doi:10.1016/j.wasman.2020.04.055 PMID:32408101

Gomonov, K., Ratner, S., Lazanyuk, I., & Revinova, S. (2021). Clustering of EU countries by the level of circular economy: An object-oriented approach. *Sustainability, 13*(13), 1–20. doi:10.3390u13137158

Hu, C.-F., Wang, H.-F., & Liu, T. (2022). Measuring efficiency of a recycling production system with imprecise data. *Numerical Algebra. Control and Optimization, 12*(1), 79–91. doi:10.3934/naco.2021052

İçöz, C., & Er, F. (2020). A statistical analysis of municipal waste treatment types in European countries. *Environmental Research & Technology, 3*(3), 113–118. doi:10.35208/ert.769634

Jacques, J., & Preda, C. (2014). Model-Based clustering for multivariate functional data. *Computational Statistics & Data Analysis, 71*, 92–106. doi:10.1016/j.csda.2012.12.004

Jansen, L. (2003). The challenge of sustainable development. *Journal of Cleaner Production, 11*(3), 231–245. doi:10.1016/S0959-6526(02)00073-2

Kneip, A., Park, B. U., & Simar, L. (1998). A note on the convergence of nonparametric DEA estimators for production efficiency scores. *Econometric Theory, 14*(6), 783–793. doi:10.1017/S0266466698146042

Lange, K. (1995). A gradient algorithm locally equivalent to the EM algorithm. *Journal of the Royal Statistical Society. Series B. Methodological, 57*(2), 425–437. doi:10.1111/j.2517-6161.1995.tb02037.x

Lu, W. (2020). Improved k-means clustering algorithm for big data mining under hadoop parallel framework. *Journal of Grid Computing, 18*(3), 239–250. doi:10.100710723-019-09503-0

Luptacik, M. (Ed.). (2010). *Mathematical Optimization and Economic Analysis*. Springer. doi:10.1007/978-0-387-89552-9

Maitra, R., Melnykov, V., & Lahiri, S. N. (2012). Bootstrapping for significance of compact clusters in multidimensional datasets. *Journal of the American Statistical Association, 107*(497), 378–392. doi:10.1080/01621459.2011.646935

Majumdar, J., Naraseeyappa, S., & Ankalaki, S. (2017). Analysis of agriculture data using data mining techniques: Application of big data. *Journal of Big Data, 4*(1), 1–15. doi:10.118640537-017-0077-4

Matsumoto, K., Makridou, G., & Doumpos, M. (2020). Evaluating environmental performance using data envelopment analysis: The case of European Countries. *Journal of Cleaner Production, 272*, 1–13. doi:10.1016/j.jclepro.2020.122637

Mesjasz-Lech, A. (2014). Municipal waste management in context of sustainable urban development. *Procedia: Social and Behavioral Sciences, 151*, 244–245. doi:10.1016/j.sbspro.2014.10.023

Nawrin, S., Rahman, M. R., & Akhter, S. (2017). Exploring k-means with internal validity indexes for data clustering in traffic management system. *International Journal of Advanced Computer Science and Applications, 8*(3), 264–272. doi:10.14569/IJACSA.2017.080337

Öner, Y., & Bulut, H. (2021). A robust EM clustering approach: ROBEM. *Communications in Statistics. Theory and Methods*, *50*(19), 4587–4605. doi:10.1080/03610926.2020.1722840

Pal, N. R., & Biswas, J. (1997). Cluster validation using graph theoretic concepts. *Pattern Recognition*, *30*(4), 847–857. doi:10.1016/S0031-3203(96)00127-6

Ramasamy, S., & Nirmala, K. (2020). Disease prediction in data mining using association rule mining and keyword-based clustering algorithms. *International Journal of Computers and Applications*, *42*(1), 1–8. doi:10.1080/1206212X.2017.1396415

Rios, A. M., & Picazo-Tadeo, A. J. (2021). Measuring environmental performance in the treatment of municipal solid waste: The case of the European Union-28. *Ecological Indicators*, *123*, 1–10. doi:10.1016/j.ecolind.2020.107328

Rogers, P. P., Jalal, K. F., & Boyd, J. A. (Eds.). (2008). An introduction to sustainable development. Earthscan (Glen Educational Foundation).

Sachs, J. D. (Ed.). (2015). *The age of sustainable development*. Columbia University Press. doi:10.7312ach17314

Shen, J., Chang, S. I., Lee, E. S., Deng, Y., & Brown, S. J. (2005). Determination of cluster number in clustering microarray data. *Applied Mathematics and Computation*, *169*(2), 1172–1185. doi:10.1016/j.amc.2004.10.076

Söküt Açar, T., & Ayman Öz, N. (2020). The determination of optimal cluster number by silhouette index at clustering of the European Union Member Countries and candidate Turkey by waste indicators. *Pamukkale University Journal of Engineering Sciences*, *26*(3), 481–487. doi:10.5505/pajes.2019.49932

Thanassoulis, E. (Ed.). (2001). *Introduction to the theory and application of data envelopment analysis: a foundation text with integrated software*. Springer. doi:10.1007/978-1-4615-1407-7

Tone, K. (2001). On returns to scale under weight restrictions in data envelopment analysis. *Journal of Productivity Analysis*, *16*(1), 31–47. doi:10.1023/A:1011147118637

Tone, K., Toloo, M., & Izadikhah, M. (2020). A modified slacks-based measure of efficiency in data envelopment analysis. *European Journal of Operational Research*, *287*(2), 560–571. doi:10.1016/j.ejor.2020.04.019

Towards A Green Economy in Europe. (2013). EU Environmental Policy Targets and Objectives 2010-2050. European Environment Agency, Report 8.

United Nations. (n.d.a). https://www.un.org/en/observances/world-population-day

United Nations. (n.d.b). https://sdgs.un.org/goals

United Nations. (n.d.c). https://www.un.org/development/desa/disabilities/envision2030-goal12.html

Worldometer. (n.d.). https://www.worldometers.info/world-population/

Xu, L., & Jordan, M. I. (1996). On convergence properties of the EM algorithm for gaussian mixtures. *Neural Computation*, *8*(1), 129–151. doi:10.1162/neco.1996.8.1.129

Yang, M.-S., Chang-Chien, S.-J., & Hung, W.-L. (2017). Learning-based EM clustering for data on the unit hypersphere with application to exoplanet data. *Applied Soft Computing*, *60*, 101–114. doi:10.1016/j.asoc.2017.06.037

Yang, M.-S., Lai, C.-Y., & Lin, C.-Y. (2012). A robust EM clustering algorithm for gaussian mixture models. *Pattern Recognition*, *45*(11), 3950–3961. doi:10.1016/j.patcog.2012.04.031

Zhou, H., Yang, Y., Chen, Y., & Zhu, J. (2018). data envelopment analysis application in sustainability: the origins, development and future directions. *European Journal of Operational Research*, *264*, 1-16.

Zhou, S., & Xu, Z. (2018). A novel internal validity index based on the cluster centre and the nearest neighbour cluster. *Applied Soft Computing*, *71*, 78–88. doi:10.1016/j.asoc.2018.06.033

ADDITIONAL READING

Altın, F. G. (2022). Handbook of research on advances and applications of fuzzy sets and logic. Broumi, S. (Ed.), A fuzzy data envelopment analysis-based performance assessment of European Union Countries' waste management practices (pp. 29-49). IGI Global.

Castillo-Gimenez, J., Montanes, A., & Picazo-Tadeo, A. J. (2019). Performance and convergence in municipal waste treatment in the European Union. *Waste Management (New York, N.Y.)*, *85*, 222–231. doi:10.1016/j.wasman.2018.12.025 PMID:30803576

Di Maria, F., Sisani, F., Contini, S., Ghosh, S. K., & Mersky, R. L. (2020). Is the policy of the European Union in waste management sustainable? An assessment of the Italian context. *Waste Management (New York, N.Y.)*, *103*, 437–448. doi:10.1016/j.wasman.2020.01.005 PMID:31952025

Khan, S., Anjum, R., Raza, S. T., Bazai, N. A., & Ihtisham, M. (2022). Technologies for municipal solid waste management: Current status, challenges, and future perspectives. *Chemosphere*, *288*(1), 1–12. doi:10.1016/j.chemosphere.2021.132403 PMID:34624349

Minelgaite, A., & Liobikiene, G. (2019). Waste problem in European Union and its influence on waste management behaviours. *The Science of the Total Environment*, *667*, 86–93. doi:10.1016/j.scitotenv.2019.02.313 PMID:30826684

Özer, B., & Erses Yay, A. S. (2021). Comparative life cycle analysis of municipal waste management systems: Kırklareli/Turkey case study. *Environmental Science and Pollution Research International*, *28*(45), 63867–63877. doi:10.100711356-020-12247-0 PMID:33492597

Pires, A., & Martinho, G. (2019). Waste hierarchy index for circular economy in waste management. *Waste Management (New York, N.Y.)*, *95*, 298–305. doi:10.1016/j.wasman.2019.06.014 PMID:31351615

Smol, M., Duda, J., Czaplicka-Kotas, A., & Szoldrowska, D. (2020). Transformation towards circular economy (CE) in municipal waste management system: Model solutions for Poland. *Sustainability*, *12*(11), 1–25. doi:10.3390u12114561

Struk, M., & Boda, M. (2022). Factors influencing performance in municipal solid waste management – A case study of Czech municipalities. *Waste Management (New York, N.Y.), 139*, 227–249. doi:10.1016/j.wasman.2021.09.022 PMID:34979352

Vyas, S., Prajapati, P., Shah, A. V., & Varjani, S. (2022). Municipal solid waste management: Dynamics, risk assessment, ecological influence, advancements, constraints and perspectives. *The Science of the Total Environment, 814*, 1–13. doi:10.1016/j.scitotenv.2021.152802 PMID:34982993

KEY TERMS AND DEFINITIONS

Landfill: It is the conventional landfill of solid waste that is widely used as it is the least costly of disposal methods such as incineration or composting.

Municipal Waste: It consists of originating from households or similar in content or structure commercial, industrial and institutional wastes collected by or on behalf of municipalities.

Recovery of Waste: It is defined as any waste management process that results in a particular product that has an economic or ecological benefit for a waste material.

Recycling of Waste: It refers to the process of collecting and processing wastes that can be reused and transformed into new products.

Sustainable Development: It is the optimum consumption of natural resources by establishing a balance between human and nature and ensuring economic development by taking into account the needs of future generations.

Waste Disposal: It refers to any non-recovery operation or the destroying of waste generated from agricultural, domestic, and industrial products.

Waste Treatment: It refers to the necessary practices to minimize the negative effects of waste on the environment.

Chapter 2
An Economic Assessment of Implementing Dominica's Nationally Determined Contributions

Don Charles
University of the West Indies, Trinidad and Tobago

ABSTRACT

The Government of the Commonwealth of Dominica has joined with the international community to pursue the reduction in its country's GHG emissions. Its Nationally Determined Contributions (NDCs) highlighted its intention to reduce emissions in the energy sector by 98.6%; the transport sector by 16.9%; the commercial, residential, agriculture, forestry, and fishing sector by 8.1%; and the solid waste sector by 78.6%. The achievement of the NDCs is conditional upon the country receiving timely access to international climate change financing, technology development, and capacity building support. However, climate finance is shrouded with bureaucracy and is very difficult for developing countries to access. This study assesses the economic impacts of implementing the NDCs in Dominica. It quantifies the cost of implementing the NDCs in the identified sectors, identifies the risks and critical factors to be considered in the implementation of the NDCs, and provides recommendations for the mobilization of finance to facilitate the NDCs implementation.

INTRODUCTION

The twenty-first Conference of the Parties (COP 21) held in 2015, was a landmark achievement in the international climate change negotiations. 195 Country Parties to the United Nations Framework Convention on Climate Change (UNFCCC) agreed to adopt a common position to take collective action to limit temperature rise to well below 20C above pre-industrial levels and to pursue efforts to limit the temperature increase even further to 1.50C by the end of the century.

DOI: 10.4018/978-1-6684-4829-8.ch002

The Paris Agreement arising out of the COP 21, seeks to catalyze global greenhouse gas (GHG) emissions reductions through a system of nationally determined country-level emissions reduction targets, also referred to as the Nationally Determined Contributions (NDCs).

The present NDCs submitted by countries are insufficient to achieve the 20C limit on global temperature rise. The current NDCs are estimated to collectively result in a global temperature rise ranging between 2.90C to 3.40C by 2100. Achieving the Paris Agreement's goals will require an emissions peak as soon as possible, followed by sharp reductions in GHG emissions.

All countries are expected to review their progress in achieving their NDCs, a process referred to as the Global Stocktake, and submit increasingly ambitious NDCs. This process is supposed to occur every 5 years, with the first opportunity to do so was officially scheduled for 2020.

The Commonwealth of Dominica, the Caribbean Community (CARICOM) Member State that is situated 150 N and 610W and is bordered by the French territories of Guadeloupe and Martinique, also submitted NDCs to the UNFCCC. Its NDCs outlines the government's intention to progressively reduce total gross greenhouse gas (GHG) emissions below 2014 levels at the following reduction rates: 17.9% by 2020; 39.2% by 2025; and 44.7% by 2030 (UNFCCC 2015b). By 2030, the country intends to achieve the reduction in total emission per sector as follows:

- energy industries – 98.6% (principally from harnessing of geothermal resources);
- transport – 16.9%;
- manufacturing and construction – 8.8%;
- commercial/institutional, residential, agriculture, forestry, fishing – 8.1%; and
- solid waste – 78.6%.

As the first Global Stocktake period has arrived, several countries are considering the review of the NDCs with a view for updating to one that reflects raised mitigation and adaptation ambition. It is expected that the implementation of the NDCs will impact many sectors of Dominica's economy. However, the full extent of the impact is presently unknown.

The main objective of this study is to assess the economic impacts of implementing the NDCs in Dominica. More specifically, this study seeks to undertake the following:

- quantify the cost of implementing the NDCs in various sectors;
- identify the risks associated with achieving the NDC targets; and
- provide recommendations for the mobilization of finance to facilitate the NDCs implementation.

The rest of this study is structured as follows. The second section reviews the literature on the evolution of the climate change policy action to the development of NDCs. The third section reviews the climate change problem in Dominica. The fourth section explains the methodology for estimating the costs and the economic impact of implanting the NDCs. The fifth section will present the results. The sixth section provides a discussion about the critical factors and risks that should be considered in the implementation of the NDCs. The seventh section considers the perspective of the residents. The eighth section furnishes some policy recommendations to mobilize finance and facilitate the implementation of the NDCs. The seventh section concludes this study.

LITERATURE REVIEW ON THE NATIONALLY DETERMINED CONTRIBUTIONS

In 1990, the Intergovernmental Panel on Climate Change (IPCC) published its first assessment report, which recognized that human-induced climate change was a real threat to countries and that international cooperation to tackle its consequences. This report influenced 166 countries to meet in 1992 to form the United Nations Framework Convention on Climate Change as the first international treaty and organization to address climate change (Schipper 2006; Bulkely & Newell 2015; UNFCCC 2015a).

The second major milestone was first Conference of the Parties (COP 1) which was held in Berlin, Germany in 1995. This conference sought to introduce the concept of Quantifiable Emissions Limitations and Reduction Obligations (QELROs) which would, in turn, set GHG emissions reduction targets for countries. This was eventually followed up 2 years later at the third Conference of the Parties (COP 3), held in Kyoto, Japan. At COP 3, countries agreed to the Kyoto Protocol, in which 38 industrial countries would be required to reduce their GHG emissions. Although the agreement had 150 signatories, only 38 industrialized countries agreed to legally binding emissions reduction commitments. This asymmetry in commitments was due to the principle of "common by differentiated responsibility". The CBDR principle recognizes that all nations are responsible for addressing the climate change problem, yet not all are equally responsible (Yu and Zhu, 2015).

Under the Kyoto Protocol, the 38 industrialized countries were categorized as Annex I, and had QELROs. The remaining signatories to the Kyoto Protocol were classified as non-Annex I countries, and had no QELROs, which were legally binding GHG emission reduction targets.

The "Annex classification" created a problem as the Annex I (developed) countries eventually became annoyed that the more industrialized non-Annex I countries had no GHG emission reduction commitments. More specifically, the United States (US), an Annex I country, was disgruntled that China, a non-Annex I country, had no GHG emission reduction commitments. The GHG emission reduction commitments were a big issue since imposing various measures to reduce GHGs would result in higher costs of production. This in turn would make several businesses less price competitive in international trade. China, in contrast, had no GHG emission reduction commitments, and thus was being perceived by the US as being given a free pass to pollute the environment while competitively competing in international trade. As a result of the Annex classification deadlock, the US withdrew from the Kyoto Protocol in 2001.

The next notable milestone was the fifteenth Conference of the Parties (COP 15) held in Copenhagen, Denmark in 2009. This conference was supposed to facilitate a successor agreement to the Kyoto Protocol, but it broke down due to disagreement between the Annex I and non-Annex I countries. At the seventeenth Conference of Parties (COP 17) in Durban, South Africa, in 2011, parties approved an extension of the Kyoto Protocol. This was signed in 2012 as the "Doha Amendment to the Kyoto Protocol". Another major outcome of COP 17, was the agreement by Parties to negotiate a successor agreement to the Kyoto Protocol by 2015 (CFR, 2013).

Several things changed between 2011 and 2015.

- First, was the change in China's position on climate change. This was driven by the occurrence of record-setting heatwaves, deadly flash floods, and typhoons have raised public concern about climate change risk in China, subsequently causing the Chinese government to adopt a proactive stance on climate change (Jing 2018).
- Second, the US government, under the Obama administration, managed to persuade the BASIC group of countries (Brazil, South Africa, India, and China) to consider adopting voluntary GHG

emission reduction commitments based on their country's circumstances. China committed to reducing the carbon intensity by 40% of 2005 levels by 2020, while India set a target of ranging from 20% to 25% over the same period (Siddiqi 2011).

- Third, the nineteenth Conference of the Parties (COP 19), in Warsaw, Poland, introduced the concept of voluntary GHG emission reduction commitments, referred to as Intended Nationally Determined Contributions (INDCs). The INDCs were to be used to collectively stimulate GHG emission reduction action by all countries. COP 19 also requested that the INDCs be communicated "well in advance" of the twenty first Conference of the Parties (COP 21) in Paris, France. Voluntary GHG emission reduction commitments were very important criteria for the US since its Congress was opposed to any legally binding climate change agreement (Clémençon, 2016).

Finally, in 2015, the international community successfully negotiated the Paris Agreement. They abandoned the idea that the major countries should be forced to reduce their GHG emissions, in favor of voluntary commitments.

The Paris Agreement was based on the overall target of limiting global temperature rise to well below 20 C by the end of the century. Upon the ratification of the Paris Agreement, countries' INDCs became their Nationally Determined Contributions.

As the Paris agreement is relatively new, there is an absence of research on the economic impact of implementing the NDCs. The economic impact of the NDC implementation is very important as it will affect several productive sectors in each implementing country. Furthermore, there will be a cost attached to the NDC implementation. While developed countries such as the US, Canada, and European countries are likely to have considered the economic impacts in their crafting of NDCs, Caribbean SIDS facing limitations in technical capacity are unlikely to have considered these assessments. Dominica, a CARICOM Member State with geothermal energy potential, has identified the energy, transport, forestry, and waste sectors as areas for NDC implementation, but has not conducted an economic impact assessment. This study intends to fill this gap.

The next section will review the climate change problem in Dominica.

CLIMATE CHANGE IN DOMINICA

Dominica's climate can be described as tropical maritime. A dry season occurs from December to May, and a rainy season occurs from June to November. The annual rainfall total is approximately 10,000mm (400 inches), but it becomes higher during the rainy season and lower in the dry season. The island also resides within the Atlantic hurricane belt, causing it to be frequently impacted by hurricanes (UNFCCC, 2015b). The most recent hurricane to impact the country was Maria in 2017, and it caused US $1.3 billion in damages and loss, which was approximately 226% of the GDP (GCoD, 2018). Table 1 displays the tropical cyclones and hurricanes1 that have been recorded to have affected Dominica from the 1900s to 2021.

The country is also vulnerable to sea level rise, a slow onset event, which threatens coastal communities (such as Concorde, Hatton Garden, Marigot, Wesley, Woodfordhill, Calibishie, Hampstead, Bense, Anse de Mai, and Anse SolDat) with coastal erosion, increased flood risk and in some areas permanent loss of land. This threat is a serious issue for the country given that out of a population of approximately 72,000 people, 90% reside in coastal villages (USDS, 2010).

Table 1. Tropical cyclones and hurricanes having affected Dominica (1930 – 2021)

Year	Disaster Subtype	Event Name
1930	Tropical cyclone	
1963	Tropical cyclone	Edith
1970	Tropical cyclone	
1979	Tropical cyclone	David
1979	Tropical cyclone	Frederick
1980	Tropical cyclone	Allen
1984	Tropical cyclone	Klaus
1989	Tropical cyclone	Hugo
1995	Tropical cyclone	Marilyn
1995	Tropical cyclone	Luis
1999	Tropical cyclone	Lenny
2007	Tropical cyclone	Dean
2011	Tropical cyclone	Tropical storm Ophelia
2015	Tropical cyclone	Hurricane Erika
2017	Tropical cyclone	Hurricane Maria

With regards to the economic impacts of climate change, two sectors immediately draw attention. The first is the agriculture sector, which historically was the main export sector for the economy. However, the erosion of trade preferences to the European market from the 1990s to the 2010s caused many people to exit the farming industry. In 1995 following the conclusion of the Uruguay Round of the General Agreement on Tariffs and Trade (GATT) the World Trade Organization (WTO) was formed. The WTO created rules on non-discrimination, namely most favored nation (MFN) and national treatment, to encourage fair trade. MFN treatment requires equal treatment amongst all members of the WTO. National treatment required that the products of locals and foreigners be treated equally (VanGrasstek, 2013). The application of these two principles to the African, Caribbean and Pacific (ACP) countries' trade with the European Union (EU) would require both regions to engage in reciprocated preferential trade.

In 1996, Several multinational banana corporations (such as Dole and Chiquita), operating in Latin American countries, filed complaints against the ACP countries' trade with the EU under the Lomé agreements. They argued that the Lome agreements were discriminatory and granted the ACP countries and unfair trading advantage against the EU's other trading partners. The WTO eventually ruled in favor of the Latin American countries. Subsequently, the EU and the ACP countries were granted a waiver on the reciprocity to continue into Contonou, Lomé's successor, but were required to negotiate a new reciprocated preferential trade agreement by the expiration of Contonou (Greenaway & Milner, 2003).

The preferences were also eroded as the banana protocol and its corresponding preferential prices were also dismantled. The Latin American stakeholders also challenged the banana protocol, and after the WTO ruled that it was discriminatory, the EU moved towards ending the system. Additionally, the hurricanes and tropical cyclones that impacted the country (Marilyn, Luis, Lenny, Dean, Orphelia, Erika, and Maria) in the aforementioned period wiped out the banana crops (the main export crop) and destroyed infrastructure networks such as roads and storage facilities.

Table 2. Dominica's GHG emissions (Gg) 1994 - 2005

	1994	2000	2001	2002	2003	2004	2005
I. Energy	72.6	106	118	113	111	111	119
A. Fuel Combustion	0.0	106	118	113	111	111	119
1. Energy Industries	20.2	34.9	40.3	33.1	36.1	34.1	41.8
2. Manufacturing and Construction	4.1	9.1	10.1	10.5	13.2	11.6	11.0
3. Transport	37.4	47.1	53.7	55.4	46.8	42.9	46.8
4. Other Sectors	10.8	14.7	13.8	14.3	15.0	22.3	19.5
a. Commercial	7.33	10.0	9.43	9.91	10.7	18.4	15.4
b. Residential	3.41	4.07	3.67	3.59	3.42	3.08	3.29
c. Agriculture	0.100	0.631	0.721	0.721	0.766	0.811	0.766
II. Land-Use Change and Forestry							
Changes in Forestry & Other Woody Biomass Stocks (Removals)	-355	-198	-198	-193	-192	-190	-188
Forest and Grassland Conversion	26.5	60.4	60.4	60.4	60.4	60.4	60.4
Abandonment of Managed Lands	-43.7	0	0	0	0	0	0
CO_2 Emissions & Removals from soils	0.37	0	0	0	0	0	0

The second industry is the fishing sector. Many former farmers shifted to the fishing industry to earn an income. Dominica's fishing industry is artisan, with all the fish being caught for local consumption. The fishermen generally sell their fish to the public at the landing sites. However, in the aftermath of a tropical storm or hurricane the fisheries storage infrastructure is usually damaged and immediately unusable. This results in wastage, and the loss of income for the small scale fishermen. Therefore, the small-scale entrepreneurs' livelihoods are under threat.

Dominica's contribution to climate change is close to nil, as its anthropogenic GHG emissions are very low, and the country's vast forestry resources are covering 63% of the total land area, and are removing the emissions produced mainly from the electricity generation sector. Table 2 displays the GHG emissions by sector.

Dominica has no hydrocarbon reserves, and thus it must import all its fuels for energy production. Dominica Electricity Services Limited (DOMLEC) is the sole electric utility company in the country, with an electrical generating capacity of 23.8 megawatts (MW) and a peak demand of 17.2 MW. The total demand for electricity in Dominica is 103.60 million kWh per year. The per capita demand is 1,439 kWh per year (World Data 2021a). The average cost of generating electricity in Dominica is US $0.39 per kilowatt/hour (kWh), which is higher than the Caribbean regional average of US $0.33/kWh (ETI 2015).

Dominica, being a volcanic island, has a potential to generate electricity from geothermal sources. The Roseau Valley area has the potential for 120 MW (DNO 2021). However, the entire country has a geothermal potential of more than 1390 MW (Koona et al. 2020). There is also potential from energy generation from other renewable sources such as solar, wind, and biomass resources.

Indeed, Dominica is vulnerable to climate change, and its NDCs represent a courageous step along with the international community to implement climate mitigation action. The next section will review the methodology for estimating the economic cost of implementing the NDCs.

Methodology for Estimating the Cost of the Implementing the NDCs

Economic impact analysis is a methodology that estimates the effect of an event on the economy in a specified area. It is useful in showing the impact of a policy, investment, project or event on stakeholders' jobs, income and costs, and the wider economy (Crompton et al., 2001).

Notably, each economic impact analysis needs to be tailored to suit the research problem at hand, and measure the potential economic impact of one event upon other economic variables.

The economic impact of NDC measures in Dominica is essentially the quantification of the cost of the measures to reduce the GHG emissions in the energy sector by 98.6%, in the transport sector by 16.9%, in the manufacturing and construction sector by 8.8%, in the forestry sector by 8.1%, and in the solid waste sector by 78.9%.2

The economic impact analysis will be done by estimating the economic impact of the actions required to reduce the GHG emissions in the energy, transport, forestry, and solid waste sectors.

Energy

To achieve the reduction in the GHG different interventions will be required. In the area of energy, to facilitate the reduction of the emissions, it would require the fuel switching from electricity generation from diesel sources to geothermal sources. This fuel switching would require a cost for the national geothermal power generation company in Dominica – the Dominica Geothermal Development Company Ltd. (DGDC).

The Government of the Commonwealth of Dominica (GoCD) has expressed interest in developing its geothermal energy potential. In fact, the GoCD has secured funding from the World Bank to develop a 7 MW geothermal facility at the cost of US $40 million (DGDC 2018; Richter 2018).3 This US $40 million represents the upfront capital costs.

A loan amortization schedule is prepared for the geothermal energy project. The time period for loan for the project is assumed to be 10 years. The frequency of payments is assumed to be quarterly, as the government collects revenues on a quarterly basis and so, quarterly payments would better align to its revenue flows than monthly payments. The discount rate for the project is assumed to be 8%, which is in line with the commercial bank deposit rate in the country.

Transport

The GHG emissions of the transport sector can be reduced through the fuel switching of the bus fleet from gasoline and diesel to compressed natural gas (CNG). This strategy can reduce the GHG emissions produced by the existing vehicles. However, Dominica presently does not import CNG. It does not have natural gas import facilities, nor CNG gas stations.

The cost of a floating storage and regasification unit (FSRU) can range between US $250 million to US $400 million depending on the regasification capacity and onboard storage. FSRUs have been developed more recently and offer substantial cost-savings relative to land-based liquified natural gas (LNG) regasification terminals (USDOE 2018). A CNG fueling station can cost between US $1.2 million to US $1.8 million (USDOE 2014; WTS 2015).4

The 4 main private bus companies in Dominica are Bávaro Express, Caribe Tours, Metro Tours and Terra Bus. They all have comfortable 52-seat buses with air-conditioning. Retrofitting the bus fleet from

Table 3. Motor vehicles registered in Dominica 2002 to 2020

Year	Number of Motor Vehicles Licensed During the Year (1)							
	Private Cars (2)	Taxis	Buses	Motor Cycles (3)	Trucks (4)	Sport Utility Vehicles SUV's	Tractors (5)	Total
2002	5,413	228	1593	656	2,315	2,156	71	12,432
2003	5,470	243	1693	775	2,414	2,110	77	12,782
2004	5,705	258	1838	851	2,525	2,330	80	13,587
2005	6,284	316	2145	933	2,748	2,602	84	15,112
2006	6,624	336	2239	1,020	2,918	2,709	85	15,931
2007	9,216	338	1298	254	2,217	3,735	24	17,082
2008	9,144	320	1334	375	2,233	3,753	21	17,180
2009	8,995	312	1356	317	2,217	3,563	34	16,794
2010	9,151	328	1385	328	2,212	3,402	35	16,841
2011	9,136	305	1375	348	2,160	3,299	34	16,657
2012	9,659	353	1359	407	2,300	3,609	41	17,728
2013	9,776	304	1,456	332	2,243	3,303	34	17,448
2014	9,287	298	1,414	249	1,997	3,035	28	16,302
2015	10,135	339	1,534	265	1,959	3,021	33	17,286
2016	10,752	309	1613	256	1961	2,846	30	17,767
2017	11,943	298	1,681	354	2,112	2,676	45	19,109
2018	11,256	244	1,554	339	2,044	2,054	56	17,491
2019	12,463	264	1,547	351	2,124	1,965	61	18,775
2020	12,457	235	1,445	325	2,081	1,706	62	18,311

diesel-fuel to CNG-fuel carries a cost. Table 3 shows the number of registered vehicles in Dominica over the 2002 to 2020 period.

The total number of buses that were registered in Dominica over the 2002 to 2020 period was 29,859. Therefore, there should be at least 29,859 buses in Dominica. The cost of converting a bus from diesel-fuel to CNG-fuel is unknown. However, the cost of converting a bus from diesel to CNG in Trinidad and Tobago (T&T), another CARICOM Member State, is TT$10,000 or US $1,500. This cost can be used as a proxy for the average cost of retrofitting a bus from diesel-fuel to CNG-fuel in Dominica. The total cost of retrofitting the bus fleet can be the product of the number of buses and the cost of retrofitting. This should be US $1,500 * 29,859 = US $44,788,500.

Forestry

In the forestry sector, CO2 emissions can be reduced through reforestation, and the conservation of forestry resources. As previously mentioned, Dominica forestry resources cover approximately 63% of the country's total land area, and can sequester as much as 188 gg of GHG emissions in a year.5 Therefore, utilizing of the forestry sector to remove GHG emissions would involve the conservation of the forestry resources.

Dominica's forestry resources are under threat as deforestation and unsustainable land use practices have resulted in the loss of 10% of its forest cover (approximately 5,000 ha) over the 1990 to 2010 period. Recognizing the importance of the forestry resources, the GoCD joined the Reduced Emissions from Deforestation and Forest Degradation plus (REDD+) where it agreed to pursue voluntary action to conserve its forestry resources to reduce GHG emissions (GoCD 2018). Approximately 20% of the forested area is designated as protected areas. However, conservation activities such as the enforcing of regulation to prevent deforestation, patrols, reforestation and afforestation, etc. are required to effectively conserve the forestry resources.

The Dominica Forestry, Wildlife and Parks Division, which falls under the jurisdiction of the Ministry of Blue and Green Economy, Agriculture and National Food Security is responsible for the conservation of the forest resources. Financial resources for forestry conservation in Dominica come largely from the government's recurring budget, which covers the cost of salaries and operating expenses, as well as some conservation projects.6 There were 23 forestry officers in Dominica in 2008 (Mongabay 2021). This would effectively result in each forestry officer patrolling 2,173 ha of forest area.7 A more manageable feat would be if each forestry officer is required to patrol only 500 ha of forest area. This would require 100 forestry officers.8

The wage of a forestry officer could be US $24,000 per annum. Therefore, the cost of hiring an additional 80 forestry officers is US $1,920,000 per annum.9

Solid Waste

In the solid waste sector, carbon dioxide is generated through the incineration of municipal solid waste. However, carbon dioxide and methane emissions from landfills can be reduced through the implementation of sustainable landfills.

Moreover, energy from waste (EfW) can be generated from the capturing of the methane gas and turning it into electricity. The sustainable landfill can produce energy from waste, through processing in 4 stages: i) the sorting and drying of the feed material; ii) the thermal cracking of the carbonaceous material in the gasifier to produce syngas; iii) the processing of the syngas to remove pollutants; and iv) the application of an energy recovery system such as a gas engine (Seo et al. 2017).

This sustainable landfill will carry a cost. A sustainable landfill with a 35 MW plasma gasifier EfW plant can cost US $127.528 million (Singh et al. 2009).

The next section presents the estimated economic impact of implementing the NDCs in Dominica.

Results and the Economic Impact of Implementing the NDCs

Energy

A loan amortization schedule is prepared for the geothermal energy project. The schedule is displayed in Table 4.

Based on the assumption for the discount rate, the loan duration, and the frequency of payout, the cost of implementing the geothermal energy project will be US $1,462,230 per quarter. The total interest that would be paid is US $18,489,196.48, and the effective interest rate for the loan would be 46%. The summary statistics of the loan is presented in Table 5.

Table 4. Loan amortization schedule for the geothermal project

PmtNo.	Beginning Balance	Scheduled Payment	Total Payment	Principal	Interest	Ending Balance	Cumulative Interest
1	$40,000,000.00	$1,462,230	$1,462,230	$662,229.91	$800,000.00	$39,337,770.09	$800,000.00
2	$39,337,770.09	$1,462,230	$2,924,460	$675,474.51	$786,755.40	$38,662,295.58	$1,586,755.40
3	$38662295.58	$1,462,230	$4,386,690	$688,984.00	$773,245.91	$37,973,311.58	$2,360,001.31
4	$37973311.58	$1,462,230	$5,848,920	$702,763.68	$759,466.23	$37,270,547.90	$3,119,467.54
5	$37270547.9	$1,462,230	$7,311,150	$716,818.95	$745,410.96	$36,553,728.94	$3,864,878.50
6	$36553728.94	$1,462,230	$8,773,379	$731,155.33	$731,074.58	$35,822,573.61	$4,595,953.08
7	$35822573.61	$1,462,230	$10,235,609	$745,778.44	$716,451.47	$35,076,795.17	$5,312,404.55
8	$35076795.17	$1,462,230	$11,697,839	$760,694.01	$701,535.90	$34,316,101.16	$6,013,940.46
9	$34316101.16	$1,462,230	$13,160,069	$775,907.89	$686,322.02	$33,540,193.27	$6,700,262.48
10	$33540193.27	$1,462,230	$14,622,299	$791,426.05	$670,803.87	$32,748,767.23	$7,371,066.35
11	$32748767.23	$1,462,230	$16,084,529	$807,254.57	$654,975.34	$31,941,512.66	$8,026,041.69
12	$31941512.66	$1,462,230	$17,546,759	$823,399.66	$638,830.25	$31,118,113.00	$8,664,871.94
13	$31,118,113	$1,462,230	$19,008,989	$839,867.65	$622,362.26	$30,278,245.35	$9,287,234.20
14	$30278245.35	$1,462,230	$20,471,219	$856,665.00	$605,564.91	$29,421,580.34	$9,892,799.11
15	$29421580.34	$1,462,230	$21,933,449	$873,798.31	$588,431.61	$28,547,782.04	$10,481,230.72
16	$28547782.04	$1,462,230	$23,395,679	$891,274.27	$570,955.64	$27,656,507.77	$11,052,186.36
17	$27656507.77	$1,462,230	$24,857,909	$909,099.76	$553,130.16	$26,747,408.01	$11,605,316.51
18	$26747408.01	$1,462,230	$26,320,138	$927,281.75	$534,948.16	$25,820,126.26	$12,140,264.67
19	$25820126.26	$1,462,230	$27,782,368	$945,827.39	$516,402.53	$24,874,298.87	$12,656,667.20
20	$24874298.87	$1,462,230	$29,244,598	$964,743.93	$497,485.98	$23,909,554.94	$13,154,153.18
21	$23909554.94	$1,462,230	$30,706,828	$984,038.81	$478,191.10	$22,925,516.13	$13,632,344.28
22	$22925516.13	$1,462,230	$32,169,058	$1,003,719.59	$458,510.32	$21,921,796.54	$14,090,854.60
23	$21921796.54	$1,462,230	$33,631,288	$1,023,793.98	$438,435.93	$20,898,002.56	$14,529,290.53
24	$20898002.56	$1,462,230	$35,093,518	$1,044,269.86	$417,960.05	$19,853,732.69	$14,947,250.58
25	$19853732.69	$1,462,230	$36,555,748	$1,065,155.26	$397,074.65	$18,788,577.44	$15,344,325.23
26	$18788577.44	$1,462,230	$38,017,978	$1,086,458.36	$375,771.55	$17,702,119.07	$15,720,096.78
27	$17702119.07	$1,462,230	$39,480,208	$1,108,187.53	$354,042.38	$16,593,931.54	$16,074,139.16
28	$16593931.54	$1,462,230	$40,942,438	$1,130,351.28	$331,878.63	$15,463,580.26	$16,406,017.79
29	$15463580.26	$1,462,230	$42,404,667	$1,152,958.31	$309,271.61	$14,310,621.96	$16,715,289.40
30	$14310621.96	$1,462,230	$43,866,897	$1,176,017.47	$286,212.44	$13,134,604.48	$17,001,501.84
31	$13134604.48	$1,462,230	$45,329,127	$1,199,537.82	$262,692.09	$11,935,066.66	$17,264,193.93
32	$11935066.66	$1,462,230	$46,791,357	$1,223,528.58	$238,701.33	$10,711,538.08	$17,502,895.26
33	$10711538.08	$1,462,230	$48,253,587	$1,247,999.15	$214,230.76	$9,463,538.93	$17,717,126.02
34	$9463538.931	$1,462,230	$49,715,817	$1,272,959.13	$189,270.78	$8,190,579.80	$17,906,396.80
35	$8190579.798	$1,462,230	$51,178,047	$1,298,418.32	$163,811.60	$6,892,161.48	$18,070,208.40
36	$6892161.482	$1,462,230	$52,640,277	$1,324,386.68	$137,843.23	$5,567,774.80	$18,208,051.63
37	$5567774.8	$1,462,230	$54,102,507	$1,350,874.42	$111,355.50	$4,216,900.38	$18,319,407.12
38	$4216900.384	$1,462,230	$55,564,737	$1,377,891.90	$84,338.01	$2,839,008.48	$18,403,745.13
39	$2839008.479	$1,462,230	$57,026,967	$1,405,449.74	$56,780.17	$1,433,558.74	$18,460,525.30
40	$1433558.737	$1,462,230	$58,489,196	$1,433,558.74	$28,671.17	$0.00	$18,489,196.48

Table 5. Loan summary statistics

Scheduled Payment	$1,462,230
Scheduled Number of Payments	40
Total Interest	$18,489,196.48
Loan Amount	$40,000,000.00
Total Repaid	$58,489,196.00
Annual Interest Rate	8%
Quarterly Interest Rate	2%
Loan Period in Years	10
Number of Payments Per Year	4
Effective Interest Rate	46%

The geothermal energy project will carry benefits as it would result in cost savings relative to a diesel fired electricity facility. This is because diesel fuel will not have to be imported to run the facility. While the operational cost for the facility is unknown at this point, a geothermal plant typically carries an operational cost ranging from US $0.01 to US $0.03 per kWh (GTO 2021). In comparison, the operational cost of the diesel fired facility is US $0.39 per kWh. Therefore, the cost savings from the geothermal energy facility will be at least US $0.36 per kWh. When multiplied by the load of 45,990,000 KWh,10 it can generate cost savings of US $16,556,400 per year.

Transport

The total cost of accessing the CNG through a FSRU and the CNG station could be:
US $400 million + US $1.8 million = US $401.8 million. This cost will be incurred by the government as it would be responsible for the sourcing of the LNG.

The total cost of retrofitting the bus fleet to CNG should be US $1,500 * 29,859 = US $44,788,500. This is the total cost that could be borne by the private sector bus companies.

The fuel switching from diesel to CNG will carry cost savings for the operators of the buses. In Dominica, the fuel cost at the pump fluctuates as the fuel is imported. The fuel can cost as high as EC$15.44 (US $5.71) per liter (DNO 2011).11 While there is presently no CNG used in Dominica, the CNG cost at the pump in T&T is TT$1 per liter. Given that the natural gas would have to be imported, a cost US $1 per liter can be assumed as the CNG cost at the pump. Therefore, the use of CNG will generate cost savings of US $4.71 per liter for the consumer. On average, a motor vehicle can travel 12.5 km per liter. If the average bus travels 20,000 km per year, then the cost savings for the bus would be 20,000 * 4.71 = US $94,200. The savings from the entire fleet would be US $94,200 * 29,859 = US $2,812,717,800.

For the private sector, the benefits derived from fuel switching outweigh the costs. The US $44,788,500 retrofitting costs is an upfront fixed cost that must be incurred in order to access the potential benefits. The operational costs are so low that the retrofitting project will generate cost savings.

In comparison, the total cost of procuring a FSRU and developing at least 1 CNG station is relatively high. If the social benefit, which includes the benefit of the private sector bus companies, is considered, then the CNG project would be beneficial for the economy as a whole.

Apart from fuel switching to CNG, the GHG emissions can be addressed through the utilization of hybrid vehicles. Hybrid vehicles use gasoline as a fuel. However, these vehicles are more fuel efficient than traditional gasoline powered vehicles. They burn less gasoline fuel during commuting and thus produce less GHG emissions. Therefore, the Government of the Commonwealth of Dominica can implement a policy to encourage the importation of hybrid vehicles.

Dominica's NDC highlights a target of reducing GHG emissions from the transport sector by 16.9%. Ideally, this could be achieved by reducing the GHG emissions from all vehicles by 16.9%. Table 3 shows the number of registered vehicles in Dominica over the 2002 to 2020 period. The total number of registered vehicles in the country over the 2002 to 2020 period is 314,615. 16.9% of 314,615 is 53,170. Therefore, the target should be to remove the absolute emissions produced by 53,170 vehicles.

Forestry

The cost of hiring additional forestry officers to provide sufficient conservation of the forest resources in Dominica is estimated to be US $1,920,000 per annum.

Solid Waste

The cost of creating a sustainable landfill with a 35 MW plasma gasifier to transform the carbon dioxide and methane emissions into energy can be approximately US $127.528 million. If loan financing is used under the same assumptions of an 8% discount rate, quarterly frequency of payments, and a 10-year maturity, it would result in an annuity payment of US $4,661,881.41 per quarter or US $18,647,525.62 per annum.12

However, a waste to energy facility can generate revenue. Assuming a tippage fee13 of US $20 per ton, and a waste use of 21,000 tons. According to De Cub et al. (2008) Dominica produces at least 21,000 tons of municipal solid waste annually. Per annum, approximately US $420,000 in revenue will be generated.

Summary of the Economic Impact of Implementing the NDCs

Summarizing the above mentioned, the energy sector may cost US $40 million which equates to US $1,462,230 per quarter and US $5,848,920 per annum; the transport sector can cost US $401.8 million for a FSRU and a CNG station, but as little as US $0 if hybrid vehicles are encouraged; the forestry sector can cost US $1,920,000 per annum to hire an additional 80 forestry officers; and the solid waste sector can cost US $127.528 million for the development of a EfW facility, which equates to US $4,661,881.41 per quarter or US $18,647,525.62 per annum in annuity payments. The implementation of these NDCs will be affected by various factors, which will be reviewed in the next section.

Risks Critical Factors to Consider for Achieving the NDCs

There are several factors which the Government of the Commonwealth of Dominica should consider in the implementation of the NDCs. The following subsections will review these factors in each of the NDC sectors.

Risks and Critical Factors to Consider for the Energy NDC

In the energy sector, as previously mentioned, the supply of electricity in Dominica is 23.8 MW. Of this, approximately 28.6% (which equates to 6.8 MW) is generated from renewable sources, namely hydro, wind, and solar energy sources. The country has 3 hydroelectric plants, with capacities of 1.3 MW, 1.8 MW, and 3.5 MW respectively. The remaining 0.2 MW of the renewable energy is generated from wind and solar sources (ETI, 2015).

With a supply of 23.8 MW and a demand of 17.2 MW, Dominica has a potential for an excess supply of 6.6 MW. Thus, this is based on the assumption that the plant will run at 100% capacity for the entire year. In reality, the plant will occasionally shut down during the year for occasional maintenance, and there will be transmissions and distribution losses. Therefore, the capacity factor and the load factor will be less than 100%.14 The transmission and distribution loss for the electricity sector in Dominica is approximately 8.2% of the total electricity production, which equates to 1.95 MW. To match the demand, the state-owned electricity company, DOMLEC, could operate its plants at less than full capacity.

The development of a 7 MW geothermal plant has the potential to add to the existing electricity generation capacity in Dominica to 30.8 MW.15 To avoid the excess supply problem, DOMLEC would be required to reduce its production capacity by at least 7 MW.

Additionally, DGDC is likely to perform as an independent power producer (IPP) for the geothermal energy, while DOMLEC would act as the distributor of the energy to residential and industrial consumers. Given these dynamics, DGDC would require a power purchase agreement (PPA) with DOMLEC for the supply of the bulk power. For the project feasibility, the PPA would be required to cover the full loan amortization period, otherwise the project financiers would be reluctant to provide the financing. Therefore, the PPA would be for a duration of at least 10 years.

Another critical aspect of the geothermal energy project is the price of the power or the feed-in-tariff (FIT). DGDC would require a FIT that would allow them to make at least a normal profit over the loan amortization period. This would mean that the FIT would allow for the recovery of the total US $40,000,000 principal, and US $18,489,196.48 cumulative interest. Since the scheduled payment is US $1,462,230, the FIT should allow for the generation of at least the aforementioned amount in revenue. Thus, the FIT should be at least US $0.127 per KWh.16

Any FIT below US $0.127 per KWh would not allow the IPP (DGDC) to recover the costs required to service its loan obligations. Since, the DGDC would be seeking to make a profit, it would desire a FIT higher than US $0.127 per KWh. If DOMLEC desires to pay a lower FIT than US $0.127 per KWh, then DGDC would need to obtain a loan amortization duration for more than 10 years.

Risks and Critical Factors to Consider for the Transport NDC

In the transportation sector, there are several risk factors which may hinder the widespread fuel switching to CNG. First, the country does not have the infrastructure to import LNG. The upfront costs of procuring a FSRU is relatively high and the government may be reluctant to use equity and loan financing to achieve this NDC target.

Second, the imported LNG would have to be regasified and then compressed to form CNG. However, the country may lack the physical capital, technical skills, and the financing capital to develop the CNG infrastructure locally.

The upfront capital costs of accessing the CNG through a FSRU and the CNG station is relatively high. The cheaper option for the GoCD is to pursue encouraging the import of hybrid electric vehicles. In fact, a government can encourage the importation of hybrid electric vehicles at little to no cost.

For example, the government can prohibit the importation of traditional gasoline and diesel fueled vehicles that are older than 3 years, while allowing the importation of hybrid vehicles that are up to 6 years. Since older hybrid vehicles would be cheaper than the 3-year-old traditional fueled vehicles, it would encourage the importation of the hybrid vehicles. Gradually, the cheaper prices should incentivize consumers to switch from traditional gasoline and diesel vehicles to hybrid vehicles. When this is done, it will gradually result in the decreased emissions from the transport sector.

Risks and Critical Factors to Consider for the Forestry NDC

Dominica has approximately 50,000 ha of natural forest, woodland and bush (CCA, 1991). Additionally, 20% of the forested areas are designated as protected areas. Despite this, the country faces deforestation. Therefore, the forestry NDC can be achieved through the conservation of the forest resources. This can be done through the employment of more forestry officers.

Alternatively, the forest resources can be conserved through the implementation for reforestation projects. In this regard, these reforestation projects should be implemented by local non-government organizations (NGOs), and community based organizations (CBOs). This argument is made because the NGOs and CBOs may contain stakeholders that are genuinely interested in conserving the forested areas and may diligently perform the conservation activities.

In fact, the Forestry, Wildlife and Parks Division can consider an option where they can implement a project to fund the conservation activities of the NGOs and CBOs. The performance of the NGOs and CBOs can be compared to a baseline scenario where they were offered no funding. This would highlight the effectiveness of the NGOs and CBOs in performing forestry conservation in Dominica.

Risks and Critical Factors to Consider for the Solid Waste NDC

In the solid waste sector, the NDC can be achieved through the development of sustainable landfills to capture methane gas and convert it into electricity. When delving into EfW, the decision maker should consider a holistic approach which takes into consideration the sources of the waste, the types of waste produced, the selection of the best option for the conversion of the waste to energy, and the final energy carrier or by-products outcome.

Solid waste can be converted to energy through the use of different technologies, namely i) incineration; ii) gasification; iii) pyrolysis; iv) mechanical biological treatment; and v) anaerobic digestion (Cantrell et al., 2008; Chen et al., 2014; Makarichi et al., 2018; Seo et al., 2017). Out of the aforementioned technologies, incineration is the only one that should not be used to implement the NDC as it does not produce syngas17, which is used as the feedstock to generate energy.

Another factor to consider is the volume of organic solid waste. The sustainable landfill must consistently generate enough organic solid waste to produce syngas required for electricity generation. If insufficient organic waste is generated, then gasification, pyrolysis, and anaerobic digestion may not be economically feasible. Instead, the better option would be the utilization of an incineration EfW plant. If an incineration EfW plant is selected, this would produce electricity, but it would not capture the methane emissions which is the target of the NDC.

The third factor is the high upfront suck cost of a EfW plant. The US $127.528 million cost of a EfW plant is greater than the US $40 million cost of the geothermal energy plant. This is a high suck cost. If the absolute value of the capital cost is considered, the geothermal energy facility appears to be the most attractive option.18

The fifth factor is the present deficiency in the policy framework for the development of waste to energy. The limitations are summarized as follows:

- There are no fiscal incentives in the country to encourage the development of waste to energy;
- The policy and legal framework do not link waste management, environmental conservation, and sustainable energy use;
- There is a lack of current data on waste generation and composition in Dominica;
- There is a weak implementation of the regulatory framework on littering and the disposal of solid waste; and
- There is a need for sustainable funding to mobilize the required funds to cover the capital costs.

The next section will review the perspective of the residents.

Residents Perspective

Energy

As previously mentioned, the cost of the energy sector NDC may be as high as US $40 million. Quarterly, this equates to US $1,462,230. However, these costs are capital costs that will be incurred by the government, not the citizens of Dominica. The citizens of Dominica will be charged a residential rate based on the kWh of electricity that they consume.

ETI (2020) notes that the residential rate for electricity in Dominica was US $0.21/ kWh up to 50kWh in 2020. The residential rate then increases to US $0.25/ kWh after 50kWh. The per capita electricity consumption in Dominica is 1,439 kWh per annum. This equates to US $302.19 per annum (EC $815.91 per annum). This is certainly affordable for the residents of Dominica.

If the Government of Dominica pursues the geothermal energy project, it will bear the capital costs of the project, while the residents will be required to pay their monthly utility bills. Given a population of 71,991 people in 2020, this can allow DOMLEC to generate US $21,754,960.29 in revenue, which is more than the US $5,848,920 per year in the potential loan amortization.

Transport

In the transport sector, if a FSRU and a CNG station are pursued, it can cost the government up to US $401.8 million. This capital cost will not be incurred by the consumer. The cheaper option for the government would be to encourage the import and use of hybrid vehicles. Foreign used hybrid vehicles are sold at the same retail price as traditional gasoline-fueled vehicles. Therefore, the fuel switching from traditional gasoline-fueled vehicles to hybrid vehicles can be achieved at no additional costs.

Forestry

The cost of hiring additional forestry workers to achieve the forestry NDC can be as high as US $1,920,000 per annum. However, this cost will be borne by the government, and covered through its annual recurrent expenditure. Therefore, the citizens of Dominica will not directly bear this cost.

Solid Waste

The cost of developing an EfW facility to achieve the NDC target can be as high as US $127.528 million. This works out to be US $4,661,881.41 per quarter or US $18,647,525.62 per annum in annuity payments. This cost will be directly borne by the government and not the citizens of Dominica.

A critical factor to the implementation of the NDCs is the financing. The next section discusses the sustainable financing options for the NDC implementation.

Recommendations to Mobilize Financing for the NDCs Implementation

Several instruments have been identified in the conservation finance literature, and can be used to finance the achievement of the NDCs in Dominica. Some of the popular conservation finance instruments include green bonds, payments for ecosystem services, debt-for-nature swaps, and donor financing (McFarland 2021).

Green Bonds

A green bond is a debt instrument which is used to finance green projects. More specifically, a green bond is a bond, which has the typical elements of a normal bond such as the principle, coupon, and the time to maturity. However, a green bond differs from a normal bond as the monies raised are used to fund specifically environmental projects or climate change action projects (Hachenberg & Schiereck, 2018). Contrary to that, a normal bond can be used to finance any activity by a government. Since governments indicate from the onset that green bonds will be used to finance green/ environmental projects, investors that are interested in environmental conservation may purchase the green bonds and help the government raise the required funds.

Green bonds can be used by Dominica to mobilize financing to fund its NDCs. In fact, green bonds are being marketed as an appropriate policy tool for countries to mobilize financing to achieve their GHG emission reduction objectives in line with their Paris Agreement commitments (Peri, 2019). However, as green bonds are debt instruments, they essentially mean that countries are borrowing to fund their climate change action projects. Subsequently, this study argues that a green bond is not the optimal choice for Dominica as the bond will increase the debt of the country.

Payments for Ecosystem Services

Payment for ecosystem services (PES) refers to payment frameworks in which the beneficiaries of ecosystem services provide payments to encourage the continued provision of ecosystem services. Ecosystem services refer to the services that are provided by the environment, but are not traded in financial markets. Therefore, the ecosystem services do not have a traditional market value, but have a real value since they are useful to human well-being.

PES follows the "beneficiary pays principle," which is analogous to the "polluter pays principle." The latter of which forces commercial enterprises to internalize their negative externalities. In contrast, the "beneficiary pays principle" allows the beneficiaries to voluntarily make payments to encourage environmental conservation. A notable example is the PES program that is implemented by the Protected Areas Conservation Trust (PACT) in Belize.

PACT was established in January 1996 as a statutory board with a mandate to encourage and promote the conservation of the natural and cultural resources of Belize (PACT, 2011). From its inception, PACT generated revenue mainly from a US $3.75 conservation fee paid by overnight foreign visitors. The total revenue accruing from the conservation fee typically accounts for over 50% of PACT's total revenue. PACT also generates revenue from a 15% commission from cruise ship passenger head tax, 20% fee on concessions within protected areas, 20% of all recreation-related licenses, as well as donations from individuals and corporations (PACT, 2011; 2017). This blend of financing allowed PACT to be self-sufficient.

In the context of Dominica, a similar approach can be applied. Rather than using the government's present fiscal budget to finance all of its NDCs, the PES approach can be utilized to mobilize the NDC finance. This can be done through the charging of a small conservation fee, perhaps around US $20, for each tourist that travels to the country. Dominica recorded approximately 322,000 tourist arrivals in 2019 (World Data, 2021b). This conservation fee would generate US $6,440,000, which is greater than the US $5,848,920 per annum in annuity payments required to finance the geothermal energy plant.

Debt-for-Nature Swaps

Another funding mechanism which can be used to fund the NDCs is a debt-for-nature swap. A debt-for-nature swap is a financial mechanism whereby, in exchange for some degree of debt forgiveness or write-off, the debtor country agrees to allocate its debt service payments towards environmental conservation projects. In the context of climate change, a debt-for-climate swap is a debt-for-nature swap where the debt service payments are allocated towards climate change adaptation and mitigation projects.

In 2020, the debt of Dominica was US $607 million, which accounted for 108.68% of its GDP (CE 2021). In 2019, the year before the COVID-19 pandemic, the debt was US $549 million, and 94.69% of its GDP, which is still high. McLean & Charles (2018) note that public debt in developing countries becomes difficult to manage when the debt to GDP ratio exceeds 40%. Additionally, McLean & Charles (ibid) study found that the Caribbean economies' debt had a negative and linear effect on economic growth.

The high debt of Dominica, as well as its negative effect upon growth, provides a justification for the utilization of the debt-for-climate swap instrument. In fact, a debt-for-climate swap could be used to finance Dominica's NDCs.

To implement the debt-for-climate swap, Dominica should first win consensus on its use from the international community. It can lobby alongside with its neighboring Caribbean SIDS as well as the Alliance of Small Island States (AOSIS) for the use of a debt-for-climate swap instrument. Once consensus is gained, the next step would be to convince a climate finance donor to provide the funds for the debt swap. Alternatively expressed, the climate finance donor would purchase the debt from the creditors at a haircut, and the debtor government would be required to allocate its debt service payments to a trust fund which will pool the finance for the climate action projects. Of course, all stakeholders[19] must sign

Table 6. Climate finance funds

Fund	Adaptation only	Mitigation only	Both Adaptation and Mitigation
Global Environment Facility		x	
Least Developed Countries Fund	x		
Special Climate Change Fund			x
Adaptation Fund	x		
Green Climate Fund			x
Clean Technology Fund		x	
The Strategic Climate Fund			
Pilot Program for Climate Resilience	x		
Forest Investment Program		x	
Scaling-Up Renewable Energy in Low Income Countries Program		x	

a memorandum of agreement (MOA) which states their willingness and terms through which they will participate in the debt-for-climate swap.

This idea is not farfetched since the Commonwealth and the Economic Commission for Latin America and the Caribbean (ECLAC) have both lobbied for the use of debt-for-climate swaps as a policy instrument to address the twin challenge of climate change and debt in the Caribbean (Mitchell, 2015; McLean & Charles, 2018).

Donor Financing

As mentioned before, the Commonwealth of Dominica's NDCs are contingent upon the receipt of international support for both finance as well as technology. In this regard, the GoCD can consider donor grant finance to fund its NDCs. There are several international funds dedicated towards financing climate change action in countries.

The climate finance funds under the UNFCCC are the Global Environment Facility (GEF), the Least Developed Countries Fund (LDCF), the Special Climate Change Fund (SCCF), the Adaptation Fund (AF), and the Green Climate Fund (GCF) (Amerasinghe et al., 2017).

The climate finance funds outside of the UNFCCC are the Clean Technology Fund (CTF), and the Strategic Climate Fund (SCF). The SCF encompasses three further climate finance programs, namely, the Pilot Program for Climate Resilience (PPCR), the Forest Investment Program (FIP) and the Scaling-Up Renewable Energy in Low Income Countries Program (SREP) (Amerasinghe et al., 2017).

Some funds were set up to fund adaptation projects only, some mitigation only, and others both adaptation and mitigation. Table 6 provides an overview of the type of projects funded by each fund.

Each of the climate finance funds has different requirements for access. In each case, an organization representing the country must first gain accreditation. The accreditation process is rigorous, and few organizations manage to pass the entire process. This effectively results in climate finance being inaccessible for developing countries (Oxfam, 2018).

Despite this challenge, four institutions in the CARICOM region have managed to gain accreditation from the GCF. They are the:

1. Caribbean Community Climate Change Center (CCCCC);
2. Caribbean Development Bank (CDB);
3. Department of the Environment – Antigua and Barbuda; and
4. PACT – Belize.

The CCCCC has accreditation for small projects of the basic fiduciary standards, and environmental and social risk in Category B. The CDB has accreditation for small projects, all fiduciary standards, and environmental and social risk in Category A. The Department of the Environment in Antigua and Barbuda has accreditation for small projects, for all fiduciary standards, and environmental and social risk in Category B. PACT has accreditation for the basic, project management and grant award fiduciary standards, and environmental and social risk in Category C (GCF 2019; Charles 2019).

Therefore, there is hope that the GoCD can gain accreditation for one of its institutions to act as the focal point to access climate finance to fund the NDCs.

The next section concludes this chapter.

CONCLUSION

Dominica's Nationally Determined Contributions highlighted the government intention to reduce emissions in the energy sector by 98.6%, the transport sector by 16.9%, the commercial/institutional, residential, agriculture, forestry, and fishing sector by 8.1%, and the solid waste sector by 78.6%. These are ambitious goals, and would require the implementation of several projects to achieve the targets.

The energy sector NDC can be achieved through the development of the country's geothermal energy potential. The government presently has plans for the development of a 7 MW geothermal energy facility. The GoCD is considering obtaining a US $40 million loan to finance the implementation of this project. Based on the assumptions of an 8% discount rate, quarterly payments, and a 10-year maturity, the annuity payment would be US $1,462,230 per quarter and US $5,848,920 per annum. The total interest that would be paid over the 10 years would be US $18,489,196.48 resulting in an effective interest rate of 46%.

The transport sector NDC can be achieved through the encouragement of fuel switching from diesel to CNG, or through the promotion of the importing of hybrid vehicles. The fuel switching option is costlier as Dominica does not have reserves of natural gas and must import it. Additionally, the country presently does not have any infrastructure to import natural gas or create CNG. The cost of procuring the required infrastructure is US $401.8 million.

In comparison, the cost of promoting the importation of hybrid vehicles can be as low as US $0 for the government. This is because only a policy adjustment is required to affect the importation of vehicles.

The forestry NDC can be achieved through the conservation of Dominica's forests. This can be done through the employment of additional forestry officers, which can cost approximately US $1,920,000 per annum. Alternatively, the government can partner with NGOs and CBOs and fund their conservation activities. This will be beneficial to both parties, the government can achieve its target of forestry conservation and the sequestering of GHG emissions, while the NGOs and CBOs will get finance to fund their activities.

The solid waste NDC can be achieved through the implementation of an energy from waste facility to capture the methane emissions. The cost of an EfW facility with 35 MW capacity can be US $127.528 million, which equates to US $4,661,881.41 per quarter or US $18,647,525.62 per annum in annuity payments.

Collectively, the cost of the NDCs can be US $5,848,920 + US $1,920,000 + US $18,647,525.62 = US $26,416,445.62 per annum. Therefore, the economic impact of implementing the NDCs be US $26,416,445.62 per annum for the Government of the Commonwealth of Dominica.

Indeed, this is a high cost for the government. The aforementioned costs are capital costs that will not be directly borne by the citizens of Dominica. The only direct cost that the citizens may incur is the cost of electricity. However, at the utility tariff rate of US $0.21/ kWh, and an average consumption of 1,439 kWh per annum, the average utility bill will be US $302.19 per annum. This is certainly affordable for the citizens of Dominica.

There are several options for the sustainable financing of these costs. The conservation finance literature has identified green bonds, debt for nature swaps, payment for ecosystem services, and donor financing as possible financing avenues.

Bond financing can be used, but it would result in the increased debt of Dominica. A debt for climate swap may be a more attractive option, especially since the country's debt to GDP ratio was 108.68% in 2020.

The payment for ecosystem services approach is also an attractive financing mechanism. In fact, the GoCD can follow the example of PACT and impose a small conservation fee, perhaps around US $20, for each tourist that travels to the country. Given that approximately 322,000 tourists visited Dominica in 2019, this conservation fee could generate US $6,440,000 in revenue. This is less than the total cost of implementing the NDCs of US $26,416,445.62. However, it is more than the US $5,848,920 per annum in annuity payments required to finance the geothermal energy plant. If the EfW project is not pursued, the conservation fee could generate sufficient revenue.

Donor financing is also an attractive financing option. However, the rigorous accreditation process of the climate finance funds makes them difficult to access by developing countries. Given that four organizations in the CARICOM region have managed to gain some form of accreditation, it highlights that accreditation is not an impossible feat. Therefore, the GoCD can be encouraged to pursue accreditation from the climate finance funds to access finance for its NDCs.

REFERENCES

Amerasinghe, N., Thwaites, J., Larsen, G., & Ballesteros, A. (2017). *The Future of the Funds: Exploring the Architecture of Multilateral Climate Finance*. World Resources Institute.

Bulkely, H., & Newell, P. (2015). *Governing Climate Change*. Routledge. doi:10.4324/9781315758237

Cantrell, K. B., Ducey, T., Ro, S. K., & Hunt, P. G. (2008). Livestock Waste-to-Bioenergy Generation Opportunities. *Bioresource Technology*, 99(17), 7941–7953. doi:10.1016/j.biortech.2008.02.061 PMID:18485701

CCA (Caribbean Conservation Association). (1991). *Dominica Country Environmental Profile*. http://www.irf.org/wp-content/uploads/2015/10/DominicaEnvironmentalProfile.pdf

CDA (Claude Davis and Associates). (2010). *Greenhouse Gas Mitigation Assessment for Dominica, Final Report*. https://unfccc.int/sites/default/files/dominica_mitigation_assessment_final_report%5B1%5D.pdf

CE (Country Economy). (2021). *Dominica National Debt*. https://countryeconomy.com/national-debt/dominica

CFR (Council on Foreign Relations). (2013). *The Global Climate Change Regime*. https://www.cfr.org/report/global-climate-change-regime

Charles, D. (2019). Untapping the Potential of the Green Climate Fund in Transforming the Caribbean Community Member States' Renewable Energy Nationally Determined Contributions into Action. In *Towards Climate Action in the Caribbean Community*. Cambridge Scholars Publishing.

Chen, D., Yin, L., Wang, H., & He, P. (2014). Pyrolysis Technologies for Municipal Solid Waste: A Review. *Waste Management (New York, N.Y.)*, 34(12), 2466–2486. doi:10.1016/j.wasman.2014.08.004 PMID:25256662

Clémençon, R. (2016). The Two Sides of the Paris Climate Agreement: Dismal Failure or Historic Breakthrough? *Journal of Environment & Development*, 25(1), 3–24. doi:10.1177/1070496516631362

CPP (Check Petrol Price). (2021). *Welcome to CheckPetrolPrice.com*. http://www.checkpetrolprice.com/global-gasbuddy/Diesel-price-in-Dominica.php

Crompton, J. L., Lee, S., & Shuster, T. J. (2001). A Guide for Undertaking Economic Impact Studies: The Springfest Example. *Journal of Travel Research*, 40(1), 79–87. doi:10.1177/004728750104000110

De Cuba, K., Burgos, F., Contreras-Lisperguer, R., & Penny, R. (2008). *Limits and Potential of Waste-To-Energy Systems in the Caribbean*. Organization of American States.

DGDC (Dominica Geothermal Development Company Ltd.). (2018). *Geothermal project to get funding from World Bank*. https://www.geodominica.dm/news/geothermal-project-to-get-funding-from-world-bank/

DNO. (2011). *Government Has No Control Over Gas Prices – PM*. https://dominicanewsonline.com/news/homepage/news/business/government-has-no-control-over-gas-prices-pm/

DNO. (2021). *Dominica has potential to generate over 120 megawatts of electricity – Vince Henderson*. https://dominicanewsonline.com/news/homepage/news/dominica-has-potential-to-generate-over-120-megawatts-of-electricity-vince-henderson/

Dominica, C. S. O. (Central Statistical Office of Dominica). (2021). *Motor Vehicles Licensed During the Year 2002 to 2020*. https://stats.gov.dm/subjects/transport/motor-vehicles-licensed-during-the-year-2002-to-2017/

EM-DAT. (2021). *The International Disaster Database*. https://www.emdat.be/

ETI (Energy Transition Initiative). (2015). *Energy Snapshot – Dominica*. https://www.nrel.gov/docs/fy15osti/62704.pdf

ETI (Energy Transition Initiative). (2020). *Dominica: Energy Snapshot*. Accessed March 22, 2022. https://www.energy.gov/sites/default/files/2020/09/f79/ETI-Energy-Snapshot-Dominica_FY20.pdf

GCF (Green Climate Fund). (2019). *Accredited Entity Directory*. https://www.greenclimate.fund/how-we-work/tools/entity-directory.

GCoD (Government of the Commonwealth of Dominica). (2018). *National Resilience Development Strategy Dominica 2030*. https://observatorioplanificacion.cepal.org/sites/default/files/plan/files/Dominica%202030The%20National%20Resilience%20Development%20Strategy.pdf

Greenaway, D., & Milner, C. (2003). *A Grim REPA?* Leverhulme Centre for Research on Globalisation Economic Policy, University of Nottingham.

GTO (Geothermal Technologies Office). (2021). *Geothermal FAQs*. https://www.energy.gov/eere/geothermal/geothermal-faqs

Hachenberg, B., & Schiereck, D. (2018). Are Green Bonds Priced Differently from Conventional Bonds? *Journal of Asset Management*, *19*(6), 371–383. doi:10.105741260-018-0088-5

Jing, L. (2018). *Does the Chinese Public Care About Climate Change?* https://www.chinadialogue.net/article/show/single/en/10831-Does-the-Chinese-public-care-about-climate-change-

Koona, R. K., Marshallb, S., Mornac, D., McCallumd, R., & Ashtinee, M. (2020). A Review of Caribbean Geothermal Energy Resource Potential. *West Indian Journal of Engineering*, *42*(2), 37–43.

Makarichi, L., Jutidamrongphan, W., & Techato, K. (2018). The Evolution of Waste-To-Energy Incineration: A Review. *Renewable & Sustainable Energy Reviews*, *91*, 812–821. doi:10.1016/j.rser.2018.04.088

Martin, C., Elges, L., & Norwsworthy, B. (2014). *Protecting Climate Finance: An Anti-Corruption Assessment of the Global Environment Facility's Least Developed Countries Fund & Special Climate Change Fund*. https://www.academia.edu/30853719/PROTECTING_CLIMATE_FINANCE

McFarland, B. J. (2021). Blue Bonds and Seascape Bonds. In B. J. McFarlan (Ed.), *Conservation of Tropical Coral Reefs* (pp. 621–648). Palgrave Macmillan. doi:10.1007/978-3-030-57012-5_15

McLean, S., & Charles, D. (2018). Caribbean Development Report: A Perusal of Public Debt in the Caribbean and its Impact on Economic Growth. *ECLAC – Studies and Perspectives Series*, *70*, 1-39.

Mitchell, T. (2015). *Debt Swaps for Climate Change Adaptation and Mitigation: A Commonwealth Proposal*. Commonwealth Secretariat.

Mongabay. (2021). *Dominica Forest Information and Data*. https://rainforests.mongabay.com/deforestation/2000/Dominica.htm

Oxfam. (2018). *Climate Finance Shadow Report 2018*. https://d1tn3vj7xz9fdh.cloudfront.net/s3fs-public/file_attachments/bp-climate-finance-shadow-report-030518-en.pdf

PACT (Protected Areas Conservation Trust). (2011). *Strategic Plan 2011 - 2016*. https://issuu.com/pact.belize/docs/strategicplan1116

PACT (Protected Areas Conservation Trust). (2017). *Home*. https://www.pactbelize.org/

Peri, M. (2019). The Green Advantage: Exploring the Convenience of Issuing Green Bonds. *Journal of Cleaner Production*, *219*, 127–135. doi:10.1016/j.jclepro.2019.02.022

Richter, A. (2018). *World Bank Has Approved Funding of $27 Million for the Development of a 7 MW Geothermal Power Plant in the Commonwealth of Dominica in the Caribbean*. https://www.thinkgeoenergy.com/world-bank-approves-27m-for-7-mw-dominica-geothermal-project-caribbean/

Schipper, E. L. F. (2006). Conceptual History of Adaptation in the UNFCCC Process. *Review of European Community & International Environmental Law, 15*(1), 82–92. doi:10.1111/j.1467-9388.2006.00501.x

Seo, Y.-C., Alam, T., & Yang, W.-S. (2017). Gasification of Municipal Solid Waste. In Y. Yun (Ed.), *Gasification for Low-grade Feedstock* (pp. 115–136). Intechopen.

Siddiqi, T. (2011). China and India: More Cooperation than Competition in Energy and Climate Change. *Journal of International Affairs, 64*(2), 73–90.

Singh, K., Kelly, S. O., & Sastry, M. K. S. (2009). Municipal Solid Waste to Energy: An Economic and Environmental Assessment for Application in Trinidad and Tobago. *The Journal of the Association of Professional Engineers of Trinidad and Tobago, 38*(1), 42–49.

UNFCCC (United Nations Framework Convention on Climate Change). (2015a). *Background on the UNFCCC: The International Response to Climate Change*. Accessed December 13, 2015. https://unfccc.int/essential_background/items/6031.php

UNFCCC (United Nations Framework Convention on Climate Change). (2015b). *Intended Nationally Determined Contribution (INDC) of the Commonwealth of Dominica*. https://www4.unfccc.int/sites/ndc-staging/PublishedDocuments/Dominica%20First/Commonwealth%20of%20Dominica-%20Intended%20Nationally%20Determined%20Contributions%20(INDC).pdf

USDOE (United States Department of Energy). (2014). Costs Associated with Compressed Natural Gas Vehicle Fueling Infrastructure. United States Department of Energy.

USDOE. (2018). *Global LNG Fundamentals*. United States Department of Energy.

USDS (United States Department of State). (2010). *Background Note: Dominica*. https://www.state.gov/r/pa/ei/bgn/2295.htm

VanGrasstek, C. (2013). *The History and Future of the World Trade Organization*. World Trade Organization. doi:10.30875/14b6987e-en

World Data. (2021a). *Energy Consumption in Dominica*. https://www.worlddata.info/america/dominica/energy-consumption.php

World Data. (2021b). *Tourism in Dominica*. https://www.worlddata.info/america/dominica/tourism.php

WTS (Waste Today Staff). (2015). *A Look at CNG Station Project Costs*. https://www.wastetodaymagazine.com/article/rew0315-cng-fueling-costs/

WTS (Waste Today Staff). (2015). *A Look at CNG Station Project Costs*. https://www.wastetodaymagazine.com/article/rew0315-cng-fueling-costs/

ENDNOTES

1. A hurricane is a very strong tropical cyclone. However, not all tropical cyclones reach the strength of a hurricane.
2. For the purposes of this study, the construction sector is not considered as Dominica has not signaled any approach to reduce their emissions in that sector.
3. The total capital cost of the geothermal facility is US $40 million. However, the World Bank agreed to provide US $27 million in funding for the project (Richter 2018).
4. The total cost of developing a CNG fueling station depends on several factors, including the fuel demand from the fleet and other users, the fleet's applications and duty cycles, site conditions, the complexity of equipment installation, and permitting processes. These factors can cause a variation in the capital costs (USDOE 2014). Nevertheless, the range of US $1.2 million to US $1.8 million is provided as a guide for the costs.
5. The Reduced Emissions from Deforestation and Forest Degradation plus (REDD+) strategy for Dominica also indicated that Dominica's forest will continue to sequester at least 100 gg of GHG emissions annually between 2018 and 2030 (GoCD 2018).
6. The Ministry of Blue and Green Economy, Agriculture and National Food Security was allocated EC$37,166,122 or 8.70% of the budget in the fiscal year 2020-2021.
7. The total forested area in Dominica is approximately 50,000 ha. Therefore, 50,000/23 = 2,173 ha per forestry officer.
8. To estimate the number of forestry officers required, the total forested area is divided by the optimal area to be patrolled by 1 officer. Therefore, 50,000 / 500 = 100 forestry officers.
9. Notably, there are other recurrent administrative costs involved in providing forestry conservation services. However, these costs are not included as they are assumed to be already covered by the Forestry, Wildlife and Parks Division. Therefore, the US $1,920,000 represents the additional costs required to provide forestry conservation services.
10. 7 MW * 0.75 capacity factor * 24 hours * 365 days = 45,990 MWh per year. 45,9990 MWh = 45,990,000 KWh.
11. The price of diesel is EC$15.44 (US $5.71) per liter (DNO 2011). Notably, the price of diesel at the pump fluctuates as the fuel is imported and its price fluctuates in international markets.
12. This was derived by dividing the principle of US $127.528 million by the Present Value Interest Factor Annuity (PVIFA) of 27.35547924.
13. The tippage fee is the fee charged by the EfW plant. The EfW charges a tippage fee as they produce electricity from waste and sell the electricity to an electricity distribution company. In some countries, a feed-in-tariff is used in place of the tippage cost.
14. The capacity factor is the ratio of a power plant's average production to its rated capability. The load factor is the ratio of a power plant's average load to its peak load.
15. If the full supply of electricity is produced, the excess supply will grow by an additional 7 MW to 13.6 MW.
16. The US $1,462,230 * 4 months = US $5,848,920 is the payment for the year. A 7 MW plant with a 75% capacity factor should generate (7 MW * 0.75 capacity factor * 24 hours * 365 days = 45,990 MWh per year). 45,9990 MWh = 45,990,000 KWh. Therefore, the FIT should be US $5,848,920 / 45,990,000 = US $0.127 per KWh.

[17] Synthesis gas, also referred to as syngas, is a mix of hydrocarbon molecules such as hydrogen, methane, carbon monoxide, carbon dioxide, water vapour, and condensable compounds.

[18] The US $127 million EfW plant with 35 MW capacity per unit cost is US $3.6 million per MW. The US $40 million geothermal plant with 7 MW capacity per unit cost is US $5.7 million per MW. Therefore, the EfW may seem cheaper on a per unit basis.

[19] A debt-for-climate swap would involve several actors, namely the creditors, the debtor (the GoCD), and a climate finance donor.

Chapter 3
Climate Change:
The New Normal Management

Ezatollah Karami
Shiraz University, Iran

Marzieh Keshavarz
https://orcid.org/0000-0002-0284-5635
Payame Noor University, Iran

ABSTRACT

Climatic and anthropogenic changes have affected water availability, food security, poverty, and migration in many parts of the world. While there will be no return to the normal climate and living, in arid and semi-arid regions, climate change has reached a point that can be best described as a new normal. Since the past cannot provide adequate guidance for the future, to manage the new normal, past knowledge should be unlearned. New normal management is a non-linear, complex, and non-deterministic behavior that considers the non-routine and uncertain features of climate change. The potential application of new normal management was assessed through in-depth review of the literature and three cases in Iran. The findings revealed significant differences between the crisis and new normal management practices and consequences. Accordingly, new normal management is a promising approach in facing climate change. To integrate new normal management into practice, political will, mobilization of resources, unlearning and relearning, and multilateral coordination are required.

INTRODUCTION: CLIMATE CHANGE AS A NEW NORMAL

The atmospheric concentrations of carbon dioxide, methane and nitrous oxide have increased to unprecedented levels during the past 800,000 years (IPCC, 2013). Such an increase in the level of greenhouse gases has significantly affected normal weather events like daily temperatures. As a result, the world is warming at a rate three times that of the 1880s (IPCC, 2018). Also, global warming has significantly enhanced the frequency and intensity of high-temperature extremes (Figure 1), i.e., heatwaves, in some parts of the world, e.g., western and central United States, West Africa, and Australia (IPCC, 2021).

DOI: 10.4018/978-1-6684-4829-8.ch003

Climate Change

Figure 1. Number of natural global disasters by year. (Own representation based on EMDAT, 2021).

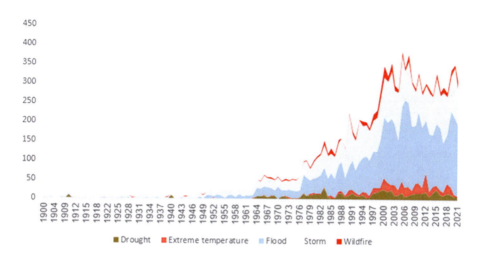

Heatwaves have caused more deaths for humans than any other extreme event or natural disaster. However, dealing with extreme heatwaves is very difficult in our changing world.

While changes in the temperature have already influenced a diverse set of physical and biological systems, the global average surface temperature is projected to rise about 2.1 to 5.7°C up by 2100, compared with the pre-industrial level (IPCC, 2021). Global warming is expected to be associated with the change of seasonality, increase in the average, maximum and minimum seasonal temperatures, increase in the frequency and intensity of heatwaves, reduction of soil moisture, and worsening of water shortage in the arid and semi-arid regions of the world (IPCC, 2021; Karimi et al., 2018).

Global warming has also led to a change in precipitation patterns. Large multi-year oscillations in precipitation have been more frequent and extreme after the late 1960s relative to the preceding decades (Figure 1), and the precipitation patterns across some regions have settled around a new norm (IPPC, 2013). The shift in the precipitation patterns declares that we can no longer rely on historical data to predict reliable rainfall patterns for our future. Moreover, the number of catastrophic events, such as floods, storms, and droughts, have significantly increased over the past four decades (Figure 1). Also, there is a considerable body of evidence, which suggests that almost every storm, drought, flood, wildfire, and heatwave are more destructive than the previous ones (Figure 2). This condition can be described as a transition to a new normal, at least from the precipitation perspective.

Warming-induced drying over the northern mid-high latitudes and ENSO-induced precipitation with drier conditions over most regions of Africa, Southeast Asia, eastern Australia, southwest United States, and southern Europe have increased the areas under drought by about 8% in the 21st century (Dai, 2013, Haile et al., 2020). Many regions, i.e., Australia, India, Indonesia, Philippines, Iran, Brazil, some parts of eastern and southern Africa, Central America, and the United States, have experienced their worst droughts in decades, which have affected hundreds of millions of people. Climate change is projected to increase the frequency and severity of droughts in arid and semi-arid regions, e.g., the central and southern United States, southern Australia, Southeast Asia, and southern Africa, by the end of this century. Also, the nature of droughts is expected to change, with a greater frequency of long lasted and extreme droughts and less frequent short-term and moderate to severe droughts (IPCC, 2021).

Figure 2. Extent of damages (1 million 2016 US$): 1964-2020 (own reperesentation on EMDAT, 2021)*

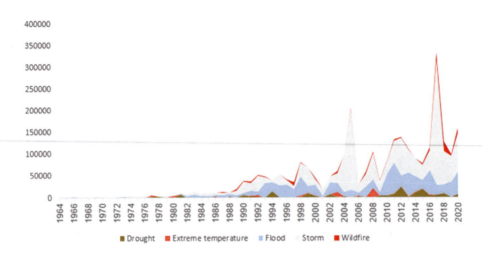

While there will be no return to the previous conditions, there is no time to minimize the climate damages in the short term under the current rate of climate change, population growth, and economic trends. Indeed, starting to harness the rapid and severe emissions would make only a tiny difference by 2050, and the great dividends would only emerge around 2100 (Arnell et al., 2013). Moreover, 15 to 40% of the global warming impacts caused by past emissions would continue for the next 1000 years. Consequently, rather than trying to recover from the past, both developed and developing countries need to accept the new normal features of the climate. This new normal condition requires adopting some coherent approaches that can explain where we have failed, where the challenges lie, and what can be done to increase resilience to the impacts of climate change and support mitigation activities.

CURRENT AND POTENTIAL IMPACTS OF CLIMATE CHANGE

Recent climate trends have directly and, or indirectly affected all-natural and human systems worldwide. However, climate change impacts on water availability, food production, and food security are more noticeable.

Climate Change Impacts on Water Availability

Approximately one-fifth of the global population has suffered from water scarcity, and about two-thirds of the world's population has faced moderate to severe water shortage for at least one month of the year (Mekonnen & Hoekstra, 2016). In arid and semi-arid regions, water resources are already stressed due to climate change and anthropogenic forces (Zarei et al., 2020). As evidence, more than 20% of water supplies are entirely or nearly dried up (Tzanakakis et al., 2020; Wreford et al., 2010), including the Colorado River in the western United States, the Yellow River in northern China, the Aral Sea in central Asia and Kaftar, Bakhtegan and Tashk lakes in southern Iran. Also, groundwater resources have little room for maneuver to expand irrigated areas in many countries, including India, Pakistan, the United

States, Iran, China, and Mexico. Moreover, in many river basins, e.g., the Ganges basin in India and Bangladesh, the Limpopo basin in Southern Africa, the Mekong basin in Southeast Asia, the Zayandeh-Rud basin in Iran, and the Murray-Darling basin in Australia, water availability and consumption are countercyclical, i.e., water consumption is highest when the lower volume of water is available for use (Mekonnen & Hoekstra, 2016; Miyan, 2015). The growth in demand for scarce resources in arid and semi-arid regions will continue to exacerbate water stress, even in the absence of climate change.

Climate change is also expected to reduce renewable surface and groundwater resources and cause significant drying in already water-stressed areas, such as the southwestern United States and Australia, the Middle East, Southeast Asia, and South Africa (CSIRO & Bureau of Meteorology, 2015; Georgakakos et al., 2014). However, in most developing countries, climate change will have an overall modest influence on future water scarcity, relative to the other drivers, including population growth, urbanization, agricultural growth, and land-use change (Georgakakos et al., 2014; Nazari et al., 2018). Therefore, even if annual rainfall increases in some water-stressed regions, it will not be promising to mitigate shortages if an increasing water demand is not curbed.

As indicated in Table 1, the number of people living under increased water resource stress will significantly rise in 2055 in northern, central and southern Africa, central Asia, western, central and eastern Europe, and central and southern America, compared with 2025. However, areas with a considerable decrease in water resource stress will be concentrated in the south and east of Asia (Table 1). The increased water availability "may not be beneficial in practice because the increases tend to come during the wet seasons and the extra water may not alleviate dry season problems" (Arnell, 2004) due to the lack of infrastructure to capture and manage this water.

Climate Change Impacts on Food Production and Food Security

Over the past 55 years, crop production has increased in various regions of the world (Table 2) due to the expansion of harvested areas or the intensification of agricultural systems. Despite significant enhancement of crop yields over five decades, climate disruptions in conjunction with socio-economic factors have generally reduced yields and nutritional value of staple cereals in arid and semi-arid regions and, also, low latitude areas, e.g., Jordan, Iran, Iraq, Syria, and Malawi. The impacts of climate disruptions are more pronounced on rainfed agriculture and fodder availability, compared with irrigated food grains (Karimi et al., 2018). Drought also has led to a significant deficiency of fodder in arid regions of Africa, India, Afghanistan, Morocco, and Mongolia, which has resulted in high mortality or livestock migration in search of fodder and drinking water.

Climate change is expected to progressively increase the annual variation in crop yields (Table 2) and livestock production in many areas (IPCC, 2018). In high latitudes, climate change may increase crop yields, although the yields may be low due to poor soil fertility and water scarcity in some regions. However, in low latitudes, it is projected to reduce crop yields progressively (Table 2). Negative impacts on crop yield may be relatively negligible up to the 2050s, but they are projected to become significantly worse by the mid-century and beyond (Table 2). These projected negative impacts will occur in the context of rapidly rising food demand. Climate change is expected to reduce fodder quality, decline in dairy production, reduce animal weight gain and exacerbate disease incidence among livestock (Porter et al., 2014).

Knowledge of the climate change impacts on crop yields and livestock production alone is not sufficient to understand food security impacts. Food security results from a complex set of natural and

Table 1. Number of people (millions) living with increased or decreased water stress for selected regions21

Regions	Scenarios of climate change plus market-oriented high growth and convergence, free trade				Scenarios of climate change plus economic growth and convergence, high environmental consciousness and technological development			
	Increased water stress		Decreased water stress		Increased water stress		Decreased water stress	
	2025	2055	2025	2055	2025	2055	2025	2055
Northern Africa	86	218	107	3	48	138	115	129
Western Africa	0	23	23	67	15	23	17	73
Middle Africa	26	65	0	0	26	36	0	0
Eastern Africa	35	13	15	35	25	100	20	19
Southern Africa	44	56	0	0	46	66	0	0
Western Asia	32	149	59	153	28	123	54	145
Central Asia	0	7	2	0	2	6	6	0
Southern Asia	186	136	207	1530	6	125	1166	1530
South-east Asia	0	0	6	6	0	0	6	6
Western Europe	110	183	12	0	38	140	2	6
Central Europe	29	80	0	0	33	59	0	0
Eastern Europe	2	15	0	0	3	7	0	0
Canada	6	7	0	0	0	7	0	0
United States	3	85	39	6	12	37	21	0
Meso-America	27	33	2	0	1	34	3	0
South America	1	46	5	19	29	46	5	6

Notes: All reported results use the HadCM3 climate model and the IPCC SRES A1 and B1 storylines. Increased/reduced water stress is defined by a change to *per capita* water availability to below/above the threshold of 1000 m^3 *per capita* per year, respectively.

social factors in which food production plays only one (albeit important) part. Most people in the world currently have enough access to food. However, in 2020 nearly 811 million people suffered from severe food insecurity, and about 2.37 billion people did not have access to adequate food (FAO et al., 2021). The vast majority of food-insecure people live in regions where agricultural productivity, knowledge, access to services, and productive resources remain low and poverty and vulnerability remain high. Table 3 points to the low food affordability and availability in the regions affected by natural hazards. This means that low resilient families did not have the capacity to smooth consumption in the face of extreme hazards. Any increase in climate extremes would exacerbate the vulnerability of food-insecure people unless significant barriers to food security can be addressed. However, all aspects of food security are projected to be influenced by climate change (IPCC, 2018).

Food security depends on food production and availability, food prices and consumption patterns, food utilization and nutrition, and the overall stability of the system (Keshavarz, 2021). Therefore, food insecurity risks are generally more significant in low-latitude areas, low-income countries, and those regions where food demand is expected to increase and the capacity to adapt remains low. Climate change is projected to increase food prices and, therefore, its affordability. While it is likely that climate change will increase food insecurity, the magnitude of the climate impacts is likely to be small in comparison with the impacts of socio-economic and technological trends, including changes in institutions and policies.

Table 2. Long-term trends in food production and projected impacts in global regions and sub-regions under future climate change scenarios

Regions/ sub-regions	harvested area change[†] (%)	Yield change[†] (%)	Yield impact (%)	Scenario[††]
World	Maize: +78.06 Rice: +38.52 Wheat: +7.79	Maize: +190.38 Rice: +148.04 Wheat: +212.7	Irrigated maize: -4, -7 Rainfed maize: -2, -12 Irrigated rice: -9.5, -12 Rainfed rice: -1, +0.07 Irrigated wheat: -10, -13 Rainfed wheat: -4, -10	A1B CSIRO, MIROC 2050
Africa, all regions	Wheat: +20.03 Maize: +136.79 Sorghum: +131.2 Millet: +75.29	Wheat: +275.42 Maize: +84.53 Sorghum: +20.71 Millet: +18.15	Wheat: -17 Maize: -5 Sorghum: -15 Millet: -10	2050
Eastern Africa	Maize: +187.67 Beans: +313.26	Maize: +73.61 Beans: +143.37	Maize: -3.1 to +15.0, -8.6 to +17.8 Beans: -1.5 to +21.8, -18.1 to +23.7	A1FI; B1 2030, 2050 HadCM3; ECHam4
Western Asia	Barley: -8.21 Wheat: + 1.18	Barley: +116.95 Wheat: +224.07	Barley: -8, +5 Wheat: -20, +18	-20%, +20% precipitation
Southern Asia	Maize: +119.55 Sorghum: -68.52	Maize: +174.07 Sorghum: +74.93	Maize: -16 Sorghum: -11 Net cereal production: -4, -10	+3°C 2050
Eastern Asia	Maize: -91.70 Rice: +3.99	Maize: +362.54 Rice: +177.16	Irrigated maize: -9.8 to -21.7 (-10.2 to -16.4) Rainfed maize: -7.9 to -27.6 (-5.6 to -18.1) Irrigated rice: -13.5 to -31.9 (+2.5 to -16.1)	+2°C -CO_2 (+CO_2)
Northern Europe, Atlantic (1), Continental (2), Mediterranean (3)	Wheat: +217.16	Wheat: +92.07	Wheat, maize, soybean: (1): -5 to +22 (2): -8 to +4 (3): -22 to 0	A2, B2 2080 HadCM3/ HIRHAM, ECHAM4/RCA3
Southern Europe, Atlantic (1), Continental (2), Mediterranean (3)	Wheat: -53.44 Maize: -42.36 Soybean: +4219.7	Wheat: +167.96 Maize: +317.69 Soybean: +339.66	Wheat, maize, soybean: (1): -26 to -7 (2): +11 to +33 (3): -27 to +5	A2, B2 2080 HadCM3/ HIRHAM, ECHAM4/RCA3
United States, Midwestern (1), Southeastern (2), Great Plains (3)	Maize: +61.39 Soybean: +773.97 Wheat: -5.56	Maize: +191.62 Soybean: +84.62 Wheat: +164.39	(1) Maize: -2.5 (-1.5) (2) Maize: -2.5 (-1.5) (1) Soybean: +1.7 (+9.1) (2) Soybean: -2.4 (+5.0) (3) Wheat: -4.4 (+2.4)	+0.8°C -CO_2 (+CO_2)
Central America	Maize: +25.35 Beans: +78.03 Rice: -19.74	Maize: +245.42 Bean: +499.43 Rice: +159.52	Maize: 0, 0, -10, -30 Bean: -4, -19, -29, -87 Rice: +3, -3, -14, -63	A2 2030, 2050, 2070, 2100
South America	Wheat: +28.89 Maize: +97.55 Soybean: +22469.5	Wheat: +183.28 Maize: +263.62 Soybeans: +153.01	Wheat: -16, -11 (+3, +3) Maize: -24, -15 (+1, 0) Soybean: -25, -14 (+14, +19)	A2, B2 2080 -CO_2 (+CO_2) PRECIS
Australia	Wheat: +87.81	Wheat: +73.40	Wheat: -15, -12	A2: Low, high plant availability Water capacity 2080 +CO_2 CCAM

[†]. Period of 1961- 2016; own representation based on FAOSTAT (2018)

[††]. - CO_2 = without CO_2 effects; +CO_2 = with CO_2 effects; CSIRO = Commonwealth Scientific and Industrial Research Organization; ECHAM4 = European Centre for Medium Range Weather Forecasts Hamburg 4; HadCM3 = Met Office Hadley Centre Climate Prediction Model 3; HIRHAM = High-Resolution Hamburg Climate Model; MIROC = Model for Interdisciplinary Research on Climate; PRECIS = Providing Regional Climates for Impact Studies; RCA3 = Rossby Centre Regional Atmospheric Model 3; Adopted from Porter et al. (2014).

Climate Change Impacts on Poverty and Migration

Climate change is expected to slow down economic growth, exacerbate poverty in most developing countries and create new poverty traps (Roy et al., 2018). In regions with high food insecurity and high inequality, e.g., Africa, wage-labor-dependent poor households that are net buyers of food are projected to be particularly affected. However, producers in food-exporting countries, such as Indonesia, the Philippines, and Thailand, would benefit from the condition (Hertel et al., 2011). Moreover, climate change is projected to increase weather risk, specifically in developing countries (Castells-Quintana et al., 2018). Highly vulnerable families can slide into chronic poverty as a result of extreme weather risks, when they are unable to rebuild their eroded assets (IPCC, 2014).

Few studies have systematically investigated migration concerning exposure to climate variability and change. However, it can be inferred from these studies that while decision making about migration had been significantly influenced by economic or other concerns in the past, there is now a good reason to perceive that environmental factors have considerably driven migration with an increase in climate variability and change, in arid and hazard-prone areas. Using a meta-regression model, Hoffmann et al. (2021) identified that regions that have been exposed to high risk of natural hazards, e.g., Latin America and the Caribbean, Sub-Saharan Africa, Middle East and North Africa, and eastern and south-eastern Asia, have experienced high levels of net migration from 1960 to 2000. Their meta-analysis suggests that migrants have tended toward leaving low productivity and climatically variable dryland and mountain ecosystems and, also, drought-prone areas to reside in more climatically resilient ecosystems.

While there is now a greater stock of migrants in the world, compared with any point in the past, climate change and environmental degradation are expected to significantly increase the displacement of people and internal migration (IPCC, 2014). Since various drivers are documented as the causes of forced migration, it is difficult to demonstrate the causal chains and links between climate change and migration. However, four possible pathways can be outlined for how climate change can affect migration. The pathways include: a) intensification of natural disasters, which often leads to short-term migration, b) increased warming, variability and anomalies of precipitation, and desertification that affect agricultural production and access to water and often cause permanent migration, c) sea-level rise, and d) competition over natural resources (Feng et al., 2010; Hoffmann et al., 2021).

ADAPTATION TO THE NEW NORMAL OF CLIMATE CHANGE

The impacts of climate change can never be decreased to zero. However, some opportunities for adaptation are present that can be mainstreamed into the existing policy programs in many cases. Historically, human societies have developed various coping strategies to deal with climatic events. However, the traditional practices that consider the regular climate may not suffice under the new normal condition, and past experiences are less reliable as a guide for managing the climate changes (Boafo et al., 2016). Therefore, more significant efforts are required to prevent the negative impacts of climate-induced hazards. However, the impacts of climate variability and change vary by region, and adaptation options have to be compatible with local geographic conditions and their socio-economic and agro-ecological characteristics. As a result, there is no single adaptation strategy that can fit all regions.

A number of the developing countries that had little capacity to adapt (e.g., the least developed countries) or were in danger of various natural hazards have pledged some financial assistance from developed

Table 3. Food security, natural resources and resilience of selected countries

Region	Country	Food security indicators[†]				Resilience[†]		
		Total food security	Affordability	Availability	Quality and safety	Total resilience	Exposure[††]	Sensitivity[†††]
Eastern Africa	Ethiopia	37.6	24.5	47.5	41.6	39.4	63.8	83.2
	Sudan	37.1	31.8	31.6	52.4	41.4	71.2	42.9
Western Africa	Chad	40.6	37.8	42.0	42.3	41.6	62.4	58.4
Sub-Saharan Africa	Angola	41.1	32.6	42.6	48.7	45.9	74.6	41.5
	Botswana	55.5	69.6	47.5	59.6	40.0	76.4	44.4
	Madagascar	40.4	36.3	41.1	39.9	47.3	77.3	78.6
	Malawi	37.3	23.6	40.9	37.1	55.9	74.2	85.4
	Mozambique	35.9	42.9	30.4	33.8	35.2	74.2	45.2
	Zambia	38.0	29.0	40.4	42.0	46.4	71.0	61.2
Middle East and North Africa	Yemen	35.7	39.3	27.6	37.4	42.1	61.3	41.3
	Syria	37.8	34.0	30.1	53.2	43.3	74.9	89.7
Central America/ Caribbean	El Salvador	59.5	65.5	59.2	63.2	45.5	69.5	67.0
	Guatemala	53.5	58.0	48.2	57.4	51.2	76.9	65.2
	Haiti	37.8	27.8	40.2	44.2	45.2	55.8	60.6
	Honduras	59.4	53.0	64.2	63.8	57.8	67.5	60.4
Asia/ Pacific	Cambodia	53.0	68.8	48.7	44.3	40.7	64.9	98.4
	Indonesia	59.2	74.9	63.7	48.5	33.0	45.5	82.7
	Laos	46.4	47.7	46.1	49.2	42.0	69.3	86.6
	Myanmar	56.7	58.9	52.2	63.0	54.7	52.9	86.8
	Pakistan	54.7	52.6	63.0	55.7	42.2	72.2	97.4

[†]. The Economist Intelligence Unit (2021)
[††]. Temperature rise, drought, flooding, and sea-level rise
[†††]. Food import dependency and dependence on natural capital

countries and have organized some adaptation plans and programs to deal with climate change. Their emphasis has been mainly devoted to the selection, design, and implementation of particular adaptation projects, such as poverty eradication, the increase of food and livelihood security, and improving agricultural productivity (Table 4). This is while a broader and more encompassing view is required. Moreover, some developing countries pay little or no attention to climate change, and their national development plans or sectoral strategies and projects are only limited to the current climatic and non-climatic risks. Even when climate change is considered, specific guidance about adapting to it is generally lacking.

In the first years following the Rio de Janeiro Conference of 1992, the developed countries perceived that adaptation was an issue for the developing countries. While the new normal features of climate and their negative consequences were rarely taken into account, it was supposed that climate change would take place gradually, and the developed countries with a greater adaptive capacity would effectively cope with the associated changes (Burton, 2011). However, high adaptive capacity does not necessarily translate into effective adaptation, and planned adaptation is imperative. Also, while the threat of climate

Table 4. Adaptation strategies

Regions	Water availability	Food production and security	Poverty
Northern Africa	- Applying supplemental irrigation of rainfed crops - Installing and maintaining wells - Constructing bunds to more effectively capture rainwater	- Conservation agriculture (e.g., Minimum tillage) - Modifying planting and fertilizing practices for crops - Providing supplemental feeding for herds/storage of animal feed - Diversifying animal species - Using different varieties (e.g., drought resistant) - Crop residue management	- Expanding and diversifying on-farm income generating activities - Selling assets - Changing cultivation techniques to decrease management costs - Developing social safety nets - Migration
Western Africa	- Water control mechanisms (Small water harvesting pits in improved yields and incomes due to improved soil moisture) - Water control mechanisms (Dry season vegetable production through irrigation to enable two crop cycles) - Construction of dams	-Changing the amount or area of land under cultivation - Using different varieties (e.g., drought resistant) - Conservation agriculture (e.g., soil protection and agroforestry) - Modifying grazing patterns for herds -Providing supplemental feeding for herds/storage of animal feed - Ensuring optimal herd size (e.g., selling of livestock) - Developing new crop and livestock varieties - Modifying planting and fertilizing practices for crops	-Commercialization of agriculture - Emerging risk transfer schemes - Developing social safety nets - Livelihood diversification - Migration - Providing access to micro-credit
Central Africa	- Water control mechanisms (Dry season vegetable production through irrigation to enable two crop cycles) - Installing and maintaining wells - Water control mechanisms (water harvesting)	- Using different varieties (e.g., drought resistant) - Diversifying crops and animals - Adjusting planting date - Modifying grazing patterns for herds -Providing supplemental feeding for herds/storage of animal feed -Changing the amount or area of land under cultivation - Ensuring optimal herd size - Build grain storage and milling facilities	- Labor migration - Selling assets - Emerging risk transfer schemes - Developing social safety nets - Livelihood diversification - Developing food security and nutrition-related safety nets
Eastern Africa	- Water control mechanisms (Earth bunds and terracing to harvest water) - Increase of water supply from groundwater pumping - Installing and maintaining wells - Establishing water distribution network for more equitable distribution - Using low-pressure drip irrigation technologies - Construction of small reservoirs	- Diversifying animal species - Conservation agriculture (e.g., soil protection and agroforestry) - Modifying grazing patterns for herds - Build grain storage and milling facilities - Developing traditional food storage technologies - Developing new or improved crop varieties) - Modifying planting and fertilizing practices for crops - Crop residue management	- Expanding and diversifying on-farm income generating activities - Expanding and diversifying off-farm employment opportunities - Establishing revolving funds - Providing access to credit - Emerging risk transfer schemes - Developing social safety nets - Migration
Southern Africa	- Water control mechanisms (Small water harvesting pits in improved yields and incomes due to the improved soil moisture) - Water control mechanisms (Dry season vegetable production through irrigation to enable two crop cycles) - Using low-pressure drip irrigation technologies - Construction of small reservoirs	- Conservation agriculture (e.g., minimum tillage) - Fertilizing practices for crops - Changing the amount or area of land under cultivation - Using different varieties (e.g., drought resistant) - Diversifying crops - Modifying grazing patterns for herds - Providing supplemental feeding for herds/storage of animal feed - Adjusting planting dates	- Commercialization of agriculture and income generation from natural resources (e.g., fuel wood) -Promote ecotourism - Creation of small, medium and microenterprises - Drought relief and agricultural subsidies - Emerging risk transfer schemes - Developing social safety nets - Livelihood diversification - Migration - Emergence of index-based insurance contracts
Northern Asia	- Supply management (e.g., improve pasture water supply)	- Improving grazing management (e.g. modifying grazing patterns for herds) - Ensuring optimal herd size - Providing supplemental feeding for herds - Diversifying crops and animal species	- Expanding access to credit - Generating alternative income - Expanding and diversifying off-farm employment opportunities - Establishing insurance system - Establishing risk fund
Western Asia	-Water control mechanisms (including irrigation and water allocation rights) - Water reuse - Increase of water supply from groundwater pumping - Demand management	- Fertilizing practices for crops - Changing cropping systems and patterns - Improving grazing management (e.g. modifying grazing patterns for herds) - Ensuring optimal herd size - Providing supplemental feeding for herds - Diversifying crops and animal species - Using different varieties (e.g., drought resistant)	- Expanding and diversifying on-farm income generating activities - Expanding and diversifying off-farm employment opportunities - Migration - Providing access to credit - Financial support to farmers
Central Asia	- Use of water-saving technology - Modernization of the existing irrigation systems - Increase of water supply from groundwater pumping	- Improving grazing management (e.g. modifying grazing patterns for herds) - Diversifying crops and animal species - Changing planting date - Crop rotation - Application of conservation tillage practices	- Labor migration - Financial support to farmers - Income redistribution
Southern Asia	-Water control mechanisms (including irrigation and water allocation rights) - Construction of reservoirs - Groundwater recharge - Increasing water productivity	- Diversifying crop varieties - Match method and timing of cultivation practices to seasonal climate - Using climate change resistant varieties	- Expanding and diversifying off-farm employment opportunities - Seasonal migration - Financial support to farmers
Eastern Asia	- Construction of reservoirs - Supplying more water resources through irrigation systems and wells -Improved management through efficiency - Water division regulation to limit water withdrawals - Improvement of water allocation policies - Establishing water permits and trade - Applying advanced water saving equipment and technologies - Implementing water price system - Water infrastructure development - Water reuse	- Changing the area of land under cultivation - Diversifying crop varieties - Application of conservation tillage practices	- Reforming economic structure - Financial support to farmers
Australia	- Flexible water allocation -Reviewing allocation rights -Supplying augmentation (water recycling, rainwater harvesting, increased storage and desalinization) - Demand management - Infrastructure upgrades - Integrated water-sensitive urban design	- Using different varieties (e.g., slower maturing, pest and drought resistant crops) - Diversifying crops - Crop residue management - Adjusting planting date - Conservation agriculture (e.g., minimum tillage)	- Establishing insurance system to transfer risk - Financial support and incentives
Western Europe	- Land management to reduce runoff - Introducing new technology such as increase of water storage capacity - Water management to improve efficiency and reduce wastage	- Changing or diversifying crop types, livestock breeds or species, and/or farm type - Introducing new technology such as improved seed and crop storage facilities - Fertilizer and pesticide management - Adjusting planting date - Implementation of early warning systems	- Establishing insurance system to transfer risk - Financial support and incentives

continued on following page

Climate Change

Table 4. Continued

Regions	Water availability	Food production and security	Poverty
Central Europe	- Applying advanced water saving equipment and technologies - Control of salt-rich groundwater level - Reducing the losses and reconversion of irrigation systems	- Ensuring optimal herd size - Implementation of early warning systems - Diversification of species in agricultural cultivations - Changing cropping systems - Crop rotations - Enhancing plant breeding activities to help the use and spreading of more resistant species	- Establishing insurance system to transfer risk - Financial support and incentives
Eastern Europe	- Provision of accurate hydro-meteorological forecast and services - Water- and energy-saving irrigation technologies - Minimization of conflicts between different end-users - Improving the water supply infrastructures' efficiency - Water quality monitoring	- Ensuring optimal herd size - Diversification of crop species - Changing cropping systems (e.g., later maturing and more fertile crops) - Change of fertilization management by utilization of organic fertilizers	- Establishing insurance system - Financial support and incentives
Northern America	- Developing integrated watershed management plans - Expanding hydrometric network - Developing and implementing a comprehensive provincial water management strategy	- Using different varieties (e.g., early maturing, pest and drought resistant crops) - Diversifying crops - Adjusting planting date	- Establishing insurance system - Financial support and incentives
Central America	- Construction of reservoirs - Increase of water supply from groundwater pumping	- Modifying planting and harvesting - Conservation agriculture (e.g., soil protection and agroforestry) - Diversifying crops - Using different varieties - Adjusting planting date - Ensuring optimal herd size - Providing supplemental feeding for herds	- Expanding and diversifying on-farm income generating activities - Establishing insurance system - Commercialization of agriculture
Southern America	- Water control mechanisms (including irrigation and water allocation rights) - Increase of water supply from groundwater pumping - Fog interception practices - Water capture - Improved management, e.g., through efficiency - Shading management system - Deficit irrigation	- Modifying planting, harvesting and fertilizing practices for crops - Using different varieties (e.g., early maturing and less water-intensive crops) - Adjusting planting date - Diversifying crops - Developing new crop and livestock varieties - Land use planning - Conservation agriculture (e.g., no-till operations and soil conservation)	- Expanding and diversifying on-farm income generating activities - Establishing insurance system - Financial support to farmers - Migration

†. Own representation based on the review of more than 50 literature.

change impacts has been recognized in developed countries, it has not led to any transformative change in actual decision making, monitoring, or preparedness efforts and subsequent support for adaptation has been low (Davidson, 2016). Meanwhile, climate change has happened faster than what was expected at the time of the Rio de Janeiro negotiations, and the greenhouse gas emissions and atmospheric concentrations have continued to rise. As a result, a fundamental movement toward planning adaptation is now observable in many developed countries, such as the United States of America, United Kingdom, Australia, Canada, and the Netherlands. However, most efforts are still in the early stage and comprehensive approach to implementing climate change adaptation is lacking.

As indicated in Table 4, various adaptive strategies have been adopted in various regions of the world. However, adaptation efforts in many regions of Asia, Africa, central and southern America tend to be isolated (Table 4). Planning for adapting to the new normal conditions and future climate change in these regions demand more structured, integrated, and flexible approaches, including making changes to the practices, policies and infrastructures that help increase the capability of managing new normal and responding to climate change. Recognizing that the past experiences are no longer sufficient guides for dealing with the current new normal and future climate change, adaptation planning is embedded in some planning processes of developed countries even though it is at a conceptual level rather than an operational one (Reisinger et al., 2014). For instance, various adaptive strategies for managing water consumption and agricultural production are adopted in Australia, Europe, and northern America (Table 4). However, implementation of specific adaptation policies, e.g., water quality concerns and diversification of land use, and also the engagement of a wider public is still limited. Furthermore, in many developed regions, e.g., North America, central and western Europe, adaptation policy is developing across all national, state, and local levels of government, but the overall pattern of response lacks clarity and cohesion (Dickinson and Burton, 2011).

In both developed and developing countries, adaptation approaches have not succeeded in the intended way, and the world has not been able to respond appropriately to an actual new normal condition. Some developed countries fare better than many developing countries in the face of adverse climate change impacts. However, the world is too connected to extricate the impacts of the current new normal conditions and near-term climate change, and developed countries can expect similar stresses on food, water, and other natural resources. Furthermore, given the recent experience of the new normal climate and the scale of the projected impacts, opportunities for adaptation are narrow (Simpson and Burpee, 2014). That is not to say that adaptation to climate change will never be achieved, but it needs much effort to overcome the adaptation challenges and incorporate adaptation into the decision making and management practices in both developed and developing countries.

The critical concerns declared in the literature regarding the ability of developing countries to adapt to climate change include a lack of comprehensive commitment to climate change adaptation, knowledge about adaptation options, access to appropriate real-time and future climate information, access to economic and natural resources, adequate infrastructure, well developed institutional capacity, accessibility of technology at various levels, accountability, financial commitments, and political will to meaningfully address climate change (Karimi et al., 2018; 2020; Keshavarz and Karami, 2018; Sharafi et al., 2020). Even in developed countries where there is an increasing awareness about climate change and an interest in comprehensive adaptation planning, there is a lack of funding, training, expertise, access to information and other resources, intra- and inter-agency coordination and mandates, as well as political opposition, competing priorities and legal obstacles hinder the engagement of the countries in adaptation (Ford and Berrang-Ford, 2011; Moser, 2011; Simpson and Burpee, 2014). Poor adaptation, overemphasizing short-term responses, or failing to project long-term consequences adequately can result in mismanagement of climate change. While mismanaging climate change in both developed and developing countries can increase the vulnerability to the present and future new normal features of climate.

CLIMATE CHANGE MANAGEMENT: CRISIS VERSUS RISK MANAGEMENT APPROACHES

The most common approach followed by both developing and developed nations is the reactive intervention relying mainly on crisis management (Wilhite, 2016). Crisis management is based on implementing actions and practices after a major catastrophic event has started and been perceived. The approach of crisis management does not try to prevent crises absolutely, and it only attempts to lessen the actual damage inflicted by a crisis (Coombs, 2015). Crisis management often results in inadequate and unsatisfactory technical and financial solutions since a) the catastrophic event is perceived as an isolated event; b) the situation is perceived to be technically, economically, and socially manageable; c) decisions are taken with little time for evaluating the optimal strategies; d) the response is costly, poorly coordinated and untimely; and e) public participation is minimal (Iglesias et al., 2007; Wilhite, 2016). Crisis management typically presents relief measures in the form of emergency assistance programs. Although emergency-response interventions provide immediate benefits for the assistance recipients, it discourages people and institutions from adopting more risk reduction practices. It is mainly because assistance may not be available to those who have employed more risk management strategies and experienced fewer damages. This emergency assistance can also foster more significant reliance on these interventions, which increases the long-term vulnerability of individuals and communities to climate change.

Some scholars have a tendency to consider crisis management as a holistic process involving the four interrelated factors a) crisis prevention, b) preparation, c) activation/response, and d) recovery and learning (e.g., Coombs, 2015; Pennington-Gray, 2017). These factors are integrated into three phases, including a) the pre-crisis phase (i.e., prevention and preparation); b) the crisis phase (i.e., activation/response); and c) the post-crisis phase (i.e., recovery and learning) (Coombs & Laufer, 2017). However, these commonly used phases are not stationary and logically identifiable as implied. Far too often, crisis prevention does not happen. Preparation, if it exists, just extends the organizational or community routines, recovery takes longer than projected and learning is minimal (Prayag, 2018). Therefore, during the last two decades, there has been an increasing shift in the focus towards a proactive or preventive approach (FAO, 2014).

The proactive approach relying mainly on "risk management" includes the measures that are prepared according to strategy planning rather than within an emergency framework. Risk management emphasizes preparedness, mitigation, prediction, and early warning (Wilhite, 2016) with appropriate planning tools and stakeholder participation. This approach utilizes predictive and diagnostic techniques for planning the required measures to prevent or minimize the climate change impacts. Such an approach includes a) scoping exercise where the context of risk assessment is established; b) risk identification; c) risk analysis where the consequences and the likelihood of climatic risks are analyzed; d) risk evaluation where adaptation or mitigation options for reduction of climate change impacts are prioritized; e) risk management or treatment where adaptation or mitigation measures are implemented in response to climate change; and f) continuous monitoring and review where the applied strategies are evaluated (Jones, 2010).

To deal with the uncertainties involved in the climate change projections and policy responses, the risk management framework is perceived as more efficient than the crisis management approach. However, management of the risks of climate change is a very complex task since the risks are uncertain, extensive, and delayed (Döll & Romero-Lankao, 2017). Moreover, the climate change risks are intertwined with many other non-climatic risks. Therefore, it is a fundamental need to characterize and quantify the uncertainty of climatic hazards and understand the societal processes determining the vulnerability of human and natural systems and diverse ways of framing risks (Pennington-Gray, 2017). Even though some developed nations, i.e., Australia, the United States of America and the Netherlands, have endorsed risk management as the leading approach for adaptation to climate change, the climate change research and assessment community had heretofore been slow in implementing the risk management approach to hedge against the potential climatic hazards and adjust the negative impacts of the new normal condition. Since risk management has not been embedded in climate change management thinking and culture, it is imperative to frame a more flexible, rigorous, reflexive, scientific, and interdisciplinary approach. It seems that new normal management is what we need to develop an appropriate capacity to manage the new normal features of climate. The argument is that traditional crisis and risk management are necessary but not sufficient in the case of the new normal condition.

NEW NORMAL MANAGEMENT: BEYOND CRISIS MANAGEMENT

Table 5 outlines the differences perceived between crisis management (CM) and new normal management (NNM) besides the possibilities for convergence of the two approaches.

Typically, CM is perceived as the management of changes that result from sudden, exceptional, and inconceivable situations and responds to these temporary crises as an isolated events (Pennington-Gray, 2017). However, NNM tends to focus on the changes that can be incremental, cumulative, or extraor-

Table 5. Comparison of crisis management and new normal management under climate change

Issue	Crisis management	New normal management
Nature of the problem	Well-known, isolated event, temporary, exceptional situation	Unknown, permanent, new normal
Time frame	Short-term focus	Short-, medium- and long-term focus
Past knowledge and experience	Learn and useful	Unlearn
Complexity	Linear, chaos, deterministic	Non-linear, highly complex, non-deterministic
Source of uncertainty	The practical inability to know the initial conditions of a system	Uncertainty is inherent in the system because of the concept of emergence
Willingness to change	Low	High
Extent of behavioral change	Low	High
Investment risk	Low	High
Goal	Make crisis bearable and tolerable until returning to the perceived past situation	Consider future opportunities, create a climate-aware society, capture emerging opportunities and strengthen resilience
Adaptive strategies	Well-known	Great challenge to determine the timing, nature and location of specific adaptive changes
Timing of adaptive response	Known	Optimal timing for adaptive responses is never obvious
Scale	Community level	Local, national and international levels
Consequences	Limited	Catastrophic
Climate change refuges	Temporary	Permanent
What can be done	Provide welfare	Mobilization, unlearn
Nature of development	Sustainable development	Alternative regenerative development
Stakeholders	Affected population	Every one
Problem solving model	Mechanistic design	Biological redesign
Planning	Demand based	Demand and supply based
Role	Solving problem	Capacity building

dinary (Prayag, 2018). While the causes and effects of new normal circumstances are unknown, NNM tends to respond to these changes in an ongoing systematic manner. Since the crisis is perceived as a low probability event, the CM approach focuses on shorter timescales and pays little or no attention to the longer-term perspectives. However, the frequency of what used to be low probability events has increased in recent decades. Hence, NNM incorporates the new normal features of climate in short-, medium- and long-term comprehensive plans and tends to terminate the hydro-illogical cycle of crisis management.

CM focuses on the existing crises and the associated strategies based on past knowledge and management strategies. However, the past knowledge and management experiences may not be adequate under the new normal condition, and may even have counter-effects. To learn about the new state of climate change, it is required to unlearn some of the normal patterns, routines, or protocols and update the existing knowledge and knowledge structures (e.g., behaviors, skills, values, or preferences) (Aledo-Ruiz et al., 2017). Unlearning can be operationalized through three different processes, including a) awareness of obsolesce rules, routines, or processes; b) relinquishing of old mistakes or partial truths that may need to be ignored; and c) relearning of new routines, rules, and patterns (Cegarra-Navarro et al., 2014).

Climate Change

CM addresses the causes and effects with linear relationships. Considering the chaos theory, CM focuses on models that assume deterministic steady-states. However, NNM emphasizes that deterministic models cannot efficiently capture the dynamic of the new normal state (Candelon et al., 2016). NNM implies that the current complicated contexts demand greater attention to reflexivity, where the causes and effects interact over time, and their feedback mechanism can lead to non-linearity. Therefore, even if we know the initial conditions, we still cannot predict the entire future. This is mainly because the strategies implemented to manage a new normal will influence the outcomes, and the possible outcome of some interactions is inherently unpredictable. Therefore, uncertainty is inherent in the complex system due to the concept of emergence.

Also, as a demand-based approach, CM tends to manage similar crises in the same ways. Moreover, individuals continue to engage in the same old behaviors rather than innovative ones. These unproductive behaviors may lead to negative consequences such as economic loss and increased vulnerability to the new normal features of climate. In contrast, the willingness to change is considerably high in NNM in order to mitigate the disaster pressure. Even for the most resilient groups or communities, when disaster shocks are absorbed, the disaster pressure changes communities substantially and sets a new normal (Overmyer, 2015). As a result, NNM most often is associated with behavioral changes to reduce the likelihood of harm. Administrators, interveners, and other individuals may experience transformation in knowledge, insights, individual and shared schemata or circumstances to respond to the disaster pressure.

As a reactionary management approach, CM ignores all risks and waits for a loss before employing countermeasures against it (Grose, 2018). While reactionary management can reduce investments risks, it is proven to be inefficient due to the depth and extent of losses that typically accompany natural crises. However, investment risk is significantly high in NNM. NNM requires considerable funds for NNM preparation and, also, development of the infrastructures that account for the increase of climate variability and change. While the risk of investment is high in this management approach, by being proactive, every loss would be less costly than when the risk is not anticipated and controlled. Moreover, the increase in investment in the new infrastructures that are stronger, more efficient, and environmentally sound (Overmyer, 2015) can help enhance resilience to unexpected dramatic risks. To increase resilience to climate change, adapting, preparing, and responding to the new normal is the main objective of NNM. This approach also takes into account other goals, including the creation of a climate-aware society, capturing the opportunities emerging in the current new normal situations, and benefiting from the future opportunities. However, the aims of CM are limiting the harm to people, property, and the environment and facilitating the overall management of a crisis to ensure a rapid and adequate response to the crisis, reduce the adverse effects on individuals and communities and make the crisis tolerable till returning to the past situation.

Although natural or man-made crises are unpredictable, they are not unexpected. Identification of the potential threats and coordination of the responses can help reduce the negative consequences of such crisis events. The strategies adopted in CM involve tackling the quickly emerging crisis with unforeseen arrival and conclusion and restricting the slowly occurring crisis that gives more time to the preparation and slower to conclude (Drennan et al., 2015). Also, a crisis is perceived as a social construct (Drennan et al., 2015). To be more specific, individuals make judgments about a crisis differently depending on their perceptions and responsibilities. Therefore, the attribution theory can be used as a guide for linking crisis events to crisis management strategies. According to this theory, willingness to adopt CM strategies is high when the crisis event is perceived as stable, and the person has high individual control or low external control. However, the great challenge facing NNM is determining the timing, nature, and location of the specific adaptive changes (Simpson & Burpee, 2014). An additional problem that arises with the new normal features of climate is that it is impossible to know what current adaptive strategies can be effective in the future or what social costs might be required in interchange for future benefits (Chorafas, 2011). While there is no rudimentary roadmap, new approaches to adaptation are required to be evaluated and monitored at the appropriate scale and timeframe. In this way, we can learn from experience and adjust the existing approaches accordingly. Moreover, the optimal timing for adaptive responses is never obvious under new normal situations. It is mainly because adaptation to new normal depends on the availability of resources, adaptation costs, individual, social, and institutional adaptive capacities, and national and international policy agendas (Simpson & Burpee, 2014). On the contrary, CM strategies are usually adopted during the acute/media-dominated phase of the crisis. However, devoting the necessary time and efforts to the pre-crisis preparation is more effective. Also, the focus of CM is often on the community level, for which pre-set crisis management plans have been developed. Nevertheless, NNM experiences are generally derived from international policy frameworks. While international policy agendas are fundamental in directing adaptive practices, policy and action at the local and national levels, where climate change impacts are being experienced, should also be considered. Therefore, to ensure effective management of the current new normal and mitigate the impacts of future climate change, great efforts and coordination are urgently needed at local, national, and international levels.

Also, it is projected that continued emissions of GHGs (IPCC, 2021) and improper management of new normal features of climate will increase the likelihood of catastrophe. If catastrophic climate events cannot be managed well, they will lead to significant harm. The significant problem is deep structural uncertainty in the science to evaluate precisely the catastrophic losses from extreme climate changes and the inability of today's management efforts to reverse these climate change impacts (Weitzman, 2009). However, based on the psychological, social-political, and technological-structural views of crisis, the consequences of a crisis can be victimization of the individuals that are physically or psychologically harmed by the crisis, destruction of social order and the commonly held values, loss of livelihood and devastation of the technological systems (Wilhite, 2016).

Moreover, the increased frequency and intensity of sudden onset crises may drive people to relocate temporarily to safer ground (Barnett & Webber, 2010). More vulnerable countries should develop and implement appropriate NNM strategies before the new normal situation reaches a crisis stage. Since catastrophic climate events develop slower than some other catastrophes, there is somewhat more chance for learning to adapt to the new normal (Weitzman, 2009). In this respect, the adaptive strategies should be prioritized with multiple advantages for people, places, and sectors and avoid increasing negative impacts on other sectors and places (Macgregor & Cowan, 2011). One strategy is the implementation of pro-climate policies at an appropriate scale and adjusting the activities that might lead to a catastrophe. Another strategy is reducing the cost of catastrophic climate events if they emerge by protecting critical infrastructures and improving early warning and emergency response capabilities (Sharafi et al., 2021). Moreover, timely relief and provision of welfare can minimize social disruption when a crisis does lead to any displacement or significant economic loss.

Also, CM focuses on applying of policies, strategies, and practices that help minimize vulnerability to crisis events and unfold crisis impacts throughout the community in the context of sustainable development. However, when people face severe environmental degradation, sustainable development is not enough, and it is required to do better than just sustaining the present situation. Therefore, NNM should focus on regenerative development. Regenerative development seeks to provide life-support services and products, and develop an economy in the ways that result in more rainforests, more fertile soils, abundant water resources, a cleaner atmosphere, and even more incredible biodiversity (Giller et al., 2021; Gosnell et al., 2019).

While CM tends to focus on crisis victims, NNM considers all actual and potentially affected populations to make hard choices in new normal situations. Also, given that timing is a vital aspect of the adaptation and mitigation of new normal features of climate (Simpson & Burpee, 2014), NNM attempts to engage numerous stakeholders in assessing the new normal situation from different angles and formulating no-regret strategies. However, in CM, a limited number of specialists intervene in the management process. Moreover, CM tends to solve real-life problems through a mechanistic approach. It is while NNM focuses on a biological approach to solve the new normal problems. Centralization of authority, formalization of procedures and practices, and specialization of functions are the main principles of mechanistic models. Whereas reduction of physical efforts and fatigues, mitigation of environmental stressors, consideration of postural factors, and increased ability to self-organize are the central tenets of the biological approach (Campion et al., 2005). While the mechanistic approach is easy to organize, it seems inefficient in situations that demand coping with rapid changes.

Also, CM tends to solve crisis-induced problems through emergency assistance programs. Although assistance programs provide immediate benefits, they can also increase the dependency of the victims on the government or relief institutes, which increases vulnerability to climate change. In contrast, NNM is not only related to sudden and gradual changes but also stability. NNM tends to enhance the capacity of all actual and potential new normal affected people and increase their resilience to the new normal situations. Capacity building demands education, training, financial resources, technological capacity, and talented human resources with expertise (Zarei et al., 2020) that are upgraded continually to meet the challenges of new normal features of climate. This is beyond the realm of CM.

MANAGING CLIMATE CHANGE: EMPIRICAL EVIDENCE

To provide some empirical evidence about the above mention theoretical issues of climate change management, examples of crisis and new normal management at micro and macro levels are provided. Since governments are slow learners and slow decision-makers, no valid macro-level case of new normal management was found to present, and only evidence for macro-level crisis management is reviewed. In the case of micro-level management of climate change, both crisis and new normal management are discussed. Farmers' management in both cases was not based on informed theoretical management concepts. Instead, they intuitively followed two different management paths to mitigate the negative consequences of climate change, which based on our conceptualization, can classify into crisis and new normal managers. In the following sub-sections, the micro-level and macro-level cases will be presented.

Micro-level Management

Data for the crisis and new normal management strategies and their consequences were collected from two extreme cases of the data reported by Keshavarz et al. (2013), which was completed by a follow-up study in November 2021.

Figure 3a. The timeline and perceived causes and effects of micro-level crisis management.

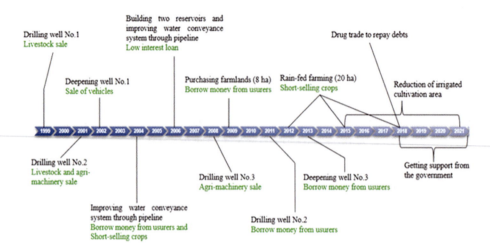

Farmer A: Crisis Management

Age: 67, Education: 8 years, Household size: 8, Farm size: 40 ha, No. of irrigation wells: 4, Main source of livelihood: on-farm income.

As a progressive farmer, he showed significant commitment and interest in farming (Figure 3). To endure the negative impacts of climate change, he adopted various strategies to extract as much water as possible (i.e., digging or deepening wells; Figure 3). However, with a crisis management mindset as well

Figure 3b.

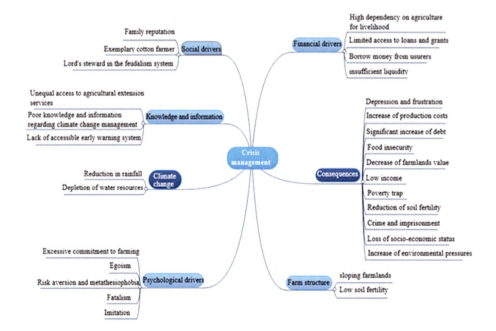

as heavy reliance on past knowledge and management strategies, the climate change impacts intensified (Figure 3). The adoption of water exploitation strategies for the second and third times indicated that he could not unlearn past knowledge and experiences even though they were ineffective. Therefore, he continued deepening or drilling boreholes by selling properties. After running out of savings, he resorted to private moneylenders to continue his utterly futile struggle to seek groundwater (Figure 3). On the other hand, the late adoption of water-saving technologies reduced the effectiveness of some adaptive strategies, such as reservoir building and improving water conveyance systems (Figure 3).

Within his management mindset, he perceived the events were isolated, temporary, and short-term, and things will be back to previous conditions soon. Therefore, he unexpectedly planned to acquire extra farmlands without enough access to water. To intensively cultivate his lands, he borrowed money from usurers and drilled and deepened his irrigation wells (Figure 3). However, he was not able to pay off his debts. Therefore, usurers were unwilling to lend him more money. Inevitably, he has reduced his irrigated cultivation area to 2 ha in the last six years, cultivating the arable lands located near the waterway. Moreover, despite precipitation deficiency, he has constantly repeated the old behavior and has cultivated about 20 ha of a rain-fed farm.

As Farmer A found himself under enormous pressure from moneylenders, he decided to sell some farmland at a depressing rate, but no one was willing to purchase it. To compensate for his past failures, he made another mistake by distributing drugs as a family business. Soon, his sons were arrested for possession of drugs (Figure 3). A series of crisis management decisions led him to dependent on government aid for survival which was socially humiliating for him.

Like a gambler, I lost all my belongings to find a drop of water'. The farmer continued 'I still taking a great risk by cultivating rain-fed crops, wishing to return to my previous magnificent life.' (November, 2021)

Farmer B: New Normal Management

Age: 65, Education: 6 years, Household size: 7, Farm size: 35 ha, No. of irrigation wells: 3, Main source of livelihood: On-farm and non-farm income.

To minimize crisis vulnerability, the government has provided numerous incentives such as technical support and meliorated access to subsidized and low-interest loans. This farmer was one of the first to use such supports (Figure 4). To increase resilience to the new situation, he has adopted several strategies, including land leveling, improvement of irrigation and water transfer systems, changing cropping patterns, and application of planting machines (Figure 4). Moreover, to reach the hard-won irrigation water sustainability, he has reduced the cultivation area before encountering severe water shortage while starting to raise beef cattle and cotton trading (Figure 4).

Like Farmer A, the government has granted him several well drilling licenses. However, based on the advice of geology and water experts (Figure 4), he decided not to dig new boreholes. This well-timed decision has reduced his production expenses to a minimum, allowing more income generation in a quasi-perpetual cycle (Figure 4). Farmer B soon realized the uncertainty of farming and the unknown, cumulative, and permanent nature of new climate conditions. Therefore, he expressed his willingness to seek water-independent sources of income. First, he started providing transportation services for one of the large industrial companies in the area. However, his contract was terminated in 2017, and he was forced to sell the cars. Meantime, his sons encouraged him to initiate a construction business in a nearby city. The risk of investment was high as a result of unbridled inflation and economic collapse in Iran.

Figure 4a. The timeline and perceived causes and effects of micro-level new normal management

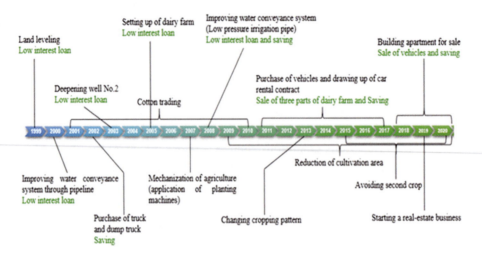

However, when the construction of his first building was finished, he had gained a substantial benefit from the deal and successfully continued this business.

Like everyone else, I would have lost all my possessions if I had not learned to go against the flow of the tide, the farmer said. (November, 2021)

Figure 4b.

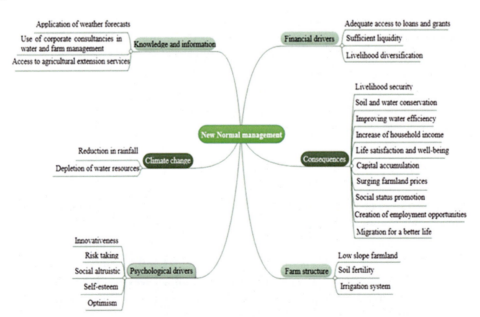

Macro-level Crisis Management: The Case of Iran's Water Bankruptcy

Three decades ago, when the first signs of water scarcity were visible, Iran's government tried to manage the "water crisis". The conceptual framework and practice of government can well be identified as crisis management. Close analysis indicates that, similar to Farmer A, the government has inappropriately approached the water crisis, except the latter is at a mega scale. Although Iran's climate change management is multifaceted, this brief analysis is focused on dam construction and water transfer as measures to mitigate climate change-induced water scarcity.

Over the past three decades, the water demand has increased because of climate variability and change, environmental degradation, population growth, economic growth, intensive irrigated agriculture, and the desire for self-sufficiency in crop production (Nazari et al., 2018). Like Farmer A, the government decision-makers have perceived the water crisis as an isolated and exceptional event. Therefore, following a mechanistic approach, they have established supply-driven water management strategies (i.e., dam construction) as a means of enhancing water availability. In 1993, there were only 28 dams in the country. However, 160 dams have been constructed during the last 28 years, increasing the total number of dams to 188 (IRNA, 2020). Despite massive investment, this management myopia has increased the water crisis and caused permanent, irreversible environmental damages and social conflicts, and unrest. Currently, the reservoirs are half-empty (i.e., 61%), and the government is incapacitated to supply sufficient drinking water in some provinces, such as Khuzestan and Sistan-Baluchestan. This human-induced water crisis has increased social conflicts among ethnic groups, leading to a new wave of political unrest in Iran.

Despite the devastating consequences of crisis management, government decision-makers could not unlearn the past. Since dams and reservoirs have proved unable to meet the ever-increasing need for water, policymakers have fostered the idea of seawater desalination and water transfer from the Persian Gulf to the water-scarce regions through pipelines with a total length of 3700 kilometers (Tehran Times, 2021). This emergency assistance project is hoped to reduce water stress in the arid and semi-arid provinces of Iran. The costs of this project are pretty huge (approximately 30.47$ thousand million; Tehran Times, 2021), and its effectiveness and environmental sustainability have been questioned by many scholars.

Regardless of investments and efforts, some experts consider the current water status of Iran as "water bankrupt". A condition in which water extraction exceeds natural rates of replenishment, with long-term environmental, economic, social, and political consequences (Collins, 2017).

THE ROAD AHEAD: MEETING THE REQUIREMENTS OF NNM

As a social construct, the NNM approach is based on at least six principles: 1) the new normal features of climate are non-routine and unknown; 2) it requires unlearning the normal patterns and routines and updating knowledge, skills, and values; 3) it demands certain preconditions to characterize new normal as a non-linear, complex and non-deterministic issue that is distinct from chaotic, linear and deterministic behaviors; 4) by intensifications of climate change and their negative impacts, the nature, degree of flexibility and adaptability of NNM will be constantly changed; 5) determining the timing, nature, and location of adaptive changes is difficult, and optimal timing for adaptation is not quite apparent; and 6) NNM requires global mobilization and unrestrained resources.

While the challenge of new normal features of climate demands robust NNM, the political will to purposefully address new normal situations and facilitate NNM is often lacking. Early warning systems and forecasts can assist in adequate and timely response to the new normal situations. However, governments in many developing countries are slow in establishing early warning systems and assessing climatic risks. In this regard, the main impediment of NNM is political reluctance to mobilize resources and funds to the adaptive responses that have long-term political benefits. Policymakers need to weigh the costs of action versus the costs of inaction. Also, to manage the new normal, integration of large-scale new normal adaptations into national investment plans is imperative. Investment plans must be consistent and equitable for all economic sectors, population groups, and regions.

Moreover, to integrate NNM into practice, new knowledge and skills are required. The governments should invest in education and training programs to train NNM teams and managers for practical management of the new normal and facilitate public response to the current and future new normal situations. Another point is that most climate change managers attempt to solve new normal problems through linear thinking. They should change their mindset and think in non-linear ways to manage the new normal. Managing the new normal challenges also requires unlearning the normal set of organizational patterns, learning to learn new knowledge and knowledge structures, and enhancement of NNM capacity. It is suggested that NNM program directors engage more in policy formation and convince policymakers to prioritize investments in NNM programs and make the transition easy. Moreover, managing the complexity of new normal situations requires skillful identification, efficient assessment, careful modification, and wise exchange of valuable technologies and adaptive responses. Therefore, the development of climate-resilient technologies, provision of timely access to context-specific climate information, and identification of appropriate entry points for climate change information is needed. The potential entry points for mainstreaming new normal adaptation and for integrating NNM into practice consist of natural resource management, land use planning, infrastructure design, new normal adaptation strategies, and environmental impact assessment. Public media and extension and advisory service providers have a vital role in disseminating responsive technologies and real-time climate change information.

Also, successful management of new normal climate change requires meaningful coordination of international, regional, national, and sectoral planners. Most climate change adaptation plans are at the national level. However, many climate change impacts cut across national boundaries and require joint data collection, analysis, monitoring, and policies. Therefore, providing an organizational structure that enhances coordination at the international and regional levels is imperative. This requires serious cultural unlearning and relearning for NNM. Furthermore, greater engagement of local communities in mainstreaming mitigation and adaptation efforts is essential.

ACKNOWLEDGMENT

We gratefully thank Prof. Donald A. Wilhite (University of Nebraska) and Prof. Sara Hughes (University of Michigan) for their critical reviews and constructive comments.

REFERENCES

Aledo-Ruiz, M. D., Ortega-Gutiérrez, J., Martinez-Caro, E., & Cegarra-Navaroo, J. G. (2017). Linking an unlearning context with firm performance through human capital. *European Research on Management and Business Economics*, *23*(1), 16–22. doi:10.1016/j.iedeen.2016.07.001

Arnell, N. W. (2004). Climate change and global water resources: SRES emissions and socio-economic scenarios. *Global Environmental Change, 14*(1), 31–52. doi:10.1016/j.gloenvcha.2003.10.006

Arnell, N. W., Lowe, J. A., Brown, S., Gosling, S. N., Gottschalk, P., Hinkel, J., Lioyd-Hughes, B., Nicholls, R. J., Osborn, T. J., Osborne, T. M., Rose, G. A., Smith, P., & Warren, R. F. (2013). A global assessment of the effects of climate policy on the impacts of climate change. *Nature Climate Change, 3*(5), 512–519. doi:10.1038/nclimate1793

Barnett, J., & Webber, M. (2010). *Accommodating migration to promote adaptation to climate change* (Policy research working paper No. 5270). The World Bank.

Boafo, Y. A., Saito, O., Jasaw, G. S., Otsuki, K., & Takeuchi, K. (2016). Provisioning ecosystem services-sharing as a coping and adaptation strategy among rural communities in Ghana's semi-arid ecosystem. *Ecosystem Services, 19*, 92–102. doi:10.1016/j.ecoser.2016.05.002

Burton, I. (2011). Adaptation to climate change: Context, status, and prospects. In J. D. Ford & L. Berrang-Ford (Eds.), *Climate Change Adaptation in Developed Nations: From theory to practice*. Springer. doi:10.1007/978-94-007-0567-8_35

Campion, M. A., Mumford, T. V., Morgeson, F. K., & Nahrgang, J. D. (2005). Work redesign: Eight obstacles and opportunities. *Human Resource Management, 44*(4), 367–390. doi:10.1002/hrm.20080

Candelon, B., Carare, A., & Miao, K. (2016). Revisiting the new normal hypothesis. *Journal of International Money and Finance, 66*(C), 5–31. doi:10.1016/j.jimonfin.2015.12.005

Castells-Quintana, D., Lopez-Uribe, M. P., & McDermott, T. K. J. (2018). Adaptation to climate change: A review through a development economics lens. *World Development, 104*, 183–196. doi:10.1016/j.worlddev.2017.11.016

Cegarra-Navarro, J. G., Wensley, A. K. P., & Sánchez Polo, M. T. (2014). A conceptual framework for unlearning in a homecare setting. *Knowledge Management Research and Practice, 12*(4), 375–386. doi:10.1057/kmrp.2013.6

Chorafas, D. N. (2011). *Sovereign debt crisis: The new normal and the newly poor*. Palgrave Macmillan. doi:10.1057/9780230307124

Collins, G. (2017). *Iran's looming water bankruptcy*. Center for Energy Studies, Rice University's Baker Institute for Public Policy.

Coombs, W. T. (2015). Ongoing Crisis Communication (4th ed.). Sage.

Coombs, W. T., & Laufer, D. (2017). Global crisis management-current research and future directions. *Journal of International Management, 24*(3), 199–203. doi:10.1016/j.intman.2017.12.003

CSIRO & Bureau of Meteorology. (2015). *Climate Change in Australia: Information for Australia's Natural Resource Management Regions*. Technical Report, CSIRO and Bureau of Meteorology.

Dai, A. (2013). Increasing drought under global warming in observations and models. *Nature Climate Change, 3*(1), 52–58. Advance online publication. doi:10.1038/nclimate1633

Davidson, D. (2016). Gaps in agricultural climate adaptation research. *Nature Climate Change*, 6(5), 433–435. doi:10.1038/nclimate3007

Dickinson, T., & Burton, I. (2011). Adaptation to climate change in Canada: A multi-level mosaic. In J. D. Ford & L. Berrang-Ford (Eds.), *Climate Change Adaptation in Developed Nations: From theory to practice*. Springer. doi:10.1007/978-94-007-0567-8_7

Döll, P., & Romero-Lankao, P. (2017). How to embrace uncertainty in participatory climate change risk management- A roadmap. *Earth's Future*, 5(1), 18–36. doi:10.1002/2016EF000411

Drennan, L., McConnell, A., & Stark, A. (2015). Risk and crisis management in the public sector. Routledge printers, Abingdon.

EMDAT. (2021). *The International Disaster Database*. Center for Research on the Epidemiology of Disasters-Cred. https://public.emdat.be/data

FAO. (2014). *Towards risk-based drought management in Europe and Central Asia*. https://www.fao.org/fileadmin/user_upload/Europe/documents/Events_2014/ECA2014/ECA_38_14_4_2_en.pdf

FAO, IFAD, UNICEF, WFP, & WHO. (2021). The State of Food Security and Nutrition in the World 2021. Transforming food systems for food security, improved nutrition and affordable healthy diets for all. FAO.

FAOSTAT. (2018). *Food and Agriculture Organization of the United Nations*. http://faostat3.fao.org/download/Q/QA/E

Feng, S., Krueger, A., & Oppenheimer, M. (2010). Linkages among climate change, crop yields and Mexico–US cross-border migration. *Proceedings of the National Academy of Sciences of the United States of America*, 107(32), 14257–14262. doi:10.1073/pnas.1002632107 PMID:20660749

Ford, J. D., & Berrang-Ford, L. (2011). Introduction. In J. D. Ford & L. Berrang-Ford (Eds.), *Climate Change Adaptation in Developed Nations: From theory to practice*. Springer. doi:10.1007/978-94-007-0567-8_1

Georgakakos, A., Fleming, P., Dettinger, M., Peters-Lidard, C., Richmond, T. C., Reckhow, K., White, K., & Yates, D. (2014). Water resources. In J. M. Melillo, T. C., Richmond, & G. W. Yohe (Eds.), Climate Change Impacts in the United States: The Third National Climate Assessment (pp. 69-112). U.S. Global Change Research Program.

Giller, K. E., Hijbeek, R., Andersson, J. A., & Sumberg, J. (2021). Regenerative agriculture: An agronomic perspective. *Outlook on Agriculture*, 50(1), 13–25. doi:10.1177/0030727021998063 PMID:33867585

Gosnell, H., Gill, N., & Voyer, M. (2019). Transformational adaptation on the farm: Processes of change and persistence in transitions to 'climate-smart' regenerative agriculture. *Global Environmental Change*, 59, 101965. doi:10.1016/j.gloenvcha.2019.101965

Grose, V. L. (2018). *Five weaknesses of enterprise risk management*. Omega Systems Group Incorporated.

Haile, G. G., Tang, Q., Hosseini-Moghari, S. M., Liu, X., Gebremicael, T. G., Leng, G., Kebede, A., Xu, X., & Yun, X. (2020). Projected impacts of climate change on drought patterns over East Africa. *Earth's Future, 8*(7), e2020EF001502.

Hertel, T. W., Burke, M. B., & Lobell, D. B. (2011). The poverty implications of climate-induced crop yield changes by 2030. *Global Environmental Change, 20*(4), 577–585. doi:10.1016/j.gloenvcha.2010.07.001

Hoffmann, R., Dimitrova, A., Muttarak, R., Cuaresma, J. C., & Peisker, J. (2020). A meta-analysis of country-level studies on environmental change and migration. *Nature Climate Change, 10*(10), 904–912. doi:10.103841558-020-0898-6

Iglesias, A., Moneo, M., & Garrote, L. (2007). Defining the planning purpose, framework and concepts. In A. Iglesias, M. Moneo, & A. López-Francos, A. (Eds.), Drought management guidelines technical annex. CIHEAM/EC MEDA Water.

IPCC. (2013). *Climate change 2013: The physical science basis.* In T. F. Stocker, D. Qin, G. K. Plattner, M. Tignor, S. K. Allen, J. Boschung, A. Nauels, Y. Xia, V. Bex, & P. M. Midgley (Eds.), Contribution of Working Group I to the Fifth Assessment Report of the Intergovernmental Panel on Climate Change. *Cambridge University Press.*

IPCC. (2014). Summary for policymakers. In C. B. Field, V. R. Barros, D. J. Dokken, K. J. Mach, M. D. Mastrandrea, T. E. Bilir, M. Chatterjee, K. L. Ebi, Y. O. Estrada, R. C. Genova, B. Girma, E. S. Kissel, A. N. Levy, S. MacCracken, P. R. Mastrandrea, & L. L. White (Eds.), Climate Change 2014: Impacts, Adaptation, and Vulnerability. Part A: Global and Sectoral Aspects (pp. 1-32). Contribution of Working Group II to the Fifth Assessment Report of the Intergovernmental Panel on Climate Change. Cambridge University Press.

IPCC. (2018). Summary for Policymakers. In V. Masson-Delmotte, P. Zhai, H. O. Pörtner, D. Roberts, J. Skea, P. R. Shukla, A. Pirani, W. Moufouma-Okia, C. Péan, R. Pidcock, S. Connors, J. B. R. Matthews, Y. Chen, X. Zhou, M. I. Gomis, E. Lonnoy, T. Maycock, M. Tignor, & T. Waterfield (Eds.), Global Warming of 1.5°C (pp. 1-32). An IPCC Special Report on the impacts of global warming of 1.5°C above pre-industrial levels and related global greenhouse gas emission pathways, in the context of strengthening the global response to the threat of climate change, sustainable development, and efforts to eradicate poverty. World Meteorological Organization.

IPCC. (2021). Summary for Policymakers. In V. Masson-Delmotte, P. Zhai, A. Pirani, S. L. Connors, C. Péan, S. Berger, N. Caud, Y. Chen, L. Goldfarb, M. I. Gomis, M. Huang, K. Leitzell, E. Lonnoy, J. B. R. Matthews, T. K. Maycock, T. Waterfield, O. Yelekçi, R. Yu, & B. Zhou (Eds.). Climate Change 2021: The Physical Science Basis. Contribution of Working Group I to the Sixth Assessment Report of the Intergovernmental Panel on Climate Change. Cambridge University Press.

IRNA. (2020). *The number of dams in the country reached 188.* Available at: https://www.irna.ir/news/84025654

Jones, R. (2010). A risk management approach to climate change adaptation. In R. A. C. Nottage, D. S. Wratt, J. F. Bornman, & K. Jones (Eds.), *Climate Change Adaptation in New Zealand: Future Scenarios and Some Sectoral Perspectives* (pp. 10–25). New Zealand Climate Change Centre.

Karimi, V., Karami, E., Karami, S., & Keshavarz, M. (2020). Adaptation to climate change through agricultural paradigm shift. *Environment, Development and Sustainability*. Advance online publication. doi:10.100710668-020-00825-8

Karimi, V., Karami, E., & Keshavarz, M. (2018). Climate change and agriculture: Impacts and adaptive responses in Iran. *Journal of Integrative Agriculture*, *17*(1), 1–15. doi:10.1016/S2095-3119(17)61794-5

Keshavarz, M. (2021). Investigating food security and food waste control of farm families under drought (A case of Kherameh County). *Quarterly Journal of Spatial Economy and Rural Development*, *34*, 83–106.

Keshavarz, M., & Karami, E. (2018). Drought and agricultural ecosystem services in developing countries. In S. Gaba, B. Smith, & E. Lichtfouse (Eds.), *Sustainable Agriculture Reviews 28: Ecology for Agriculture* (pp. 309–359). doi:10.1007/978-3-319-90309-5_9

Keshavarz, M., Karami, E., & Vanclay, F. (2013). Social experience of drought in rural Iran. *Land Use Policy*, *30*(1), 120–129. doi:10.1016/j.landusepol.2012.03.003

Macgregor, N. A., & Cowan, C. E. (2011). Government action to promote sustainable adaptation by the agriculture and land management sector in England. In J. D. Ford & L. Berrang-Ford (Eds.), *Climate Change Adaptation in Developed Nations: From theory to practice* (pp. 385–400). Springer. doi:10.1007/978-94-007-0567-8_28

Mekonnen, M. M., & Hoekstra, A. Y. (2016). Four billion people facing severe water scarcity. *Science Advances*, *2*(2), e1500323. doi:10.1126ciadv.1500323 PMID:26933676

Miyan, M. A. (2015). Droughts in Asian least developed countries: Vulnerability and sustainability. *Weather and Climate Extremes*, *7*, 8–23. doi:10.1016/j.wace.2014.06.003

Moser, S. C. (2011). Entering the period of consequences: The explosive us awakening to the need for adaptation. In J. D. Ford & L. Berrang-Ford (Eds.), *Climate Change Adaptation in Developed Nations: From theory to practice*. Springer. doi:10.1007/978-94-007-0567-8_3

Nazari, B., Liaghat, A., Akbari, M. R., & Keshavarz, M. (2018). Irrigation water management in Iran: Strategic planning for improving water use efficiency. *Agricultural Water Management*, *208*, 7–18. doi:10.1016/j.agwat.2018.06.003

Overmyer, T. C. (2015). *Urban resilience: Reframing climate change for action and advocacy* [Unpublished MSc. Thesis]. Purdue University, West Lafayette, IN, United States.

Pennington-Gray, L. (2017). Reflections to move forward: Where destination crisis management research needs to go. *Tourism Management Perspectives*, *25*, 136–139. doi:10.1016/j.tmp.2017.11.013

Porter, J. R., Xie, L., Challinor, A. J., Cochrane, K., Howden, S. M., Iqbal, M. M., Lobell, D. B., & Travasso, M. I. (2014). Food security and food production systems. In C. B. Field, V. R. Barros, D. J. Dokken, K. J. Mach, M. D. Mastrandrea, T. E. Bilir, M. Chatterjee, K. L. Ebi, Y. O. Estrada, R. C. Genova, B. Girma, E. S. Kissel, A. N. Levy, S. MacCracken, P. R. Mastrandrea, & L. L. White (Eds.), Climate Change 2014: Impacts, Adaptation, and Vulnerability. Part A: Global and Sectoral Aspects (pp. 485-535). Contribution of Working Group II to the Fifth Assessment Report of the Intergovernmental Panel on Climate Change. Cambridge University Press.

Prayag, G. (2018). Symbiotic relationship or not? Understanding resilience and crisis management in tourism. *Tourism Management Perspectives*, *25*, 133–135. doi:10.1016/j.tmp.2017.11.012

Reisinger, A., Kitching, R. L., Chiew, F., Hughes, L., Newton, P. C. D., Schuster, S. S., Tait, A., & Whetton, P. (2014). Australia. In V. R. Barros, C. B. Field, D. J. Dokken, M. D. Mastrandrea, K. J. Mach, T. E. Bilir, M. Chatterjee, K. L. Ebi, Y. O. Estrada, R. C. Genova, B. Girma, E. S. Kissel, A. N. Levy, S. MacCracken, P. R. Mastrandrea, & L. L. White (Eds.), Climate Change 2014: Impacts, Adaptation, and Vulnerability (pp. 1371-1438). Part B: Regional Aspects. Contribution of Working Group II to the Fifth Assessment Report of the Intergovernmental Panel on Climate Change. Cambridge University Press.

Roy, J., Tschakert, P., Waisman, H., Abdul Halim, S., Antwi-Agyei, P., Dasgupta, P., Hayward, B., Kanninen, M., Liverman, D., Okereke, C., Pinho, P. F., Riahi, K., & Suarez Rodriguez, A. G. (2018). Sustainable Development, Poverty Eradication and Reducing Inequalities. In V. Masson-Delmotte, P. Zhai, H. O. Pörtner, D. Roberts, J. Skea, P. R. Shukla, A. Pirani, W. Moufouma-Okia, C. Péan, R. Pidcock, S. Connors, J. B. R. Matthews, Y. Chen, X. Zhou, M. I. Gomis, E. Lonnoy, T. Maycock, M. Tignor, & T. Waterfield (Eds.), Global Warming of 1.5°C. An IPCC Special Report on the impacts of global warming of 1.5°C above pre-industrial levels and related global greenhouse gas emission pathways, in the context of strengthening the global response to the threat of climate change, sustainable development, and efforts to eradicate poverty. World Meteorological Organization.

Sharafi, L., Zarafshani, K., Keshavarz, M., Azadi, H., & Van Passel, S. (2020). Drought risk assessment: Towards drought early warning system and sustainable environment in western Iran. *Ecological Indicators*, *114*, 106276. doi:10.1016/j.ecolind.2020.106276

Sharafi, L., Zarafshani, K., Keshavarz, M., Azadi, H., & Van Passel, S. (2021). Farmers' decision to use drought early warning system in developing countries. *The Science of the Total Environment*, *758*, 142761. doi:10.1016/j.scitotenv.2020.142761 PMID:33183818

Simpson, B., & Burpee, G. (2014). *Adaptation under the new normal of climate change: The future of agricultural extension and advisory services.* USAID.

Tehran Times. (2021). *Rouhani inaugurates 2nd, 3rd phases of Persian Gulf water transfer project.* Https://www.tehrantimes.com/news/459096

The Economist Intelligence Unit. (2021). *Global Food Security Index 2020.* Author.

Tzanakakis, V. A., Paranychianakis, N. V., & Angelakis, A. N. (2020). Water Supply and Water Scarcity. *Water (Basel)*, *12*(9), 2347. doi:10.3390/w12092347

Weitzman, M. L. (2009). On modeling and interpreting the economics of catastrophic climate change. *The Review of Economics and Statistics*, *91*(1), 1–19. doi:10.1162/rest.91.1.1

Wilhite, D. (2016). *Drought- management policies and preparedness plans: Changing the paradigm from crisis to risk management. In Land Restoration.* Elsevier Inc. doi:10.1016/B978-0-12-801231-4.00007-0

Wreford, A., Moran, D., & Adger, N. (2010). *Climate Change and Agriculture: Impacts, Adaptation and Mitigation.* OECD Publications. doi:10.1787/9789264086876-en

Zarei, Z., Karami, E., & Keshavarz, M. (2020). Co-production of knowledge and adaptation to water scarcity in developing countries. *Journal of Environmental Management, 262*, 110283. doi:10.1016/j.jenvman.2020.110283 PMID:32090886

Chapter 4
Climate Change Information Sources:
Fact-Checking and Attitude

Adebowale Jeremy Adetayo
https://orcid.org/0000-0001-7869-5613
Adeleke University, Nigeria

Blessing Damilola Abata-Ebire
Federal Polytechnic, Nigeria

Yetunde Omodele Oladipo
University of Salford, UK

ABSTRACT

This study sought to investigate the relationship between climate change information sources, fact-checking, and attitude among students at Adeleke University. A descriptive survey design was adopted for the study using a questionnaire. The data were analysed using descriptive and inferential statistics. The majority of the 688 survey respondents had an unworried attitude about climate change. Students were discovered to obtain climate change information through Google, television, friends, family, Facebook, radio, YouTube, and Instagram. Students were discovered to often fact-check climate change information. Facebook, Twitter, Instagram, YouTube, radio, church/mosque, friends, religious leaders, and fact-checking have a significant relationship with climate change attitudes. The study concluded that using social media and religious aspects as a source of climate change information may associate to an unworried attitude about climate change. As a result, it suggests addressing religious concerns about climate change.

INTRODUCTION

Nigeria is the world's most populous black nation, and it is endowed with an abundance of natural re-

DOI: 10.4018/978-1-6684-4829-8.ch004

sources and features (Adetayo, 2021b). Nigeria has a tropical climate with two precipitation regimes: low precipitation in the north and high precipitation in the southwest and southeast. It has not suffered the very destructive impacts of climatic extremes such as tornadoes, hurricanes, tsunamis, and earthquakes. This, however, may be changing as a result of climate change.

The Intergovernmental Panel on Climate Change (IPCC) defines climate change as any change in climate over time, whether caused by natural variability or human action (IPCC, 2014). Following a succession of earth tremors in various sections of the nation in recent years, Nigerian seismologists have unanimously agreed that the country is no longer an earthquake-free zone (aseismic), as previously believed (Orakpo, 2017).

It has been noted that atmospheric CO_2 levels have risen from 280 to 400 parts per million during the previous century, and may reach 550 parts per million by 2050 (IPCC, 2014). The frequency and intensity of extreme droughts have increased (Diffenbaugh et al., 2017) and are expected to grow further (Lehner et al., 2017), with potentially disastrous consequences for tree mortality (Hartmann et al., 2018). Also, rainfall durations and intensities have risen, resulting in massive runoffs and flooding in several parts of Nigeria (Enete, 2014).

It is crucial to emphasise, however, that developing countries and the impoverished are more vulnerable to climate change (Adger et al., 2016). Rural farmers in Sub-Saharan Africa, including Nigeria, are projected to be especially vulnerable to climate change, owing to the confluence of poverty, poor infrastructure and technological development, and reliance on rain-fed agriculture (Adimassu & Kessler, 2016; Nelson et al., 2014). According to vulnerability assessments, Nigerian states in the north are more vulnerable to climate change than those in the south (Ignatius, 2016). The rising temperature has hampered productivity in Nigerian animal production, particularly in poultry, swine, cattle, sheep, and goats. Annual output reductions of about 15% have been recorded (Gbenga et al., 2020).

There is scientific agreement that the rise in greenhouse gases created by human activity is significantly impacting the Earth's climate. A greenhouse gas (abbreviated GHG) is a gas that absorbs and emits radiant energy in the thermal infrared spectrum. Water vapour, carbon dioxide, methane, nitrous oxide, and ozone are the principal greenhouse gases in the Earth's atmosphere. The release of these gases has been directly linked to the use of fossil fuels, which is a key source of revenue for Nigeria.

Since many Nigerians have not directly experienced the worst effects of climate change, aside from floods and other minor repercussions, they may grow up with an unworried attitude about climate change. As a result, the sources of information about climate change becomes important for most Nigerians since they would only understand the disastrous effects of climate change through other's experiences and not primarily theirs.

Nigerians, particularly students, obtain the majority of their information through social media, mainstream media, and individuals around them. These several sources of information have recently become controversial. Social media, for example, has been known to transmit a lot of false news while simultaneously containing a lot of real news. On the other hand, the news media, which should be the arbiter of true news, has been divided on the topic of climate change, with liberal media being more favourable to minimising it than conservative news media (Adetayo, 2021a). Similarly, students may obtain information about climate change through religious sources, which have likewise been divided on the matter (Bardon, 2020). As a result, there have risen a surge of fact checking among people to confirm if information they receive from different sources are facts or fake.

It has been stated that we are presently living in a post-factual relativism-dominated period (Aelst et al., 2017). This suggests that the epistemic premises of information and factual knowledge have become a source of increased scepticism and suspicion. Political fact-checking evolved as a new journalistic approach and platform in response to the uncontrolled spread of disinformation (Fridkin et al., 2015).

Fact-checkers indicate which news can be believed and which news is erroneous by rigorously analysing the statements foregrounded in political news articles concerning climate change, so leading people to make the most informed judgments. Because fact-checkers make use of message simplicity and a focus on facts, both of which are proposed characteristics of correcting information (Lewandowsky et al., 2012), it may be a successful method of correcting misinformation. As a result, this chapter seeks to determine whether where students get their information correlates with their attitudes regarding climate change, and whether fact-checking correlates with their attitudes. Also, the chapter seeks to find out it gender has a role to play as it relates to attitude of students towards climate change.

Hypotheses

1. There is a significant relationship between information sources and climate change attitude among students in Adeleke University, Nigeria;
2. There is a significant relationship between fact-checking and climate change attitude among students in Adeleke University, Nigeria.
3. There is a significant difference between female and male students' attitude towards climate change among students in Adeleke University, Nigeria;
4. There is a significant difference between female and male students' level of fact-checking among students in Adeleke University, Nigeria.

METHODS

Settings

Adeleke University is a private, faith-based educational institution on 520 acres of property in Ede, Osun State, Nigeria's southwest. The Adventist education philosophy, which emphasises healthy living, is the foundation of the school. The institution was formed in 2011 by the Springtime Development Foundation (SDF), a charitable, non-profit organisation dedicated to providing poor students with access to high-quality higher education.

Research Design

The descriptive survey design was utilised in the chapter to characterise the important aspects of the phenomena of interest. The survey research method was used for this chapter because it provided an accurate and trustworthy picture of how the independent variables (information sources and fact-checking) relate to the dependent variable (climate change attitude) among Adeleke University students in Osun State, Nigeria.

Population and Sampling Size

The study population included all Adeleke University students. A simple random sampling technique was adopted to select 704 students.

Research Instrument

A structured questionnaire was used as the study's instrument. A questionnaire is useful because it can reach a large number of people and collect data quickly and cheaply. The study instrument was validated using face validity. The questionnaire was examined by an expert in the field of information management to assess the instrument's face validity. Furthermore, several questionnaire components were modified based on suggestions/observations to guarantee an accurate response from the sample group. A few elements were changed and clarified as a result of the expert's recommendations, while others were eliminated with the same meaning.

Method of Data Collection

The questionnaire was delivered to students at Adeleke University in Osun State, Nigeria after it had been reviewed and revised. Respondents were assured that any information they provided would be kept strictly confidential and would only be used for academic research purposes. Such information was likewise not made available to a third party. The research was carried out between November 15th and December 17th, 2021. A total of 688 responses were received, for a response rate of 97.7%. As a consequence, all 688 responses were included in the analysis for this study.

Method of Data Analysis

The information gathered was organised and analysed with descriptive and inferential statistics. To answer study questions, descriptive statistics such as frequency counts, percentages, mean and standard deviation scores were employed. Inferential statistics were employed to analyse the formulated hypothesis, employing Pearson product-moment correlation for analysing hypotheses 1 and 2.

RESULTS

This section highlights the study's findings using descriptive and inferential statistics. Tables 1-5 highlight the specifics.

According to Table 1, the majority of responders were female (59.6%) in the faculty of basic and medical sciences (30.4%). Most are Christians (81.7%) within the age group of 16 - 20 (69.9%). The religion is representative of the university under consideration, which is a faith-based university owned by a Christian. Furthermore, most students in Nigerian private institutions enrol between the ages of 15 and 20, as seen by the study age group.

Climate change is a serious issue all around the world. Students in Nigeria are not ignorant of it since they get information about it. According to the study's findings, the majority of students get their climate change information from Google (86.9%), Television (73.0%), friends (69.2%), family (67.9%), Face-

Climate Change Information Sources

Table 1. Demographic information of respondents

	Frequency	Percentage
Gender		
Female	410	59.6
Male	278	40.4
Total	688	100.0
Religion		
Christianity	562	81.7
Islam	126	18.3
Total	688	100.0
Age Group		
11-15	46	6.7
16-20	479	69.6
21-25	123	17.9
26 & above	40	5.8
Total	688	100.0
Faculty		
Science	212	30.8
Engineering	71	10.3
Art	33	4.8
Basic and Medical Science	209	30.4
Business and Social Science	162	23.5
Total	688	100.0

Table 2. Information sources on climate change

Information Sources	Yes	No
Facebook	382(55.5%)	306(44.5%)
Twitter	193(28.1%)	495(71.9%)
Instagram	368(53.5%)	320(46.5%)
YouTube	371(53.9%)	317(46.1%)
TV	502(73.0%)	186(27.0%)
Radio	377(54.8%)	311(45.2%)
Newspapers	330(48.0%)	358(52.0%)
Church/Mosque	323(46.9%)	365(53.1%)
Blog	287(41.7%)	401(58.3%)
Google	598(86.9%)	90(13.1%)
Family	467(67.9%)	221(32.1%)
Friends	476(69.2%)	212(30.8%)
Religious Leaders	296(43.0%)	392(57.0%)

Table 3. Fact-checking of climate change information

Fact-checking	Always	Often	Rarely	Never	Mean	Std. D
I check the source of any information received	242(35.2%)	238(34.6%)	188(27.3%)	20(2.9%)	3.02	0.861
I crosscheck the information from other mainstream news outlets	187(27.2%)	267(38.8%)	197(28.6%)	37(5.4%)	2.88	0.871
Investigating the author's background and credibility	120(17.4%)	233(33.9%)	244(35.5%)	91(13.2%)	2.56	0.928
Verifying information received from health personnel	218(31.7%)	225(32.7%)	192(27.9%)	53(7.7%)	2.88	0.944
Paying attention to URLs for misspellings and omissions	173(25.1%)	248(36.0%)	206(29.9%)	61(8.9%)	2.77	0.925
Avoidance of websites with excessive adverts	224(32.6%)	243(35.3%)	168(24.4%)	53(7.7%)	2.93	0.935
I watch out for extreme/impossible claims	208(30.2%)	265(38.5%)	160(23.3%)	55(8.0%)	2.91	0.921
I visit fact checking websites	247(35.9%)	246(35.8%)	142(20.6%)	53(7.7%)	3.00	0.935
I read the whole article critically	250(36.3%)	220(32.0%)	180(26.2%)	38(5.5%)	2.99	0.920

book (55.5%), radio (54.8%), YouTube (53.9%), and Instagram (53.5 percent). This result demonstrates Google's popularity among students since they frequently utilise it to search for information. It is not surprising, then, that students utilise Google to get information about climate change as it is not routinely covered by local news outlets. The usage of social media platforms such as Facebook, YouTube, and Instagram to gather climate change information demonstrates the media's appeal among young people, who make up the majority of responders (Adetayo & Williams-Ilemobola, 2021; Auxier & Anderson, 2021). Twitter was the least utilised source for climate change information by respondents, confirming Twitter's decrease among adolescents (Hughes, 2021).

Table 3 shows that students frequently fact-check climate change information they receive. The rationale for the fact check is unclear, although the current political and health climate might play a factor. For several years, there has been an increase in the severity of fake news throughout the world, which has raised the sensitivity of information about various controversial subjects. This might be one of the reasons for fact-checking the source of any information received (mean=3.02). Students were also found to visit fact-checking websites regularly (mean=3.00), since many social media platforms today flag content they believe to be fake news. To prevent being misled about climate change, students frequently read the whole article critically (mean=2.99), while avoiding websites with extensive advertisements (mean=2.93). They also keep an eye out for extreme/impossible claims (mean=2.91) made by people attempting to influence behaviour. As a consequence, students were found to cross-check climate change information from other mainstream news outlets and verify information acquired from health personnel (mean=2.88), as well as pay attention to URLs for misspellings and omissions (mean=2.77). The least used approach of fact-checking by students is to explore the author's background and credibility (mean=2.56).

Table 4 reveals that students have an unconcerned attitude towards climate change. The majority of students believe that the media is often too alarmist about issues like climate change (mean=2.92). This could be as a result of the divergent views about climate change portrayed in today's politicalized news.

Climate Change Information Sources

Table 4. Attitude towards climate change

Attitude	SA	A	D	SD	Mean	Std. D
Jobs today are more important than protecting the environment	150(21.8%)	192(27.9%)	270(39.2%)	76(11.0%)	2.60	0.947
I am unwilling to make personal sacrifices for the sake of the environment	115(16.7%)	254(36.9%)	239(34.7%)	80(11.6%)	2.59	0.900
The media is often too alarmist about issues like climate change	178(25.9%)	313(45.5%)	163(23.7%)	34(4.9%)	2.92	0.830
Nature is strong enough to cope with the impact of climate change	172(25.0%)	297(43.2%)	169(24.6%)	50(7.3%)	2.86	0.876
I do not feel a moral duty to do something about climate change	104(15.1%)	232(33.7%)	261(37.9%)	91(13.2%)	2.51	0.904
Nothing I do on a daily basis contributes to the problem of climate change	130(18.9%)	257(37.4%)	229(33.3%)	72(10.5%)	2.65	0.904
There is no point in me doing anything about climate change because changes in climate cannot be reversed	129(18.8%)	241(35.0%)	221(32.1%)	97(14.1%)	2.58	0.049
It is too early to say whether climate change is really a problem	122(17.7%)	251(36.5%)	235(34.2%)	80(11.6%)	2.60	0.910
Claims that human activities are changing the climate are exaggerated	159(23.1%)	240(34.9%)	197(28.6%)	92(13.4%)	2.68	0.974

Although local news outlets in Nigeria does not regularly broadcast news about climate change, students however have regular access to social media and foreign news through their phones. They also believe that nature is strong enough to cope with the impact of climate change (mean=2.86), this further reveals the complacency about climate change among Nigerians, who don't feel the extremes effect of climate change as developed nations often do. It is not surprising that the students believe the claims that human activities are changing the climate are exaggerated (mean=2.68) since they also believe the media are alarmist about it. As a result, they are firm that nothing they do daily contributes to the problem of climate change (mean=2.65). As a nation where poverty rates are high and millions of graduates are without a job, it is not surprising that they believe that jobs today are more important than protecting the environment (mean=2.60). If the problem of employment availability in Nigeria is not sufficiently addressed, this may create further challenges for climate change mitigation. Other attitudes among students discovered in the study include: it is too early to say whether climate change is a problem, unwillingness to make personal sacrifices for the sake of the environment, no point in doing anything about climate change because climate changes cannot be reversed, and a lack of a moral obligation to do something about climate change.

Table 5 contains a summary of data analysis on the test of the significance of the relationship between information sources and climate change attitude, which demonstrates that Facebook, Twitter, Instagram, YouTube, radio, Church/Mosque, Friends and religious leaders have a significant relationship with climate change attitude. This means that using Facebook, Twitter, Instagram, YouTube, radio, Church/Mosque, Friends and religious leaders as a source of climate change information will associate with an increase in attitude towards climate change. However, Church/Mosque has the strongest relationship with a strength of 21.1%, this is followed closely by Facebook with a relationship strength of 17.0%. Other climate change information sources, such as Television, Newspapers, blog, Google and family were

Table 5. Information sources and attitude

		Attitude
Facebook	Pearson Correlation	.170**
	Sig. (2-tailed)	.000
twitter	Pearson Correlation	.095*
	Sig. (2-tailed)	.013
Instagram	Pearson Correlation	.098*
	Sig. (2-tailed)	.010
YouTube	Pearson Correlation	.092*
	Sig. (2-tailed)	.015
Television	Pearson Correlation	-.038
	Sig. (2-tailed)	.324
Radio	Pearson Correlation	.091*
	Sig. (2-tailed)	.016
Newspaper	Pearson Correlation	.051
	Sig. (2-tailed)	.180
Church/Mosque	Pearson Correlation	.211**
	Sig. (2-tailed)	.000
Blog	Pearson Correlation	.002
	Sig. (2-tailed)	.965
Google	Pearson Correlation	.005
	Sig. (2-tailed)	.895
Family	Pearson Correlation	.024
	Sig. (2-tailed)	.534
Friends	Pearson Correlation	.086*
	Sig. (2-tailed)	.024
Religious Leaders	Pearson Correlation	.196**
	Sig. (2-tailed)	.000
	N	688

shown to be unrelated to climate change attitudes. From the foregoing, the alternative hypothesis is accepted, while the null hypothesis is rejected. This suggests that there is a significant relationship between some information sources and climate change attitudes among Adeleke University students in Nigeria.

Table 6 contains a summary of data analysis on the test of the significance of the relationship between information sources and climate change attitude, which demonstrates that fact-checking has a significant relationship with climate change attitude ($r = 0.341$ N= 688, P< 0.05). This means that students that fact-check climate change information will likely have an unworried attitude towards climate change. As a result, the alternative hypothesis is accepted, while the null hypothesis is rejected. This suggests that there is a significant relationship between fact-checking and climate change attitude among Adeleke University students in Nigeria.

Table 6. Fact-checking and attitude

Correlations			
			Attitude
FackCheck	Pearson Correlation		.341**
	Sig. (2-tailed)		.000
	N		688

	Gender	N	Mean	Std. Deviation	Std. Error Mean
Attitude	Male	278	24.8741	5.75511	.34517
	Female	410	23.3951	5.11386	.25256
FackCheck	Male	278	26.9388	5.93769	.35612
	Female	410	25.2610	6.18245	.30533

Independent Samples Test						
		Levene's Test for Equality of Variances		t-test for Equality of Means		
		F	Sig.	T	df	Sig. (2-tailed)
Attitude	Equal variances assumed	5.220	.023	3.537	686	.000
FackCheck	Equal variances assumed	3.171	.075	3.549	686	.000

Table 6 in relation to attitude, indicate that t=3.537, p =0.000. This means that the stated null hypothesis was not accepted. This was as a result of the t-value (3.537) with a p-value of 0.000 at 0.05 alpha value. This means that there was a significant difference between female and male students' attitudes towards climate change. However, the male means score was higher than the female. This implies that male students are more likely to have an unconcerned attitude towards climate change than females. Furthermore, in terms of fact-checking, it was revealed that males are more likely to fact-check information about climate change than females. The result showed a significant difference between the two genders (t=3.549, p=0.000).

DISCUSSION

Climate change has been a global trend for many years, with several conferences organised by various organisations such as the United Nations. Nigeria is one of the countries that has committed to reducing climate change, with the Southwest and Southeast being less vulnerable than other sections of the country. The South-South (Niger Delta area) of Nigeria is the most susceptible owing to sea-level rise, higher precipitation, coastal erosion, and floods – all of which have resulted in the relocation of numerous populations (Matemilola et al., 2019).

However, to garner widespread support for climate change mitigation in the future, the public must support it. This has required the need to measure pupils' attitudes about climate change. Despite this, research has revealed that Nigerians are largely unaware of climate change (Galadima & Lawal, 2017). This current study found that students have an unworried attitude regarding climate change. They feel that the media is frequently too concerned about matters such as climate change and that statements that human activities are altering the environment are overblown. This is consistent with the findings of

Schmid-Petri (2017), who discovered that significant minorities still believe climate change is a hoax or that its threats are overblown.

People try to exert control over conditions that have the potential to influence their life; nevertheless, if they believe they cannot generate the desired results, they will not seek to change things (Bandura, 1977). This might be the case in this study because students believe there is little they can do to counteract climate change. They also believe that nature is resilient enough to withstand the effects of climate change. This is also consistent with previous studies' findings that believe climate change has a limited direct influence on the world (Broto & Bulkeley, 2013). Some even believe that the danger of climate change is quite low (Mead et al., 2012). As a result, many of the study's participants feel that nothing they do daily adds to the problem of climate change. This is confirmed by Dunlap (2013), who found scepticism about climate change among people.

Social media is where many students receive their information. Students in this research, on the other hand, obtained climate change information via Google, television, friends, family, Facebook, radio, and Instagram. This is supported by Asadu et al. (2018), who found information on climate change was mainly from neighbours/friends. Umegbolu's (2020) research, which discovered that radio and television were the primary sources of information on climate change, is also consistent with the study. This study was also supported by Malgwi and Joshua (2021), who found that radio, television, social media, family, and friends were the most common ways for people to learn about climate change.

In addition, YouTube was discovered to be a key source of climate change information. YouTube is the third most visited website in the world, with over four billion video views per day, more than 72 hours of video posted every minute, and beats traditional media by being both a massive source of knowledge and a producer of visual memes (Pew Research Center, 2012; Xu et al., 2015). YouTube's popularity may have led to its usage as a source of climate change information. Our findings, however, contradict those of Scott-Parker et al. (2016), who identified religious leaders as a key source of climate change information. Scott-Parker et alfindings .'s may be attributable to the fact that religious leaders are normally more trusted than any other group when it comes to climate change messages. However, in today's modern period, young people, who make up the majority of this survey, are more likely to use technology to obtain information (Adetayo & Williams-Ilemobola, 2021).

According to the study, social media platforms such as Facebook, Twitter, Instagram, YouTube, radio, Church/Mosque, Friends, and religious leaders have a significant relationship with climate change attitudes. Other studies have revealed that the media increases people's environmental awareness and is a major impact in moulding people's attitudes (Arlt et al., 2011; Carvalho, 2010). Furthermore, Nigeria is a religiously diverse nation, with Islam dominating the north and Christianity dominating the south. Because religion is such an important part of people's lives, they might have a high level of confidence in their religious leaders. This study discovered that churches, mosques, and religious leaders had a major impact on climate change attitudes. This suggests that churches, mosques, and religious leaders contribute to the respondents' unfavourable attitude regarding climate change. This puts to light the fact that climate change has impacted not just politics but even religion. As a result, climate change activists have a lot of work to do.

Males were shown to be more likely than females to have an unworried attitude regarding climate change. This might be because women are more likely to view climate change to be a significant problem, to be worried about how it will affect them personally, and to believe that big lifestyle adjustments are required to address the issue (Pew Research Center, 2015).

In today's environment, misinformation concerning climate change may be found in mainstream media (Painter & Gavin, 2015). This might be why students in this present study are found to be fact-checking. One would think that fact-checking would make students to have a worried attitudes towards climate change. However, in this study, this is not the case. The study discovered that fact-checking is associated with students' unfavourable attitudes regarding climate change. Unfortunately, addressing misinformation is considerably more difficult than merely delivering facts, especially when false information is repeated or accepted by elites (Kuklinski et al., 2000). The effectiveness of new information sent via a correction or counter-frame may be determined by perceptions of a source's legitimacy, the quality of material included inside the new/counter-frame, or the strength of individuals' initial opinions (Druckman & Lupia, 2016). It appears that the majority of respondents in the research already had that attitude regarding climate change, and fact-checking does not modify that attitude for the better, but rather makes them more pessimistic about it. For example, despite public clarifications by news organisations and fact-checking agencies, false claims like the Affordable Care Act introducing Bdeath Panels or Barack Obama not being born in the United States have been difficult to dispel (Berinsky, 2017; Uscinski & Parent, 2014).

Other research has revealed that the efforts of several fact-checkers, such as PolitiFact, The Associated Press Fact Check, and FactCheck.org, to determine the truth of political assertions can be successful at times (Lewandowsky et al., 2012). Individuals, on the other hand, tend to reject corrections that contradict their beliefs (Cobb et al., 2013). According to the research on both political polarisation and selective exposure, while self-selecting news, people reject material that contradicts their priors while seeking information that is coherent with their current ideas (Bennett & Iyengar, 2008). As a result, audiences may play a key part in constructing their biased information environment in which adherence to pre-existing ideas trumps the requirement for factual accuracy (Taber & Lodge, 2006).

CONCLUSION

The study concludes that obtaining climate change information via social media, friends, radio, and religious sources may associate to an unconcerned attitude on climate change. Students learned about climate change mostly via Google, television, friends, family, Facebook, radio, YouTube, and Instagram. The most popular source of climate change information was the Google search engine, while Facebook was the most popular social media site. Friends and relatives were deemed more valuable than churches/mosques and church leaders when seeking information about climate change. Students were shown to be unconcerned about climate change, most likely due to the modest effects of climate change in Nigeria, as compared to developed nations. The more students who acquire their climate change information from Facebook, Twitter, Instagram, YouTube, radio, Church/Mosque, Friends, and religious leaders, the less concerned they are about climate change. Fact-checking did not increase the likelihood of being concerned about climate change, but rather was associated with the likelihood of being unconcerned about it. Males were shown to be more likely than females to have an unconcerned attitude toward climate change and to be less inclined to fact-check information about it. As a result, it is proposed in the report that religious concerns regarding climate change be addressed. There should also be a shift in how social media is used to disseminate climate change information, as well as a shift in how fact-checking is done. Libraries and librarians, who have been found to be more trustworthy, may be used in fact-checking services.

The study's conclusions have far-reaching implications for social media companies, climate change experts, and the government. It will assist social media companies in revising their approach to fact-checking climate change material uploaded on their sites. It will assist climate change policymakers in determining which platforms are adverse to moving forward with climate change mitigation implementation. This research would also aid the government in taking Nigerians' attitudes on climate change seriously by attempting to reshape their views and attitudes toward it.

REFERENCES

Adetayo, A. J. (2021a). Fake News and Social Media Censorship. In R. J. Blankenship (Ed.), *Deep Fakes, Fake News, and Misinformation in Online Teaching and Learning Technologies*. IGI Global. doi:10.4018/978-1-7998-6474-5.ch004

Adetayo, A. J. (2021b). Leveraging Bring Your Own Device for Mobility of Library Reference Services: The Nigerian Perspective. *The Reference Librarian*, 62(2), 106–125. doi:10.1080/02763877.2021.1936342

Adetayo, A. J., & Williams-Ilemobola, O. (2021). Librarians' generation and social media adoption in selected academic libraries in Southwestern, Nigeria. *Library Philosophy and Practice (e-Journal)*, 4984. https://digitalcommons.unl.edu/libphilprac/4984

Adger, W. N., Huq, S., Brown, K., Declan, C., & Mike, H. (2016). Adaptation to climate change in the developing world. *Progress in Development Studies*, 3(3), 179–195. doi:10.1191/1464993403ps060oa

Adimassu, Z., & Kessler, A. (2016). Factors affecting farmers' coping and adaptation strategies to perceived trends of declining rainfall and crop productivity in the central Rift valley of Ethiopia. *Environmental Systems Research*, 5(1), 1–16. doi:10.118640068-016-0065-2

Arlt, D., Hoppe, I., & Wolling, J. (2011). Climate change and media usage: Effects on problem awareness and behavioural intentions. *The International Communication Gazette*, 73(1), 45–63. doi:10.1177/1748048510386741

Asadu, A. N., Ozioko, R. I., & Dimelu, M. U. (2018). Climate Change Information Source and Indigenous Adaptation Strategies of Cucumber Farmers in Enugu State, Nigeria. *Journal of Agricultural Extension*, 22(2), 136–146. doi:10.4314/jae.v22i2.12

Bandura, A. (1977). Self-efficacy: Toward a unifying theory of behavioral change. *Psychological Review*, 84(2), 191–215. doi:10.1037/0033-295X.84.2.191 PMID:847061

Bardon, A. (2020, September 9). *Faith and politics mix to drive evangelical Christians' climate change denial*. https://theconversation.com/faith-and-politics-mix-to-drive-evangelical-christians-climate-change-denial-143145

Bennett, W. L., & Iyengar, S. (2008). A New Era of Minimal Effects? the Changing Foundations of Political Communication. *Journal of Communication*, 58(4), 707–731. doi:10.1111/j.1460-2466.2008.00410.x

Berinsky, A. J. (2017). Rumors and Health Care Reform: Experiments in Political Misinformation. *British Journal of Political Science*, 47(2), 241–262. doi:10.1017/S0007123415000186

Broto, V. C., & Bulkeley, H. (2013). A survey of urban climate change experiments in 100 cities. *Global Environmental Change*, *23*(1), 92–102. doi:10.1016/j.gloenvcha.2012.07.005 PMID:23805029

Carvalho, A. (2010). Media(ted)discourses and climate change: A focus on political subjectivity and (dis)engagement. *Wiley Interdisciplinary Reviews: Climate Change*, *1*(2), 172–179. doi:10.1002/wcc.13

Cobb, M. D., Nyhan, B., & Reifler, J. (2013). Beliefs Don't Always Persevere: How Political Figures Are Punished When Positive Information about Them Is Discredited. *Political Psychology*, *34*(3), 307–326. doi:10.1111/j.1467-9221.2012.00935.x

Diffenbaugh, N. S., Singh, D., Mankin, J. S., Horton, D. E., Swain, D. L., Touma, D., Charland, A., Liu, Y., Haugen, M., Tsiang, M., & Rajaratnam, B. (2017). Quantifying the influence of global warming on unprecedented extreme climate events. *Proceedings of the National Academy of Sciences of the United States of America*, *114*(19), 4881–4886. doi:10.1073/pnas.1618082114 PMID:28439005

Druckman, J. N., & Lupia, A. (2016). Preference Change in Competitive Political Environments. *Annual Review of Political Science*, *19*(1), 13–31. doi:10.1146/annurev-polisci-020614-095051

Dunlap, R. E. (2013). Climate Change Skepticism and Denial: An Introduction. *The American Behavioral Scientist*, *57*(6), 691–698. doi:10.1177/0002764213477097 PMID:24098056

Enete, I. C. (2014). Impacts of climate change on agricultural production in Enugu state, Nigeria. *Journal of Earth Science & Climatic Change*, *5*(9). https://www.cabdirect.org/cabdirect/abstract/20153228679

Fridkin, K., Kenney, P. J., & Wintersieck, A. (2015). Liar, Liar, Pants on Fire: How Fact-Checking Influences Citizens' Reactions to Negative Advertising. *Political Communication*, *32*(1), 127–151. doi:10.1080/10584609.2014.914613

Galadima, A., & Lawal, A. M. (2017). Climate Change Situation in Zamfara State: Farmers' Awareness and Agricultural Implications. *Chemistry and Materials Research*, *9*(8), 7–11. https://www.iiste.org/Journals/index.php/CMR/article/view/38210

Gbenga, O., Opaluwa, H. I., Olabode, A., & Ayodele, O. J. (2020). Understanding the Effects of Climate Change on Crop and Livestock Productivity in Nigeria. *Asian Journal of Agricultural Extension. Economia e Sociologia*, *38*(3), 83–92. doi:10.9734/ajaees/2020/v38i330327

Hartmann, H., Moura, C. F., Anderegg, W. R. L., Ruehr, N. K., Salmon, Y., Allen, C. D., Arndt, S. K., Breshears, D. D., Davi, H., Galbraith, D., Ruthrof, K. X., Wunder, J., Adams, H. D., Bloemen, J., Cailleret, M., Cobb, R., Gessler, A., Grams, T. E. E., Jansen, S., ... O'Brien, M. (2018). Research frontiers for improving our understanding of drought-induced tree and forest mortality. *The New Phytologist*, *218*(1), 15–28. doi:10.1111/nph.15048 PMID:29488280

Ignatius, A. M. (2016). Rurality and climate change vulnerability in Nigeria: Assessment towards evidence based even rural development policy. *2016 Berlin Conference on Global Environmental Change*. 10.17169/REFUBIUM-21841

IPCC. (2014). *AR5 Climate Change 2014: Impacts, Adaptation, and Vulnerability*. https://www.ipcc.ch/report/ar5/wg2/

Kuklinski, J. H., Quirk, P. J., Jerit, J., Schwieder, D., & Rich, R. F. (2000). Misinformation and the Currency of Democratic Citizenship. *The Journal of Politics*, *62*(3), 790–816. doi:10.1111/0022-3816.00033

Lehner, F., Coats, S., Stocker, T. F., Pendergrass, A. G., Sanderson, B. M., Raible, C. C., & Smerdon, J. E. (2017). Projected drought risk in 1.5°C and 2°C warmer climates. *Geophysical Research Letters*, *44*(14), 7419–7428. doi:10.1002/2017GL074117

Lewandowsky, S., Ecker, U. K. H., Seifert, C. M., Schwarz, N., & Cook, J. (2012). Misinformation and Its Correction: Continued Influence and Successful Debiasing. *Psychological Science in the Public Interest*, *13*(3), 106–131. doi:10.1177/1529100612451018 PMID:26173286

Malgwi, P. G., & Joshua, W. K. (2021). Assessment of the Perception and Awareness of Climate Change and the Influence of Information Amongst Tertiary Education Students in North-East Nigeria. *Library and Information Science Digest, 14*, 14–24. https://lisdigest.org/index.php/lisd/article/view/154/139

Matemilola, S., Adedeji, O. H., Elegbede, I., & Kies, F. (2019). Mainstreaming Climate Change into the EIA Process in Nigeria: Perspectives from Projects in the Niger Delta Region. *Climate (Basel)*, *7*(2), 29. doi:10.3390/cli7020029

Mead, E., Roser-Renouf, C., Rimal, R. N., Flora, J. A., Maibach, E. W., & Leiserowitz, A. (2012). Information Seeking About Global Climate Change Among Adolescents: The Role of Risk Perceptions, Efficacy Beliefs, and Parental Influences. *Atlantic Journal of Communication*, *20*(1), 31–52. doi:10.1080/15456870.2012.637027 PMID:22866024

Nelson, G. C., Van der Mensbrugghe, D., Ahammad, H., Blanc, E., Calvin, K., Hasegawa, T., Havlik, P., Heyhoe, E., Kyle, P., Lotze-Campen, H., von Lampe, M., Mason d'Croz, D., van Meijl, H., Müller, C., Reilly, J., Robertson, R., Sands, R. D., Schmitz, C., Tabeau, A., ... Willenbockel, D. (2014). Agriculture and climate change in global scenarios: Why don't the models agree. *Agricultural Economics*, *45*(1), 85–101. doi:10.1111/agec.12091

Orakpo, E. (2017, January 4). *Earthquake in Nigeria: Measures to avert devastating impacts— Experts*. Vanguard News. https://www.vanguardngr.com/2017/01/earthquake-nigeria-measures-avert-devastating-impacts-experts/

Painter, J., & Gavin, N. T. (2015). Climate Skepticism in British Newspapers, 2007–2011. *Environmental Communication*, *10*(4), 432–452. doi:10.1080/17524032.2014.995193

Pew Research Center. (2012, July 16). *YouTube & News*. https://www.pewresearch.org/journalism/2012/07/16/youtube-news/

Pew Research Center. (2015, December 2). *Women more than men see climate change as personal threat*. https://www.pewresearch.org/fact-tank/2015/12/02/women-more-than-men-say-climate-change-will-harm-them-personally/

Schmid-Petri, H. (2017). Do Conservative Media Provide a Forum for Skeptical Voices? The Link Between Ideology and the Coverage of Climate Change in British, German, and Swiss Newspapers. *Environmental Communication*, *11*(4), 554–567. doi:10.1080/17524032.2017.1280518

Scott-Parker, B., Nunn, P. D., Mulgrew, K., Hine, D., Marks, A., Mahar, D., & Tiko, L. (2016). Pacific Islanders' understanding of climate change: Where do they source information and to what extent do they trust it? *Regional Environmental Change*, *17*(4), 1005–1015. doi:10.100710113-016-1001-8

Taber, C. S., & Lodge, M. (2006). Motivated Skepticism in the Evaluation of Political Beliefs. *American Journal of Political Science*, *50*(3), 755–769. doi:10.1111/j.1540-5907.2006.00214.x

Umegbolu, E. (2020). Awarness and Knowledge of Health Implications of Climate Change in Oji River Lga of Enugu State, Southeast Nigeria. *European Journal Pharmaceutical and Medical Research*, *7*(7), 204–210. https://www.researchgate.net/profile/Emmanuel-Umegbolu-2/publication/342611614_AWARENESS_AND_KNOWLEDGE_OF_HEALTH_IMPLICATIONS_OF_CLIMATE_CHANGE_IN_OJI_RIVER_LGA_OF_ENUGU_STATE_SOUTHEAST_NIGERIA/links/5efcea56a6fdcc4ca4411954/AWARENESS-AND-KNOWLEDGE-OF-HEALTH-IMPLICATIONS-OF-CLIMATE-CHANGE-IN-OJI-RIVER-LGA-OF-ENUGU-STATE-SOUTHEAST-NIGERIA.pdf

Uscinski, J. E., & Parent, J. M. (2014). *American conspiracy theories*. Oxford University Press., doi:10.1093/acprof:oso/9780199351800.001.0001

Van Aelst, P., Strömbäck, J., Aalberg, T., Esser, F., de Vreese, C., Matthes, J., Hopmann, D., Salgado, S., Hubé, N., Stępińska, A., Papathanassopoulos, S., Berganza, R., Legnante, G., Reinemann, C., Sheafer, T., & Stanyer, J. (2017). Political communication in a high-choice media environment: A challenge for democracy? *Annals of the International Communication Association*, *41*(1), 3–27. doi:10.1080/23808985.2017.1288551

Xu, W. W., Park, J. Y., & Park, H. W. (2015). The networked cultural diffusion of Korean wave. *Online Information Review*, *39*(1), 43–60. doi:10.1108/OIR-07-2014-0160

ADDITIONAL READING

Bolin, J. L., & Hamilton, L. C. (2018). The news you choose: News media preferences amplify views on climate change. *Environmental Politics*, *27*(3), 455–476. doi:10.1080/09644016.2018.1423909

Ejaz, W., Ittefaq, M., & Arif, M. (2022). Understanding influences, misinformation, and fact checking concerning climate-change journalism in Pakistan. *Journalism Practice*, *16*(2-3), 404–424. doi:10.1080/17512786.2021.1972029

Popoola, O. O., Yusuf, S. F. G., & Monde, N. (2020). Information sources and constraints to climate change adaptation amongst smallholder farmers in Amathole District Municipality, Eastern Cape Province, South Africa. *Sustainability*, *12*(14), 5846. doi:10.3390u12145846

Rosenthal, S. (2022). Information sources, perceived personal experience, and climate change beliefs. *Journal of Environmental Psychology*, *81*, 101796. doi:10.1016/j.jenvp.2022.101796

KEY TERMS AND DEFINITIONS

Attitude: It is the sum of a person's inclinations and sentiments, prejudice or bias, preconceived beliefs, thoughts, fears, dangers, and other topics. It entails a complicated arrangement of evaluative ideas, sentiments, and proclivities toward certain acts.

Climate Change: This refers to long-term changes in temperatures and weather patterns triggered by natural causes such as fluctuations in the solar cycle or human activities, especially the use of fossil fuels such as coal, oil, and gas.

Fact-Checking: Fact-checking is the practise of verifying factual information in order to increase the validity and accuracy of reporting.

Information Source: It is a person, object, or location from which information originates, emerges, or is received.

Chapter 5
Deployment and Optimization of Virtual Power Plants and Microgrids:
An Opportunity for the Energetic Transition in Algeria

Abdelmadjid Recioui
https://orcid.org/0000-0001-9028-3910
University of Boumerdes, Algeria

ABSTRACT

In Algeria, the government trend is towards the adoption of clean sources in electrical energy production due to the imoprtant solar and wind potential that is available in the desert. Virtual power plants (VPP) are modular designed entities based on software communication technologies which efficiently integrate, organize, and manage decentralized generation, storage, and consumption through a smart energy management system (EMS). VPP can only be created if there is a market to sell its power and services whereas microgrids (MG) can be created anywhere and are not market dependent. In this chapter, a description of VPP and MG is presented. The key components are described, and a comparative study is done to assess which option to adopt for the Algerian context. A review about the deployment experiences, trends, and operation optimization is presented. The aim is to assess how to deploy them in a collaborative way to fit the Algerian future energy sector perspective.

INTRODUCTION

Algeria's energy sector is transforming to become progressively reliant on the electricity produced from renewable energy sources (RES). This shift is due to RES benefits in terms of greenhouse gas emissions, sustainability and also the depletion of fossil sources. As evidence of this transformation, the 100% renewable energy project in 2050 is under investigation (Hasni et al., 2021). The objective of

DOI: 10.4018/978-1-6684-4829-8.ch005

30% share has already to be respected in accordance with Algeria's committments in Paris Agreement (CEREFE, 2020).

Electricity production from wind and solar energy is not stable and intermittent which completely modifies the traditional model of an electricity system in which the demand is the main fluctuating factor. The distributed generation of electricity is supplied on lower voltage levels and the installation of RES power plants drastically leads to the decentralization of electricity generation. This coexistence between the traditional power plants and the distriduted generation affects the distribution grids and results in two-way power flow which stress on the the overall power system (Recioui and Bentarzi, 2021). There are developments to be made on the current grid in Algeria in terms of an increasingly deployment of automated grid systems (Mohamed and Hanane, 2021). The renewable energy sources tarnsform the passive energy consumers to prosumers that own renewables (on their rooftop) individually or as part of a community-driven energy campaign (Gui and MacGill, 2018).

Virtual Power Plant (VPP) constitute an increasingly deployed smart grid application that combines distributed energy resources (DER) (e.g., distributed generation, controllable loads and energy storage systems) in a collaborative pattern (Asmus, 2010). Distributed Energy Resources describes the vast array of small- and large-scale energy technologies owned by consumers and businesses and even governments. DERs include mature technologies such as distributed generation (e.g., solar PV and wind), demand response, electric vehicle fast chargers (not only cars but the electrification of buses, rail, ports and fleets), energy storage (batteries, thermal storage), microgrids (which facilitate renewable generation, storage and grid resiliency), energy efficiency and smart appliances. These resources all connect to the grid at the distribution level. And as the penetration of DERs increase, the distribution network changes from a single-source radial network to a multi-source grid, meaning utilities must maintain voltage limits, watch for short circuit interruption limits and maintain the interconnected network's stability, as DERs make the whole process more complex and challenging. On the upside, the grid can be transformed to be more reliable and efficient. Using a VPP, a community could potentially manage the way community-generated energy is used (Verkade and Höffken, 2019) and/ or take part and gain revenues from energy trade, network support and balancing services (El Bakari and Kling, 2012; Klaassen and Van der Laan, 2019). DERs have to deal with some challenges when they are connected to the grid. These include: How to manage voltage levels, how to predict supply and demand, How DERs can be connected to grids cheaper and faster while reducing costs and operating within the technical limits of the power system and, How to overcome the technical and commercial challenges of managing a grid integrating DERs (Pudjianto et al., 2007).

VIRTUAL POWER PLANTS VERSUS MICROGRIDS

Microgrids

A micro-grid is a small-sized version of the complete grid system where a form of electricity generation, storage, distribution and consumption exist within clearly defined electrical boundaries. A micro-grid could be stand alone or a grid-connected, with a common point of coupling. A microgrid comprises one or more DER assets such as solar panels, wind turbines, diesel or gas generators, combined heat and power (CHP) units, etc. that produce electrical power (generation) co-located with the energy loads (consumption). They often include energy storage systems that have two opposite modes, acting either

Figure 1. A VPP illustration

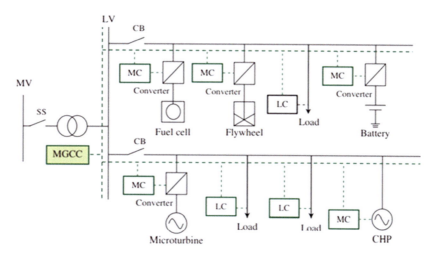

as a load (when charging) or a generator (when discharging). Using sophisticated microgrid controllers, microgrids can be connected directly to the grid but can be dynamically islanded as needed, or disconnected from the centralized power grid without causing power interruption to those site loads supported by the microgrid. They are local, intelligent, and independent (Hirsch, Parag and Guerrero, 2018).

Virtual Power Plants

Virtual power plants are software-based control and optimization of a network of generation and demand side storage. Virtual power plants are a decentralized, scaled collection or portfolio of power generating units such as the DER assets described above, or an aggregation of flexible power consumers willing to provide strategic load reductions when needed. VPPs can be considered a cloud-based distributed power plant that brings together heterogeneous DER in order to enhance electrical power generation, as well as trade it in the electricity market. To make the system more efficient or reliable for distribution facilities and users, battery storage systems are employed in conjunction with traditional solar panels, wind turbines and other power plants. These units are interconnected, and their produced power is dispatched smartly through a central control room with the aim of balancing load between power generation and power consumption of the networked systems traded on the energy exchange market. The objective of a VPP is to relieve the load on the grid by smartly distributing the power generated by the individual units during periods of peak load. Additionally, the combined power generation and power consumption of the networked units in the VPP can trade on the energy exchange. The use of VPPs can also help maximize the renewable content within microgrids while reducing operational costs and emissions, focusing on grid stability (Naval and Yusta, 2021).

Key Differences Between Microgrids and VPP

- Micro-grids can be both grid-connected or off-grid systems, VPP's are always grid connect systems.

- Micro-grids can 'isolate' themselves, allowing them to function independently from the grid. VPPs cannot since they are a combination of resources using mostly exisiting grid infrastructure, so when the grid is down, a VPP is unable to deliver power.
- Micro-grids are dependent upon inverters and smart switches, VPP's are dependent upon smart meters and other associated technology.
- Micro-grids have physical electrical boundaries, VPP's don't because the geographic area, combination of participating resources i.e batteries, and the customers can be changed.
- Microgrids functionally require some capacity for local storage such as battery systems. VPPs can function with or without the presence of a storage system.
- Microgrids are normally traded only in the form of retail distribution (i.e., focused on end-user supply), while the VPPs are traded on a wholesale market.
- Microgrids currently face legal and political obstacles, while VPPs operate with fewer restrictions or legal challenges.

Overall, micro-grids and VPPs offer similar reliability, resiliency and economy for the end user and distribution utility. It really comes down to scale. The goal is to deliver benefits to stakeholders within a defined network boundary or a large area, even including the grid as a whole. Microgrids, virtual power plants, and other distributed energy systems offer a variety of advantages and tradeoffs. Choosing an optimized solution is a complex task, as specific site requirements and energy needs vary significantly: one size (or solution) does not fit all.

VPP AND MICROGRIDS: A REVIEW

Microgrids and virtual power plants (VPPs) are two remarkable solutions for reliable supply of electricity in a power system. Since these structures include distributed energy resources (DERs), scheduling of these resources is then very important (Mancarella, 2012; Mancarella, 2014). Given the lack of current standards, a variety of microgrid and VPP models are therefore on the increase. Some of them focus on their reliability, while other models concentrate on the max- imization of the economic opportunities by selling the excess energy services to the larger networks (Pedrasa, Spooner, and MacGill, 2011; Su, 2012).

Some literature reviews have been published on the topic of microgrids and VPP concepts by focusing on DERs to overcome concerns in power systems. Some features of microgrids are investigated in (Liang and Zhuang, 2014), and a literature review on the stochastic modeling and optimization tools for a microgrid is provided. The description of microgrid design principles considering the operational concepts and requirements arising from participation in active network management is presented in (Palizban et al., 2014). The paper proposes the application of IEC/ISO 62264 standards to microgrids and VPPs, along with a review of micro- grids, including advanced control techniques, energy storage systems, and market participation. (Palizban et al., 2014) describes the developed operational concepts of microgrids that have an impact on their participation in active network management for achieving targets. The paper addresses the principles behind island-detection methods, black-start operation, fault management, and pro- tection systems along with a review of power quality. Since the concept of smart controlled DER merits consideration, reference (Nikonowicz and Milewski, 2012) reviews some VPP ideas and gives a general overview of VPP. (Aghaei et al., 2016) reviews the challenges and the problems caused by charging/discharging of plug in electric vehicles and investigates their capabilities as a solution to integrate the renewable energy sources and DR programs in power systems. In (Aien et al., 2016), a review on uncertainty modeling methods for power system studies is given that makes sense about the strengths and weakness of these methods. The literature review in (Basak et al., 2016) reveals that the integration of DERs, operation, control, power quality issues, and stability of microgrid system should be explored to implement microgrid successfully in real power scenario. Reference (Behera et al., 2015) gives an idea about different optimization techniques, their advantage and disadvantage with respect to a wind farm. The main objective of (Bouzid et al., 2015) is to give a state-of-the-art description for distributed power generation systems based on renewable energy and to explore the power converters connected in parallel to grids which are dis tinguished by their contribution to the formation of the grid vol tage and frequency. The study performed in (Dhillon et al., 2014) aimed to review the basics of wind energy and pumped storage plant system along with their current status, applications, and challenges in volved in their operation under deregulated market and optimization techniques used in the scheduling. (Fathima and Palanisamy, 2015) reviewed the concept of hybrid renewable energy systems and application of optimization tools and techniques to microgrids. In this reference, a framework of diverse objectives has been outlined for which optimization approaches were applied to empower the microgrid. A review of modeling and applications of renewable energy generation and storage sources is also presented. In (Gamarra and Guerrero, 2015), the technical literature on optimization techniques applied to microgrid planning is reviewed and the guidelines for innovative planning methodologies focused on economic feasibility are defined. Also, some trending techniques and new microgrid plannin gap- proaches are pointed out. Reference (Haider et al., 2016) presents an overview of the literature on residential DR systems; load-scheduling techniques; and the latest technology that supports residential DR applications. Furthermore, challenges are highlighted and analyzed to become relevant research topics with regard to the residential DR of smart grid. The literature review shows that most DR schemes suffer from an externality problem that involves the effect of high-level customer consumption on the price rates of other customers, especially during peak period. In (Hare et al., 2016), a review of different failure modes occurring in various microgrid components is presented and also a review on various fault diagnosis approaches available in the technical literature is provided. Reference (Kaur et al., 2016) presents a literature review on microgrid central controller. The evolution and advancement of microgrid central controller technology is explored and presented in a compact form. The classification of microgrid central controllers based on the outcomes found in the process of review is proposed. The role of central controller in the domains of microgrid protection, stability, and power quality are also summarized. In Khan et al. (2016), a review of existing optimization objectives, constraints, solution approaches, and tools used in microgrid energy management is presented. The contribution of (Romo and Micheloud, 2015) is to apply the literature review to the power quality problems and to test the mina real distribution system that has plugin electric vehicles and photovoltaic panels. The results show that a coordinated delay charge mode reduces loading on transformers at peak hours and improves voltage regulation. Additionally, it is shown that photovoltaic panels introduce a power factor reduction during day time in the main feeder. Reference (Siddaiah and Saini, 2016) offers a review of the research work carried out in planning, configurations, modeling, and optimization techniques of hybrid renewable energy systems for off-grid applications. This paper presents a review of various mathematical models proposed by different researchers. These models have been developed based on objective functions, economics, and reliability studies involving design parameters. Reference (Vandoorn et al., 2013) provides a survey of control strategies for the converter interfaces of the DERs and shows detailed figures of the control schemes. In (Yang et al., 2015), scheduling methods are reviewed and categorized based on their computational techniques to integrate plugin electric vehicles and then, various existing approaches covering analytical scheduling are surveyed. Reference (Yanine et al., 2014) proposes such strategies for integrating hybrid micro-generation power systems into the grid through home o- static control a same anstoreconcile power supply and DR management. These strategies can be designed and implemented in the microgrid supervisory control system for the purpose of eliciting energy efficiency and thriftiness in consumers to build energy sustainability in the system. Topologies and control strategies of multi-functional grid-connected inverters are reviewed in (Zeng et al., 2013), and detailed explanation, comparison, and discussion on multi- functional grid-connected inverters are achieved. The study in (Nosratabadi et al., 2017) reviews the scheduling of distributed energy resources according to different aspects, such as modeling techniques, reliability, environmental impact, and uncertainties. This review is based on the comparison of microgrids and VPPs. Reference (Yavuz et al., 2019) presents the different types of VPPs and their characteristics, communication technologies, and optimization and prediction algorithms. References (Mashhour and Moghaddas-Tafreshi, 2009; Zhang et al., 2019) provide an overview of microgrid and VPP operations. In (Ghavidel et al., 2016), the authors describe the components (generation resources, storage, and flexible loads) that compose a VPP. Articles (Cheng et al., 2019; Lv et al., 2017) provide an overview of VPP composition and the optimization of its energy resources. The authors of (Nikonowicz and Milewski, 2012) present arguments regarding the structure and control methods of VPPs. Other papers focus on the analysis of tools for the design and assessment of renewable energy systems. In References (Bahramara et al., 2016; Abbaszadeh et al, 2020), the authors propose the modeling and technical-economic optimization of hybrid renewable energy systems by using HOMER software, a powerful tool for the design of renewable energy sites.

The VPP contains numerous generation and distribution units to supply and meet the load demand. Therefore, an optimization strategy is required in VPP to handle the smooth operations between the sources and the consumers. Several optimization strategies, models, and algorithms are proposed in the literature and utilized in different applications (Podder et al., 2020).

In (Arslan and Karasan, 2013), an energy management considering model that minimizes system cost based on linear modeling of VPP has been presented. An analysis of cost and emissions impacts has been done and a PHEV-penetration network has been introduced. In addition, a sensitivity analysis of cost and emissions of DER and gasoline has been performed. In (Zamani et al., 2016), a probabilistic method using Electrical and thermal energy resources to minimize cost and maximize profit has been presented. The work considered Point Estimate Model (PEM) and Energy and Reverse Scheduling method to deal with uncertainties. The authors in (Mnatsakanyan and Kennedy, 2015) proposed A novel demand response model to ensure maximum profit with minimum cost. The work focused essentially on modeling a fair billing system that does not depend on forecast values of future demand. The proposed billing system depends on customers' behavior and demand response. In (Liu et al., 2018), a combined interval and deterministic optimization approach has been presented for better risk management. The deterministic optimization with less computation time concerns system profit and handles uncertainties. In (Caldon et al., 2004), non-linear minimization techniques were considered to minimize production cost with feasible power distribution to consumers. The work considered cost function in the power distribution system and performs the test with and without Combined Heat Power. In (Raab, 2011), A direct hierarchical and distributed control approach applied an EV management module (EVMM) that minimizes cost has been proposed. It introduces three control strategies: direct, hierarchical, and distributed control strategy. The control directly controls power flow with the distributed control based on the market price. In (You et al., 2009), a market-based VPP model has been peresented for the optimal power distribution in the electricity market. It considers the general bidding and price signal scenarios for system operation. In (Salmani et al., 2009), a multi-objective optimization problem for optimal operation of VPP focusing on economic stability and maximizes profit has been sought. In (Tascikaraoglu et al., 2014), an adaptive load dispatch and forecasting strategy for Solar, wind, hydrogen, and thermal power system have been done. The forecasting algorithm predicts future values and the load dispatch algorithm sustains the optimal operation. In (Almadhor, 2019), a generation driver control strategy that introduces mesh network infrastructure has been proposed for optimal power distribution from generation units to loads. The objective was to reduce transmission loss with emphasis on load management. The authors in (El Bakari and Kling, 2012) proposed a VPPoptimization scheme that considers both centralized and decentralized operation. They analyzed the irregularity and uncertainties in the energy market and considered demand-side management. In (Iacobucci et al., 2016), an optimization model for SAEV and VPP for Solar, wind, and dispatch generation units has been considered. The optimal operation between SAEV and VPP with microgrids that saves power and minimizes cost is searched considering weather conditions and transportation system. In (Ju et al., 2019), a multi-objective robust scheduling model for a WPP, PV and gas storage tank to maximizes profit and minimizes risks has been proposed. The system considers power to the gas energy conversion process and demand response and uncertainties. A three-stage algorithm for optimal solution: payoff table, fuzzy linearization, and rough weight calculation has been presented. In (Kahlen et al., 2018), a mixed rental-trading strategy to maximize profit and minimize system cost has been proposed. An optimal charging and discharging platform has been peresented with prediction of future demand response pattern. In (Qiu et al., 2017), a two-stage optimal scheduling model to first maximize profit of the DA market, and second minimize system costing in the RT market. In (Hernández, 2013), a multi-agent system (MAS) has been proposed for optimal modeling of VPP. The forecasting algorithm is based on ANN and it predicts future demands and customer behavior considering different level of MAS for optimal operation. The authors in (Gong et al., 2011) proposed fuzzy multiple objective optimizationalgorihms for system stability, maximum profit and minimum cost. A two-step interactive satisfactory optimization method is employed to maintain the priority level as well as an acceptable level with an optimal power management considering priority requirement. In (Dimeas and Hatziargyriou, 2007), an agent-based control strategy for PV generator, wind turbine, battery, diesel generator and CHP plant has been presented. The authors introduced three-level MAS: field level, management level, and enterprise-level to facilitate VPP control and operation with focus on economic efficacy. In (Yousaf et al., 2017), to maximize system profit, an intelligent auto-control System (IAS) for optimal power distribution between sources and loads has been presented. The scheme reduces transmission loss and sustains system stability and reliability. In (Skarvelis-Kazakos et al., 2013), a high accuracy multi-agent system (MAS) based on hierarchical control structure to control carbon emission considering the environmental efficacy has been proposed.

REVIEW ABOUT LEADING VPP DEPLOYMENT EXPERIENCES

Virtual power plant market is growing rapidly. As from the prediction, its market is going to rise from 191.5 million from 2016 to 1,187.5 million by 2023 (Inkwood Research., 2018). Major drivers are shift from centralized generation to distributed, wide integration of renewable energy sources and rising demand for cloud-based energy management systems. The total global VPP capacity in 2019 was just below 4 GW, and this is forecast to grow to 35 GW by 2028, a compound annual growth rate (CAGR) of 27.1%.

Figure 2. Total VPP Market Forecast in $m and MW
Source: Guidehouse Total VPP Market Forecast in $m and MW

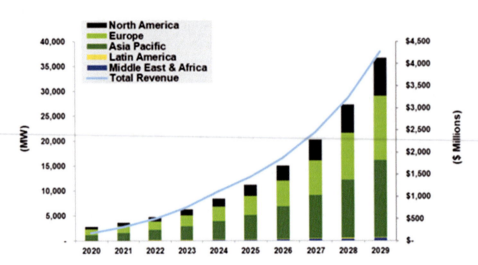

South Australia: In South Australia

AGL Energy reports they are aggregating the battery systems of 1,000 households with rooftop solar PV in Adelaide to create a virtual power plant (VPP) totaling 5 megawatts in capacity. The energy retailer is working with software provider Enbala to optimize DER coordination and maximize the overall benefits across the utility customer, wholesale markets (energy and ancillary) and network service value streams. AGL's VPP could be the largest in the world sometime in 2021; it is among the most sophisticated due to the variety of vendor assets involved (three different kinds of batteries) and grid services rendered. Almost 25 percent of homes within AGL's service territory (1.7 million prosumers) feature rooftop solar PV systems, illustrating the scope of potential assets available to include in VPPs in Australia, one of the world's hotspots when it comes to DER and VPP innovation. For some lessons learned from the many VPP demonstration projects happening in Australia please see AEMO's excellent website.

Jeju Island in South Korea Aiming to Be Carbon-free Paradise

In South Korea, Jeju Island is building a state-of-the-art VPP exploring the nexus between smart transportation, smart electricity services and smart grid upgrades. South Korea since 2009 has been pursuing aggressive advances in smart grid technology with Jeju Island serving as a pilot. Jeju Island's plans to rely upon 100 percent carbon-free energy for electricity and transportation by 2030 aiming to become the world's 'first carbon-free island'. The PEV population is forecast to reach 371,000 by 2030, supported by 75,513 charging stations, a substantial potential grid resource. EVs and their ability to provide grid support will be managed by recognizing the state of charge of the mobile batteries while optimizing the market-responsive behavior of the EV aggregators. The lessons learned from these tests will help this EV fleet become an integral part of the project's strategy to address the variability of wind and solar energy as all fossil fuels are phased out.

Japan Plans Largest VPP In the World

Japan remains short on capacity following the Fukushima nuclear plant accident in 2011, *and* has a long, long history exploring DERs. As a result, it is developing the largest behind-the-meter (BTM) virtual power plant (VPP) in the world. The VPP will aggregate approximately 10,000 DERs, initially focussing on batteries, but with a growing scope to include solar, EVs and smart home thermostats, and will also be capable of being transformed into a microgrid. The project will provide residential consumers with batteries at no upfront cost, and in the event of a power outage, will act as an emergency power source for the resident. In exchange, customers will allow the company to access the batteries in market capacity transactions. Japan and South Korea both pledged to become carbon neutral by 2050 in October, 2020. Japan is the world's third-largest economy and fifth-biggest emitter. The Republic of Korea is the world's 11th largest economy and sixth largest exporter.

England's Most Advanced VPP

In Cornwall, England, Centrica and sonnen have installed a network of 100 domestic batteries to form the UK's most advanced VPP. Cornwall's local energy market will deliver up to 585 kilowatt-hours of battery capacity and up to 169 kilowatts of solar power to local grid company Western Power Distribution and to the national transmission system operator, National Grid. The residential storage assets will be aggregated through a platform provided by Shell-owned (and Avolta Energy partner) energy storage company sonnen. Cornwall's local energy market will deliver up to 585 kilowatt-hours of battery capacity and up to 169 kilowatts of solar power to local grid company Western Power Distribution and to the national transmission system operator, National Grid. The residential storage assets will be aggregated through a platform provided by Shell-owned (and Avolta Energy partner) energy storage company sonnen.

RE IN ALGERIAN ELECTRICITY SECTOR

At the end of 2013, the installed capacity of electricity generation reached 15.1 GW. This is an increase of about 18% compared to the precedent year and due to the new power plants being installed and starting operation. Between 2001 and 2013, electricity production rose from 26,250 GWh to 57,397 GWh. The main source for the production of electricity is gas with a relative percentage to the total amount produced of over 92%. Although there are other sources of electricity, namely oil and hydro-power, these play only a minor role.

Algerian Policy Regarding RE

Algeria adopted a policy of green energy and launched an ambitious renewable energy development and energy efficiency program. This vision of the Algerian government focuses on the enhancement of renewable resources such as solar energy and their utilization to diversify energy sources. This enters in the global sustainable energetic era.

The renewable energy program targets providing a local power level of 22000 MW from renewable sources in 2030. The energy efficieny program has as objective to set up an energy reduction of 63 millions of Total Primary Energy (TPE) in all sectors by 2030. This is done by introducing high performance lighting, thermal isolation and solar water-heating, clean fuels (LPG and LNG) and high-efficieny equipment. This program allows a reduction in CO_2 Emission of 193 million tonnes.

The main objective of the Algerian Renewable Energy and Energy Efficiency Development Plan is to expand usage of renewable energies and to diversify energy sources in the country. The policy aims to install 22,000 MW of power generating capacity from renewable sources between 2011 and 2030 (of which 12,000 for internal usage and 10,000 MW for export) and meet 20% of electricity generation from renewables by 2030. Renewable energy development will drive sustainable economic development of the country, increase energy security supply in Algeria and will have job-creation factor. Solar energy (both solar PV and solar thermal) is recognized by the Algerian government as a primary renewable technology to be developed. The potential for wind, biomass, geothermal and hydropower energies is comparatively very small. The Plan forecasts that solar electricity production will increase up to 37% of total national electricity production by 2030. Greater energy efficiency will be achieved by:

1. Improvement of heat insulation of buildings;
2. Development o solar water heating;
3. Promotion of co-genaration;
4. Developing colas cooling systems;
5. Conversing simple cycle power plants to combined cycle power plants, wherever possible;
6. Desalinating brackish water using renewable energy and
7. Substituting all mercury lamps by sodium lamps.

Renewable Energy Potential in Algeria

Solar Potential

Due to its geographical location and its large area, Algeria possesses the highest solar potential in the world. The sunstroke duration over the quasi-whole national territory exceeds 2000 hours annually and may attend 3900 hours at high plateaus and Sahara. The amount of energy received annually over an area of $1m^2$ is $3kWh/m^2$ in the North and exceeds $5.6 kWh/m^2$ in the South.

Wind Potential

The wind ressources in Algeria vary from one region to the other. This is due to the topographic and clymatic diversity. Indeed, Algeria is divided mainly into two distinct geographic zones. The first is the Miditerranean North having a coast of 1200 km and two moutain series: The Tellian Atlas and the Saharian Atlas between which are sandwished the plains and high plateaus. The second is the South characterized by a Saharian climate.

Figure 4 shows that the wind speed in the South is higher than the North and in particular in the South-East. The wind speed exceeds 7 m/s with values attaining 8 m/s in the region of Tamantasset (Ain Amguel).

For the North, the average wind speed is smaller. However, some micro-climates exista t coastal sites such as Oran, Bejaïa and Annaba. In the high plateaus of Tébessa, Biskra, M'sila and El bayadh, the average wind speed may reach 6 to 7 m/s.

Figure 3. Average annual direct global irradiation map in Algeria (Period 2002-2011)

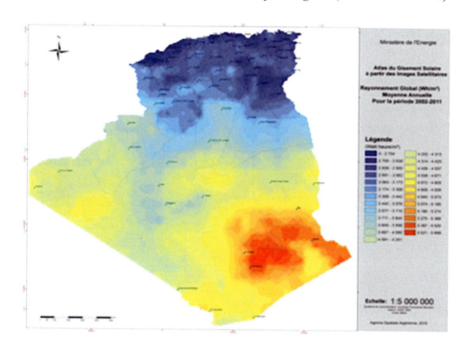

Geothermal Potential

The compilation of geological, geochemical and geophysical data allows to identify more than 200 hot sources in the North. One third of these sources have temperature values highre than 45°C. There are sources which can attain high temperature values of about 118°C at Biskra. Some srudies on the thermal gradient have allowed the discovery of three zones with a gradient exceeding 5°C/100m. These are:

- Zone of Relizane and Mascara
- Zone of Aïne Boucif and Sidi Aïssa
- Zone of Guelma and Djebel El Onk

Hydraulic Potential

The golbal precipitations over the Algerian territory are very important and are estimated at 65 billion of m^3. However, due to reduced precipitation days, concentration on limited areas, high evaporation and quick evacuation to the sea, these quanitities of precipitation are not very beneficial to the country.

Typically, the surface ressources decrease from North to south. 25 billions of ressources in m^3 turn out to be useful and renewable from which 2/3 comes from surface ressources. 103 dam zones have been identified and only 50 dams are exploited.

Figure 4. Average Annual Wind speed at 50 m (Period: 2001-2010)

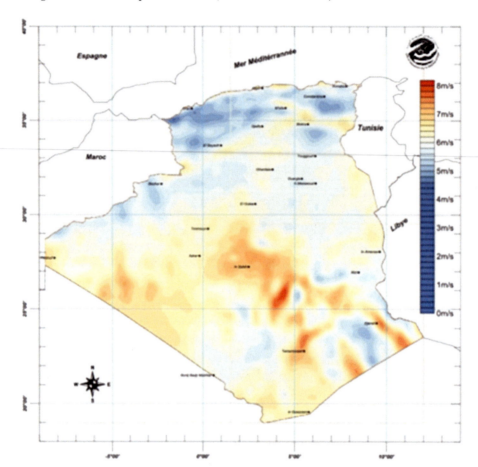

The Energy Sector Perspectives for Algeria

The Algerian sector perspectives with respect to electricity and environmental issues are summarized in table 1. As the fossil fuels are the main sources of electricity generation, Algeria has adopted a global strategy to mitigate climate change, with a view to achieving sustainable development. In this regard, Algeria established a regulatory framework, government actions and incentive measures that should be implemented over the period 2021–2030, to fight against climate warming. The proposed strategy is ambitious and aims at significantly reducing greenhouse gas (GHG) emissions. This strategy concerns mainly three kinds of GHG emissions: CO2, Methane (CH2), and Nitrous dioxide (N2O). A set of investment projects will be set up in several activity sectors, such as energy, industry, agriculture and forestry, transport, construction and the environment. The cornerstone of this strategy is the energy sector where a program to promote RE and energy efficiency has been adopted in 2011. Using national financial efforts, Algeria has set up as a perspective to reduce GHG emissions by 7% to 22%, by 2030.

The established RE program considers promoting significant investment in all RE sources (photovoltaic, concentrated solar power (CSP), geothermal, wind, biomass and cogeneration). By 2030, it is expected to produce a total of 22,000 MW by using RE, with 12,000 MW devoted to the national

Table 1. Algeria's targets and actions in the energy sector (Algeria's INDC-UNFCCC, 2015)

Targets (towards 2030)	Intended Actions (towards 2030)
Greenhouse Emission reduced by 7% to 22%	• Electricity from renewable sources up to 30% • Large deployment of High performance lighting • Generalization of Thermal Insulation buildings • Increase percentage LPG and NLG in total energy consumption • Gas flaring reduced to less than 1%

market, and 10,000 MW to exports. The production of this amount of RE implies that more than 300 billion m3 of natural gas will be saved (this quantity represents eight times the national consumption of 2014). Additionally, this implies that 348 Mt CO2 equivalents of CO2 emission could be reduced. The key achievements thus far are listed in table 2.

The VPP/MG Deployment Requirements for Algeria

Communication

Communication is one of the most important basic requirements for the operation of VPP as well of intelligent power networks. Generation units, controllable loads and all agents offering power flow flexibility and aiming on benefits from variable tariffs must use a standardized and reliable communication system (Saoud and Recioui, 2017). Hence, there is a need for a standardized communication language and easy data exchange between all parts (Buchholz et al., 2012). Table 3 lists typical standards for communication in electric power systems and that are necessary for the operation of VPP and the communication with measurement systems, market and the system operator.

Forecast-oriented Power Unit Scheduling

For setting up an optimal power unit schedule, the VPP operator needs to distinguish the individual characteristics of the available resources. This information is not only necessary for an optimal market participation, but also for enabling the system operator to make day-ahead and intra-day forecasts as well as to know the maximum controllable power for critical situations. The demand forecast for high numbers of small consumers can be realized using standardized load profiles. Bigger customers, like

Table 2. RE targets and achived installations for electricity production in Algeria (Algerian Ministry of Energy, 2019); IRENA. Data & Statistics, 2020); Bouznit et al., 2020)

Energy source	2020 Target (MW)	2018 Installed Capacity (MW)	% achieved
Photovoltaic	3000	410	13.67
Wind	1010	10	00.99
Hybrid solar Power	-	25	-
Others	515	-	-
Total	4525	445	09.83

Table 3. Communication standards that fit VPP (CEN/CENELEC/ETSI Joint Working Group, (2011; Gungor et al., 2011; Richter et al., 2018)

Application	Communication standard
Telecontrol	IEC60870, IEV 371
Substation communication	IEC61850
Distribution management	IEC61968
Energy Market Communication	IEC 62325
Metering data exchange	IEC62051/IEC 62056
Energy Management system	IEC61970/61969

industry companies with bilateral contracts, require separate measurement systems in order to meet the requirements for providing dedicated demand side potential. In contrast, a larger percentage of units in the VPP are generation units and especially volatile ones like PV plants and wind turbines. Consequently, an exact forecast for those units is indispensable from the operator 's point of view, e.g., for scheduling the secondary control reserve (Chang, 2014). The forecast for wind and PV generation can be realized with different methods. Therefore, the key factors are wind speed and solar radiation (Moskalenko, 2014).

CONCLUSION

The further the world moves in the direction of the sustainable energy landscape of the future, the more it will be able to use microgrids so that the central grid is more of a backup than a mainstay. Most importantly, this vision of the future is informed both by optimism about the power of technology to help utilities transition to renewable energy in a fast and stable way, and by a genuine awareness that such a grand transition will require coordination. Not just technical coordination, but policy and economic coordination at different levels in one geography and across borders. As has been seen in the pandemic, a crisis reveals the lack of a well-coordinated effort that involves policymakers, industry, and the public.

Algeria is making notable progress in the development of its renewable energy sector, yet challenges remain. Issues related to energy subsidies, renewable tender rules, and the financial and operational health of Sonelgaz, the state-owned utility and counterparty in renewable tenders, present barriers for some investors. Security threats, particularly in the south of the country, combined with recent political changes and protests, may also pose real challenges as well. Yet Algeria's overall trajectory toward a cleaner and more sustainable power sector is clear. While the precise path and timeline of the sector's development will unfold with time, there is clear momentum from the increasingly competitive economics of renewable power technologies, growing consumer demand, and Algeria's financial incentives to diversify its electricity sector. For investors with the appropriate risk appetite, know-how, and financial wherewithal, Algeria's clean power sector represents an interesting investment opportunity.

VPP can be useful in development of RES and increasing of their contribution to net power generation. This is extremely important considering the ever-increasing electricity demand and primary energy sources depletion. The deployment of VPP in Algeria can be part of energetic transition and has many benefits regarding environmentally friendly energy generation. The resource data and the the vast territory of Algeria revealed a large potential in terms of mainly wind and solar energy that can be exploited to build power plants. Although the renewable energy plants could be situated in a dispersed way throughout the country, some regions in the South Algeria can constitute areas where plants are concentrated and hence ideal locations of VPP. The placement of VPP on these regions may further boost the deployment of new renewable energy sources and manage the control over the electrical grid. Actions to be taken to initiate the VPP deployment in Algeria should imperatively promote domestic energy generation under the government control and attract investments for development of renewable energy plants and advanced communication technologies with reasonable Return on Investment period.

REFERENCES

Abbaszadeh, M. A., Ghourichaei, M. J., & Mohammadkhani, F. (2020). Thermo-economic feasibility of a hybrid wind turbine/PV/gas generator energy system for application in a residential complex in Tehran, Iran. *Environmental Progress & Sustainable Energy, 39*(4), 1–12. doi:10.1002/ep.13396

Aghaei, J., Nezhad, A. E., Rabiee, A., & Rahimi, E. (2016). Contribution of plug-in hybrid electric vehicles in power system uncertainty management. *Renewable & Sustainable Energy Reviews, 59*, 450–458. doi:10.1016/j.rser.2015.12.207

Aien, M., Hajebrahimi, A., & Fotuhi-Firuzabad, M. (2016). A comprehensive review on uncertainty modeling techniques in power system studies. *Renewable & Sustainable Energy Reviews, 57*, 1077–1089. doi:10.1016/j.rser.2015.12.070

Algeria's INDC-UNFCCC. (2015). *Algeria's Intended Nationally Determined Contribution (INDC) to Achieve the Objectives of the United Nations Framework Convention on Climate Change (UNFCCC).* Available online: https://www4.unfccc.int/sites/submissions/indc/Submission%20Pages/submissions.aspx

Algerian Ministry of Energy. (2019). *New Energies, Renewables and Energy Management* [Energies Nouvelles, Renouvelables et Maitrise de l'Energie.]. Algerian Ministry of Energy.

Almadhor, A. (2019, January). Intelligent control mechanism in smart micro grid with mesh networks and virtual power plant model. In 2019 16th IEEE Annual Consumer Communications & Networking Conference (CCNC) (pp. 1-6). IEEE. doi:10.1109/CCNC.2019.8651822

Arslan, O., & Karasan, O. E. (2013). Cost and emission impacts of virtual power plant formation in plug-in hybrid electric vehicle penetrated networks. *Energy, 60*, 116–124. doi:10.1016/j.energy.2013.08.039

Asmus, P. (2010). Microgrids, virtual power plants and our distributed energy future. *The Electricity Journal, 23*(10), 72–82. doi:10.1016/j.tcj.2010.11.001

Bahramara, S., Moghaddam, M. P., & Haghifam, M. R. (2016). Optimal planning of hybrid renewable energy systems using HOMER: A review. *Renewable & Sustainable Energy Reviews, 62*, 609–620. doi:10.1016/j.rser.2016.05.039

Basak, P., Chowdhury, S., Halder nee Dey, S., & Chowdhury, S. P. (2012). A literature review on integration of distributed energy resources in the perspective of control, protection and stability of microgrid. *Renewable & Sustainable Energy Reviews, 16*(8), 5545–5556. doi:10.1016/j.rser.2012.05.043

Behera, S., Sahoo, S., & Pati, B. B. (2015). A review on optimization algorithms and application to wind energy integration to grid. *Renewable & Sustainable Energy Reviews*, *48*, 214–227. doi:10.1016/j.rser.2015.03.066

Bouzid, A. M., Guerrero, J. M., Cheriti, A., Bouhamida, M., Sicard, P., & Benghanem, M. (2015). A survey on control of electric power distributed generation systems for microgrid applications. *Renewable & Sustainable Energy Reviews*, *44*, 751–766. doi:10.1016/j.rser.2015.01.016

Bouznit, M., Pablo-Romero, M. D. P., & Sánchez-Braza, A. (2020). Measures to promote renewable energy for electricity generation in Algeria. *Sustainability*, *12*(4), 1468. doi:10.3390u12041468

Buchholz, B. M., Brunner, C., Naumann, A., & Styczynski, A. (2012, July). *Applying IEC standards for communication and data management as the backbone of smart distribution*. In *2012 IEEE Power and Energy Society General Meeting*. IEEE.

Caldon, R., Patria, A. R., & Turri, R. (2004, September). Optimisation algorithm for a virtual power plant operation. In *39th International Universities Power Engineering Conference, 2004. UPEC 2004* (Vol. 3, pp. 1058-1062). IEEE.

CEN/CENELEC/ETSI Joint Working Group. (2011). *Standards for Smart Grids, Final Report*. Author.

CEREFE (2020). *Energy Transition in Algeria: Lessons, State of Play and Prospects for accelerated development of Renewable Energies, (2020 Edition)* [Transition Energétique en Algérie: Leçons, Etat des Lieux et Perspectives pour un Développement Accéléré des Energies Renouvelables, (Edition 2020)]. Commissariat aux Energies Renouvelables et à l'Efficacité Energétique, Premier Ministre, Alger.

Chang, W. Y. (2014). A literature review of wind forecasting methods. *Journal of Power and Energy Engineering*, *2*(04), 161–168. doi:10.4236/jpee.2014.24023

Cheng, L., Zhou, X., Yun, Q., Tian, L., Wang, X., & Liu, Z. (2019, November). A review on virtual power plants interactive resource characteristics and scheduling optimization. In *2019 IEEE 3rd Conference on Energy Internet and Energy System Integration (EI2)* (pp. 514-519). IEEE. 10.1109/EI247390.2019.9061780

Dhillon, J., Kumar, A., & Singal, S. K. (2014). Optimization methods applied for Wind–PSP operation and scheduling under deregulated market: A review. *Renewable & Sustainable Energy Reviews*, *30*, 682–700. doi:10.1016/j.rser.2013.11.009

Dimeas, A. L., & Hatziargyriou, N. D. (2007, November). Agent based control of virtual power plants. In *2007 International Conference on Intelligent Systems Applications to Power Systems* (pp. 1-6). IEEE. 10.1109/ISAP.2007.4441671

El Bakari, K., & Kling, W. L. (2012, May). Fitting distributed generation in future power markets through virtual power plants. In *2012 9th International Conference on the European Energy Market* (pp. 1-7). IEEE. 10.1109/EEM.2012.6254692

Fathima, A. H., & Palanisamy, K. (2015). Optimization in microgrids with hybrid energy systems–A review. *Renewable & Sustainable Energy Reviews*, *45*, 431–446. doi:10.1016/j.rser.2015.01.059

Gamarra, C., & Guerrero, J. M. (2015). Computational optimization techniques applied to microgrids planning: A review. *Renewable & Sustainable Energy Reviews, 48*, 413–424. doi:10.1016/j.rser.2015.04.025

Ghavidel, S., Li, L., Aghaei, J., Yu, T., & Zhu, J. (2016, September). A review on the virtual power plant: Components and operation systems. In 2016 IEEE international conference on power system technology (POWERCON) (pp. 1-6). IEEE.

Gong, J., Xie, D., Jiang, C., & Zhang, Y. (2011). Multiple objective compromised method for power management in virtual power plants. *Energies, 4*(4), 700–716. doi:10.3390/en4040700

Gui, E. M., & MacGill, I. (2018). Typology of future clean energy communities: An exploratory structure, opportunities, and challenges. *Energy Research & Social Science, 35*, 94–107. doi:10.1016/j.erss.2017.10.019

Gungor, V. C., Sahin, D., Kocak, T., Ergut, S., Buccella, C., Cecati, C., & Hancke, G. P. (2011). Smart grid technologies: Communication technologies and standards. *IEEE Transactions on Industrial Informatics, 7*(4), 529–539. doi:10.1109/TII.2011.2166794

Haider, H. T., See, O. H., & Elmenreich, W. (2016). A review of residential demand response of smart grid. *Renewable & Sustainable Energy Reviews, 59*, 166–178. doi:10.1016/j.rser.2016.01.016

Hare, J., Shi, X., Gupta, S., & Bazzi, A. (2016). Fault diagnostics in smart micro-grids: A survey. *Renewable & Sustainable Energy Reviews, 60*, 1114–1124. doi:10.1016/j.rser.2016.01.122

Hasni, T., Malek, R., & Zouioueche, N. (2021). *ALGERIA 100% RENEWABLE ENERGIES: Recommendations for a national energy transition strategy* [L'ALGÉRIE 100% ÉNERGIES RENOUVELABLES: Recommandations pour une stratégie nationale de transition énergétique]. Friedrich Ebert Stiftung.

Hernández, L., Baladron, C., Aguiar, J. M., Carro, B., Sanchez-Esguevillas, A., Lloret, J., Chinarro, D., Gomez-Sanz, J. J., & Cook, D. (2013). A multi-agent system architecture for smart grid management and forecasting of energy demand in virtual power plants. *IEEE Communications Magazine, 51*(1), 106–113. doi:10.1109/MCOM.2013.6400446

Hirsch, A., Parag, Y., & Guerrero, J. (2018). Microgrids: A review of technologies, key drivers, and outstanding issues. *Renewable & Sustainable Energy Reviews, 90*, 402–411. doi:10.1016/j.rser.2018.03.040

Iacobucci, R., McLellan, B., & Tezuka, T. (2018). The synergies of shared autonomous electric vehicles with renewable energy in a virtual power plant and microgrid. *Energies, 11*(8), 2016. doi:10.3390/en11082016

Inkwood Research. (2018). *Virtual Power Plant Market - Scope, Size, Share, Analysis by 2026.* Available at: https://www.inkwoodresearch.com/reports/global-virtual-power-plant-market/

IRENA, Data, & Statistics. (2020). *International Renewable Energy Agency: Abu Dhabi, United Arab Emirates.* Available online: https://www.irena.org/Statistics

Ju, L., Zhao, R., Tan, Q., Lu, Y., Tan, Q., & Wang, W. (2019). A multi-objective robust scheduling model and solution algorithm for a novel virtual power plant connected with power-to-gas and gas storage tank considering uncertainty and demand response. *Applied Energy, 250*, 1336–1355. doi:10.1016/j.apenergy.2019.05.027

Kahlen, M. T., Ketter, W., & van Dalen, J. (2018). Electric vehicle virtual power plant dilemma: Grid balancing versus customer mobility. *Production and Operations Management, 27*(11), 2054–2070. doi:10.1111/poms.12876

Kaur, A., Kaushal, J., & Basak, P. (2016). A review on microgrid central controller. *Renewable & Sustainable Energy Reviews, 55*, 338–345. doi:10.1016/j.rser.2015.10.141

Khan, A. A., Naeem, M., Iqbal, M., Qaisar, S., & Anpalagan, A. (2016). A compendium of optimization objectives, constraints, tools and algorithms for energy management in microgrids. *Renewable & Sustainable Energy Reviews, 58*, 1664–1683. doi:10.1016/j.rser.2015.12.259

Klaassen, E., & van der Laan, M. (2019). *Energy and Flexibility Services for Citizens Energy Communities*. USEF.

Liang, H., & Zhuang, W. (2014). Stochastic modeling and optimization in a microgrid: A survey. *Energies, 7*(4), 2027–2050. doi:10.3390/en7042027

Liu, Y., Li, M., Lian, H., Tang, X., Liu, C., & Jiang, C. (2018). Optimal dispatch of virtual power plant using interval and deterministic combined optimization. *International Journal of Electrical Power & Energy Systems, 102*, 235–244. doi:10.1016/j.ijepes.2018.04.011

Lv, M., Lou, S., Liu, B., Fan, Z., & Wu, Z. (2017, October). Review on power generation and bidding optimization of virtual power plant. In *2017 International Conference on Electrical Engineering and Informatics (ICELTICs)* (pp. 66-71). IEEE. 10.1109/ICELTICS.2017.8253242

Mancarella, P. (2012). Smart multi-energy grid: concepts, benefits and challenges. IEEE PES General Meeting, San Diego, CA, United States.

Mancarella, P. (2014). MES (multi-energy systems): An overview of concepts and evaluation models. *Energy, 65*, 1–17. doi:10.1016/j.energy.2013.10.041

Mashhour, E., & Moghaddas-Tafreshi, S. M. (2009, January). A review on operation of micro grids and virtual power plants in the power markets. In *2009 2nd International Conference on Adaptive Science & Technology (ICAST)* (pp. 273-277). IEEE. 10.1109/ICASTECH.2009.5409714

Mnatsakanyan, A., & Kennedy, S. W. (2014). A novel demand response model with an application for a virtual power plant. *IEEE Transactions on Smart Grid, 6*(1), 230–237. doi:10.1109/TSG.2014.2339213

Mohamed, T., & Hanane, A. (2021). The energy transition in algeria: How to prepare after oil on the horizon 2030? [La transition énergétique en Algérie: comment préparer l'après pétrole à l'horizon 2030?]. *Journal of Economic Sciences Institute, 24*(1), 1367–1382.

Moskalenko, N. (2014). *Optimal Dynamic Energy Management System in Smart Homes*. Otto-von-Guericke-Universität Magdeburg.

Naval, N., & Yusta, J. M. (2021). Virtual power plant models and electricity markets-A review. *Renewable & Sustainable Energy Reviews, 149*, 111393. doi:10.1016/j.rser.2021.111393

Nikonowicz, Ł., & Milewski, J. (2012). Virtual power plants-general review: structure, application and optimization. *Journal of Power Technologies, 92*(3).

Nosratabadi, S. M., Hooshmand, R. A., & Gholipour, E. (2017). A comprehensive review on microgrid and virtual power plant concepts employed for distributed energy resources scheduling in power systems. *Renewable & Sustainable Energy Reviews, 67*, 341–363. doi:10.1016/j.rser.2016.09.025

Palizban, O., Kauhaniemi, K., & Guerrero, J. M. (2014). Microgrids in active network management—Part I: Hierarchical control, energy storage, virtual power plants, and market participation. *Renewable & Sustainable Energy Reviews, 36*, 428–439. doi:10.1016/j.rser.2014.01.016

Pedrasa, M. A. A., Spooner, T. D., & MacGill, I. F. (2011). A novel energy service model and optimal scheduling algorithm for residential distributed energy resources. *Electric Power Systems Research, 81*(12), 2155–2163. doi:10.1016/j.epsr.2011.06.013

Podder, A. K., Islam, S., Kumar, N. M., Chand, A. A., Rao, P. N., Prasad, K. A., Logeswaran, T., & Mamun, K. A. (2020). Systematic categorization of optimization strategies for virtual power plants. *Energies, 13*(23), 6251. doi:10.3390/en13236251

Pudjianto, D., Ramsay, C., & Strbac, G. (2007). Virtual power plant and system integration of distributed energy resources. *IET Renewable Power Generation, 1*(1), 10–16. doi:10.1049/iet-rpg:20060023

Qiu, J., Meng, K., Zheng, Y., & Dong, Z. Y. (2017). Optimal scheduling of distributed energy resources as a virtual power plant in a transactive energy framework. *IET Generation, Transmission & Distribution, 11*(13), 3417–3427. doi:10.1049/iet-gtd.2017.0268

Raab, A. F., Ferdowsi, M., Karfopoulos, E., Unda, I. G., Skarvelis-Kazakos, S., Papadopoulos, P., . . . Strunz, K. (2011, September). Virtual power plant control concepts with electric vehicles. In *2011 16th International Conference on Intelligent System Applications to Power Systems* (pp. 1-6). IEEE. 10.1109/ISAP.2011.6082214

Recioui, A., & Bentarzi, H. (Eds.). (2020). *Optimizing and Measuring Smart Grid Operation and Control*. IGI Global.

Richter, A., Moskalenko, N., Hauer, I., Schröter, T., & Wolter, M. (2017, June). Technical integration of virtual power plants into German system operation. In *2017 14th International Conference on the European Energy Market (EEM)* (pp. 1-6). IEEE. 10.1109/EEM.2017.7981876

Romo, R., & Micheloud, O. (2015). Power quality of actual grids with plug-in electric vehicles in presence of renewables and microgrids. *Renewable & Sustainable Energy Reviews, 46*, 189–200. doi:10.1016/j.rser.2015.02.014

Salmani, M. A., Tafreshi, S. M., & Salmani, H. (2009). Operation optimization for a virtual power plant. *Proceedings of the 2009 IEEE PES/IAS Conference on Sustainable Alternative Energy (SAE)*, 1–6. 10.1109/SAE.2009.5534848

Saoud, A., & Recioui, A. (2017). A review on Data communication in smart grids. *Algerian Journal of Signals and Systems, 2*(3), 162–179. doi:10.51485/ajss.v2i3.42

Siddaiah, R., & Saini, R. P. (2016). A review on planning, configurations, modeling and optimization techniques of hybrid renewable energy systems for off-grid applications. *Renewable & Sustainable Energy Reviews, 58*, 376–396. doi:10.1016/j.rser.2015.12.281

Skarvelis-Kazakos, S., Rikos, E., Kolentini, E., Cipcigan, L. M., & Jenkins, N. (2013). Implementing agent-based emissions trading for controlling Virtual Power Plant emissions. *Electric Power Systems Research*, *102*, 1–7. doi:10.1016/j.epsr.2013.04.004

Su, W., & Wang, J. (2012). Energy management systems in microgrid operations. *The Electricity Journal*, *25*(8), 45–60. doi:10.1016/j.tej.2012.09.010

Tascikaraoglu, A., Erdinc, O., Uzunoglu, M., & Karakas, A. (2014). An adaptive load dispatching and forecasting strategy for a virtual power plant including renewable energy conversion units. *Applied Energy*, *119*, 445–453. doi:10.1016/j.apenergy.2014.01.020

Vandoorn, T. L., De Kooning, J. D. M., Meersman, B., & Vandevelde, L. (2013). Review of primary control strategies for islanded microgrids with power-electronic interfaces. *Renewable & Sustainable Energy Reviews*, *19*, 613–628. doi:10.1016/j.rser.2012.11.062

Verkade, N., & Höffken, J. (2019). Collective energy practices: A practice-based approach to civic energy communities and the energy system. *Sustainability*, *11*(11), 3230. doi:10.3390u11113230

Yang, Z., Li, K., & Foley, A. (2015). Computational scheduling methods for integrating plug-in electric vehicles with power systems: A review. *Renewable & Sustainable Energy Reviews*, *51*, 396–416. doi:10.1016/j.rser.2015.06.007

Yanine, F. F., Caballero, F. I., Sauma, E. E., & Córdova, F. M. (2014). Building sustainable energy systems: Homeostatic control of grid-connected microgrids, as a means to reconcile power supply and energy demand response management. *Renewable & Sustainable Energy Reviews*, *40*, 1168–1191. doi:10.1016/j.rser.2014.08.017

Yavuz, L., Önen, A., Muyeen, S. M., & Kamwa, I. (2019). Transformation of microgrid to virtual power plant–a comprehensive review. *IET Generation, Transmission & Distribution*, *13*(11), 1994–2005. doi:10.1049/iet-gtd.2018.5649

You, S., Traeholt, C., & Poulsen, B. (2009, June). A market-based virtual power plant. In *2009 International Conference on Clean Electrical Power* (pp. 460-465). IEEE.

Yousaf, W., Asghar, E., Meng, H., Songyuan, Y., & Fang, F. (2017, October). Intelligent control method of distributed generation for power sharing in virtual power plant. In *2017 IEEE International Conference on Unmanned Systems (ICUS)* (pp. 576-581). IEEE. 10.1109/ICUS.2017.8278411

Zamani, A. G., Zakariazadeh, A., & Jadid, S. (2016). Day-ahead resource scheduling of a renewable energy based virtual power plant. *Applied Energy*, *169*, 324–340. doi:10.1016/j.apenergy.2016.02.011

Zeng, Z., Yang, H., Zhao, R., & Cheng, C. (2013). Topologies and control strategies of multi-functional grid-connected inverters for power quality enhancement: A comprehensive review. *Renewable & Sustainable Energy Reviews*, *24*, 223–270. doi:10.1016/j.rser.2013.03.033

Zhang, G., Jiang, C., & Wang, X. (2019). Comprehensive review on structure and operation of virtual power plant in electrical system. *IET Generation, Transmission & Distribution*, *13*(2), 145–156. doi:10.1049/iet-gtd.2018.5880

Chapter 6
Does the Development of New Energy Vehicles Promote Carbon Neutralization?
Case Studies in China

Poshan Yu
Soochow University, China & Australian Studies Centre, Shanghai University, China

Shucai Xu
School of Vehicle and Mobility, Tsinghua University, China

Ziling Cheng
Independent Researcher, China

Michael Sampat
Independent Researcher, Canada

ABSTRACT

This chapter aims to study whether and how current practices based on the development of new energy vehicles can help promote carbon neutrality in China and in turn contribute to the improvement of global issues related to climate change. Meanwhile, this chapter will explore the role of the government in promoting the development of new energy vehicles in the aspect of sustainable development through policies. How do these institutions promote the development of new energy vehicles in various companies and different provinces in China? How will these developments in turn affect stakeholders in various sectors? This chapter will investigate the impact of China's devotion to the field of new energy vehicles on the sustainable development and low-carbon economy issue. Cases from China will be used to illustrate the improvement the new energy vehicle will make to the low-carbon economy. This chapter will provide suggestions for the government to deal with the problems that occur in the field of new energy vehicles and solve the confronted problems in the aspect of climate change.

DOI: 10.4018/978-1-6684-4829-8.ch006

INTRODUCTION

As the environmental damage caused by human economic activities intensifies, the global climate also deteriorates further. Analysis of global climate research shows that our home planet, on which we depend, is slowly warming. The causes of this change are diverse and manifest in both anthropogenic and natural factors. The impact is also obvious, and it is related to the development of nature and human society. This chapter aims to study whether and how current practices based on the development of new energy vehicles (NEVs) can help promote carbon neutrality in China in turn contribute to the improvement of global issues related to climate change. Meanwhile, this chapter will explore the role of the government in promoting the development of NEVs in the aspect of sustainable development through policies. How do these institutions promote the development of NEVs in various companies and different provinces in China? How these developments will in turn affect stakeholders in various sectors. This chapter will investigate the impact of China's devotion to the field of NEVs on the sustainable development & low-carbon economy issue. Cases from China will be used to illustrate the improvement the new energy vehicle will make to the low-carbon economy. This chapter will provide suggestions for the government to deal with the problems that occur in the field of NEVs and solve the confronted problems in the aspect of climate change.

CLIMATE CHANGE, SUSTAINABLE DEVELOPMENT AND CHINA UNDER CITESPACE ANALYSIS

Considering the current situation, slowing down global warming should be on the top of humanity's agenda. Effective actions must be taken as soon as possible to avoid further damage (1850–1900) (He et al., 2022).

Simultaneously, the global climate report 2020 issued by the World Meteorological Organization points out that the concentration of major global greenhouse gases will continue to rise in 2020, and the global average temperature is about 1.2 centigrade higher than that before industrialization (Ahmad & Hossain, 2015). Due to global warming, risks such as extreme weather disasters and the outbreak of global infectious diseases may occur, which will hinder economic development and exacerbate geopolitical tensions.

With the analysis conducted by CiteSpace, some relevance can be figured out with the keyword Climate change (Figure 1). Air quality, CO_2 emission, and CO_2 reduction are highly related to the topic. Furthermore, with the combination of the frequency of the terms, energy consumption can be involved. Hence, the following parts will investigate the impact of China's devotion to the field of NEVs on the sustainable development & low-carbon economy issue.

In response to global climate change, 175 countries around the world signed the Paris Agreement in April 2016, committing to limit the rise in global average temperatures to less than 2 degrees Celsius compared to pre-industrial times (Jin, 2021).

Afterwards, at the 75th session of the United Nations General Assembly in September 2020, Chinese President Xi Jinping announced that "carbon dioxide emissions will strive to peak by 2030, and efforts will be made to achieve carbon neutrality by 2060".

Achieving carbon peak and carbon neutral is an inherent requirement for promoting high-quality development. To achieve high-quality development, we must unswervingly implement the new devel-

Table 1. Keywords that were used with high frequency for 5 years from 2017 to 2021, analyzed by CiteSpace with the data of 2891 core journals from WOS

Visible	Count	Centra...	Year	Keywords
✔	708	0.01	2017	climate change
✔	301	0.01	2017	impact
✔	189	0.01	2017	model
✔	173	0.01	2017	climate
✔	154	0.02	2017	temperature
✔	135	0.02	2017	variability
✔	96	0.03	2017	trend
✔	90	0.02	2017	co2 emission
✔	79	0.02	2017	emission
✔	76	0.02	2017	precipitation
✔	75	0.01	2017	water
✔	73	0.03	2017	pattern
✔	68	0.02	2017	system
✔	67	0.03	2017	economic growth
✔	66	0.03	2017	management
✔	65	0.01	2017	growth
✔	62	0.02	2017	carbon
✔	58	0.02	2017	biodiversity
✔	56	0.02	2017	response
✔	56	0.02	2017	policy
✔	55	0.01	2017	china
✔	54	0.01	2017	land use
✔	54	0.01	2017	energy consumption
✔	54	0.02	2017	drought
✔	51	0.03	2017	forest
✔	50	0.02	2017	adaptation
✔	47	0.01	2017	dynamics
✔	47	0.01	2017	framework
✔	47	0.01	2017	united states
✔	46	0.02	2017	simulation
✔	46	0.03	2017	diversity
✔	46	0.02	2017	ecosystem service
✔	45	0.01	2017	carbon dioxide
✔	45	0.02	2017	performance
✔	44	0.01	2017	uncertainty
✔	44	0.02	2017	air pollution
✔	43	0.01	2017	conservation
✔	43	0.01	2017	carbon emission
✔	41	0.01	2017	mortality
✔	41	0.01	2017	cmip5
✔	41	0.02	2017	consumption
✔	39	0.02	2017	ecosystem
✔	38	0.01	2017	ocean
✔	37	0.02	2017	renewable energy
✔	36	0.07	2017	city
✔	36	0.01	2017	circulation
✔	36	0.02	2017	projection
✔	35	0.03	2017	greenhouse gas emission
✔	34	0.03	2017	nitrogen

Source: Website of Science

Figure 1. Visualization in terms of the paramount links relating to climate change terms
Source: Website of Science

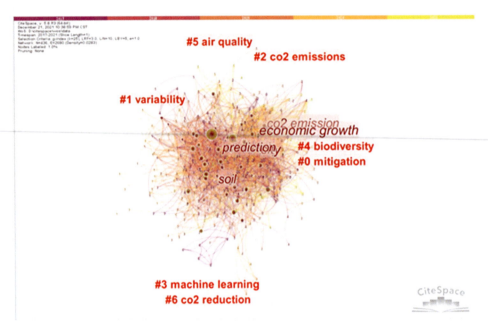

opment concept, adhere to the system concept, take the overall green transformation of economic and social development as the leader, take the green and low-carbon development of energy as the key, accelerate the formation of the industrial structure, mode of production, lifestyle and spatial pattern that conserves resources and protects the environment, and unswervingly take the ecological priority, green and low-carbon road of high-quality development (Zhang et al., 2022). In the past two years, under the stimulating effect of the "double carbon", domestic hotspot industries such as electric vehicles, photovoltaic and wind power have flourished. In the long run, the further organic combination of green transformation of the economy and society and high-quality development will certainly promote the high-end, intelligent and green transformation of traditional industries, facilitate the optimization and upgrading of the entire industrial chain, and promote the quality change, efficiency change and power change of China's economic development (Jiang et al., 2022b), thus shaping a new advantage for China to participate in international cooperation and competition. It is a new era of sustainable development that is dawning. This emerging era is more than simply a move away from fossil fuels. It is a sweeping change that touches every industry, the way people make and consume goods, and the way working-class citizens earn their living.

Under these circumstances, how to balance environmental, economic and social benefits in the context of China's development goals of carbon capping and carbon neutrality (Jin, 2021), and studying the issue of promoting NEVs has become a very urgent problem. China has become the world's largest carbon dioxide emitter and energy consumer. It has been reported that in the year 2016, nearly 30% of global CO_2 emissions came from China (The World Bank, 2021), with serious environmental problems and a high increase rate (figure 2). Therefore, a low-carbon economy has become the best choice for China's development (Shi et al., 2022). Meanwhile, as electrification becomes one of the most promising ways for low carbon transition in the transport sector (Li, Luo, & Song, 2021), the electric vehicle as a clean

Figure 2. Change in per capita territorial carbon dioxide emissions worldwide from 2000 to 2020, by selected countries (measured in %)
Source: globalcarbonatlas.org

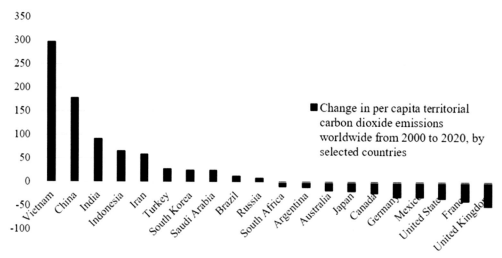

new energy transportation tool has attracted more and more attention to reach proposals of carbon peak and carbon neutrality policy (Ding et al., 2022).

Figure 3. Electric vehicle production forecast - selected countries 2023
Source: Forschungsgesellschaft Kraftfahwesen Aachen; Roland Berger

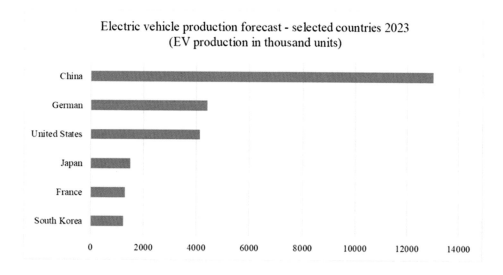

LITERATURE REVIEW

New Energy Vehicle

NEVs [including battery electric vehicles (BEVs), plug-in hybrid electric vehicles (PHEVs) and Fuel cell electric vehicles (FCEVs)], especially BEVs, promise both much higher energy efficiency, and eventual high decarbonization, if and when running on renewable electricity. can effectively reduce carbon emissions in the field of transportation (Tan, Tang, & Lin, 2018).

NEVs and especially electric cars are rapidly changing the outlook of the car industry in China, the largest vehicle market in the world (figure 3) (Hu, Javaid, & Creutzig, 2021). In the five-year plan from 2001 to 2005, the government launched the National High-Tech Research and Development (R&D) Plan for NEVs. The plan pointed out the direction of development of the NEV industry, namely the "three vertical and three horizontal" strategy (Three verticals refers to promoting the development of three types of NEVs, including hybrid electric vehicle, pure electric vehicle, and fuel cell vehicle. Three horizontal means paying attention to the R&D of three technologies, including battery, motor, and electronic control). This initiative indicates that the Chinese government used to attach great importance to the NEVs. Recently, developing the NEVs industry is regarded as one of the essential paths to achieving carbon emission reduction after China established the goal of carbon neutralization and carbon peak (Wang et al., 2021).

Carbon Peak, Carbon Neutral Target

Carbon peak refers to China's commitment to stop the growth of carbon dioxide emissions by 2030, and gradually reduce them after reaching the peak.

Carbon neutral means that an enterprise, group or individual measures the total amount of greenhouse gas emissions produced directly or indirectly within a certain period, and then offsets its carbon dioxide emissions through planting and reforestation, energy-saving and emission reduction, to achieve "zero" carbon dioxide emissions (30·60 goals).

The development of renewable energy, which is an important way to solve environmental protection, has now become a research focus around the world (Wang & Li, 2016). With the announcement that "carbon dioxide emissions will peak by 2030 and efforts will be made to achieve carbon neutrality by 2060", the peak carbon and carbon neutrality targets are beginning to remind China to take action to address the problems posed by climate change. As such, the power sector plays an important role in energy conservation and emission reduction in China, and the impact of power consumption on carbon emissions needs to be considered to reduce environmental pollution. For example, Wang et al. developed an environmental assessment model to examine the relationship between carbon emissions, population, electricity consumption and energy consumption demand (Wang & Li, 2016). By Wiryadinata et al. (2019), a methodology is proposed to identify the main technical and economic drivers influencing the implementation of carbon-neutral energy systems on university campuses.

China's emissions reduction actions will contribute greatly to the improvement of global issues related to climate change. After the outbreak of COVID-19, China has shown strong vitality in restoring social order and economic recovery. China's economy is expected to continue to grow. In addition, China's population will continue to grow until at least 2030 (Li et al., 2018). However, an adjustment in subsidy schemes and the breakout of COVID-19 appear to slow down the uptake of NEVs in the Chinese market (Hu et al., 2021). The sudden outbreak of COVID-19 reduces global NEV sales in 2020 (down 18% compared to 2019), but the long-term outlook remains unaffected and estimated production still infers that China will account for the world's largest share (figure 2). The explosion of the NEV sector in global capital markets during the pandemic is a prime example of this. While many car companies were hit hard by the health crisis in 2020, Tesla bucked the trend; its share price soared by almost 700% in the year, making it the most valuable car manufacturer in the world (Harper, 2020). Meanwhile, some Chinese companies also grasp the opportunity to develop their company.

Low Carbon Economy

An important environmental consequence of subsidies for fossil fuels is that it encourages the substitution of renewable energy, capital and labor with fossil fuels, thus hindering the low carbon transition (Li & Sun, 2018). Low-carbon economy refers to a form of economic development guided by the concept of sustainable development, through various means such as technological innovation, institutional innovation, industrial transformation and new energy development, to minimize the consumption of high-carbon energy sources such as coal and oil (Yu, Weng, & Ahuja, 2022), reduce greenhouse gas emissions and achieve a win-win situation for both economic and social development and ecological environmental protection. The benefits of a low carbon economy are clear. From improving air quality and human health, to reducing the risk of future climate change, to creating jobs and economic opportunities, the benefits of reducing carbon are innumerable (Yang et al., 2021).

Air pollution - caused by fossil fuel-based emissions - has caused respiratory illnesses and even premature death for many people who breathe in these particles (Jiang et al., 2022a). Switching to renewable energy sources such as wind power will help reduce the pollution associated with these diseases.

In addition to cleaning the air we breathe and reducing emissions and the greenhouse gas pollution that warms our planet, it could also help minimize the risk of future climate change - such as more extreme weather events, heat waves and, according to the US EPA, adverse impacts on air and water quality (Wang & Feng, 2020).

GOVERNMENT ROLES IN PROMOTING NEV TO FULFILL THE CARBON PEAK AND CARBON NEUTRAL GOAL IN CHINA

In response to global climate change, the Chinese government has developed a series of environmental policies to control and reduce greenhouse gas emissions (Y.-J. Wang et al., 2021). In 2009, the Chinese government officially announced that by 2020, carbon dioxide emissions per unit of GDP (i.e., carbon intensity) would be reduced by 40%~45% from 2005 levels. In June 2015, the Chinese government submitted to the United Nations a "national independent Contribution Document", proposing that China's CO_2 emissions will peak in 2030, when carbon intensity will be 60%~65% lower than in 2005 (Yu, Jiao, & Sampat, 2022).

On November 2, 2020, the General Office of the State Council issued the "New Energy Vehicle Industry Development Plan (2021-2035)", which proposed that by 2025, new energy vehicle sales should reach about 20% of total new vehicle sales, with pure electric vehicles accounting for about 4.32% (figure 4). According to the path of "carbon peaking" and "carbon neutral", 2030 will be the period of carbon

Figure 4. The estimated share of new vehicle sales by fuel type in China, 2025-2026 (in %)
Source: http://www.catarc.info/; Forward Intelligence (Qianzhan)

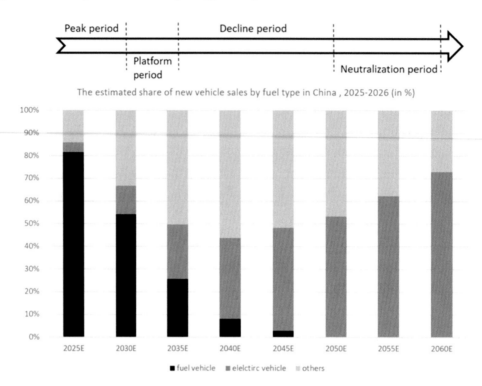

peaking and 2060 will be the year of carbon neutrality. The penetration rate of pure electric vehicles is expected to exceed 70% of the overall vehicle fleet by 2060 (Ge & Lin, 2021).

Several previous studies have confirmed this theory. Guo's study defines the key sectors driving China's energy consumption and CO2 emissions in terms of input-output relationships and demand elasticities. The findings suggest that key sectors in the Chinese economic system not only drive energy consumption and CO2 emissions in other sectors but also consume large amounts of fossil energy and

Figure 5. Share of energy-related CO2 emissions in China 2019, by sector
Source: Enerdata; Climate Transparency

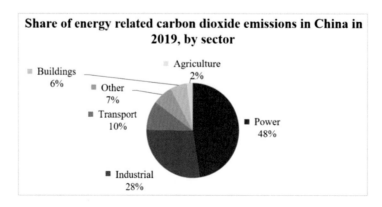

Table 2. Significant policies concerning New energy vehicle and their implications

Title	Issuing authority	Issued date	Main implications
Notice on the Pilot Demonstration and Promotion of Energy-saving and New Energy Vehicles	Ministry of Finance, Science and Technology Ministry, et al.	23/1/2009	The central finance will provide a one-off flat-rate subsidy for the purchase and use of energy-saving and NEVs by demonstration and promotion units in the relevant public service areas of the pilot cities
Notice on the Expansion of Demonstration and Promotion of Energy-saving and New Energy Vehicles in Public Service Areas	Ministry of Finance, Science and Technology Ministry, Ministry of Industry and Information Technology, National Development and Reform Commission (NDRC), et al.	7/6/2010	The Ministry of Finance advances funds for demonstration and promotion grants to pilot cities through provincial finance departments
Notice on Furthering the Pilot Project on Demonstration and Promotion of Energy-saving and New Energy Vehicles	Ministry of Finance, Science and Technology Ministry, Ministry of Industry and Information Technology, National Development and Reform Commission (NDRC), et al.	10/11/2011	1. Proposes the implementation of exemptions from restrictions such as license plate auctions, lotteries and traffic restrictions for NEVs 2. It also vigorously pursues the construction of infrastructure such as charging facilities to facilitate the use of NEVs
The State Council on the Issuance of the Energy-saving and New Energy Vehicle Industry Development Plan (2012-2020)	The State Council	9/7/2012	Defines the positioning of each energy type of vehicle
Notice the continuation of the promotion and application of New Energy Vehicles	Ministry of Finance, Science and Technology Ministry, Ministry of Industry and Information Technology, National Development and Reform Commission (NDRC), et al.	13/9/2013	For the first time, subsidies for NEVs are being rolled back, with range segments and differentiated subsidy standards for pure electric vehicles
Guidance from the General Office of the State Council on accelerating the promotion and application of New Energy Vehicles	The State Council	21/7/2014	Tax incentives for NEVs announced
Notice on the Adjustment of Financial Subsidy Policy for the Promotion and Application of New Energy Vehicles	Ministry of Finance, Science and Technology Ministry, Ministry of Industry and Information Technology, National Development and Reform Commission (NDRC), et al.	29/12/2016	In the Annex "New Energy Vehicle Promotion Subsidy Scheme and Product Technical Requirements", the mass-energy density of the power battery system for pure-electric passenger vehicles is required
Notice on Adjusting and Improving the Financial Subsidy Policy for the Promotion and Application of New Energy Vehicles	Ministry of Finance, Science and Technology Ministry, Ministry of Industry and Information Technology, National Development and Reform Commission (NDRC), et al.	2/2/2018	The subsidy standard for NEVs has been adjusted and a transition period has been set for switching between the new standards, and the implementation plan for the subsidy standard has been further refined
Regulations on the Admission of New Energy Vehicle Manufacturing Enterprises and Products	Ministry of Industry and Information Technology	19/8/2020	To further relax the access threshold, stimulate market vitality, strengthen supervision in and after the event, and promote the high-quality development of China's new energy vehicle industry

Source: www.most.gov.cn ; www.mof.gov.cn ; www.gov.cn ; www.miit.gov.cn

emit large amounts of CO2, stimulated by the demand from other sectors. These sectors should therefore receive close attention from the Chinese government, with road transport being the most important sector (Guo, Zhang, & Zhang, 2018).

In addition, the case studies show that road transport in China has experienced a wide range of development patterns and due to its dominant role in the overall transport sector, effective measures should be taken to help achieve its low carbon development.

For China, as the world's largest energy consumer, the carbon emissions in the transport sector have increased quickly, from 205 Mt in 2000 to 690 Mt in 2016 with the average annual growth rate reaching 7.4%. Meanwhile, transportation is the third large emitter of carbon dioxide (figure 5). With the continued structural change, the energy demand and carbon emission of the transport sector is likely to increase further in China. This motivates a growing focus of attention also in policymaking and within the Chinese government (Hu et al., 2021).

As a previous study showed, the transport sector is a major emitter of greenhouse gases. Four phases have been identified under the national plan: an initial pre-development phase (2001-2011) to protect nascent innovations; a take-off phase (2012-2020) to develop core technologies and expand market share; an acceleration phase (2021-2035) to strengthen this trend by encouraging key technological breakthroughs, infrastructure development and international cooperation; and a final sprint phase (2036 -) further develop the low-carbon transition of China's automotive industry (Wu, Shao, Su, & Zhang, 2021). These phases above illustrate the essential role of government in the Chinese context. Policy incentives are the important driver for the development of the NEV industry in the early stages.

As one of the strategic emerging industries, the NEV industry receives strong support from the Chinese government. Policies about the NEVs are collected in table 2, showing the changing attitude towards the NEVs throughout the past decade. Previous studies suggested that different countries should choose rational policies that were suitable for the development of the NEV industry, and if the government uses financial policies blindly, there are hidden pitfalls (X. Wang et al., 2021). In other words, the policy issued by the government imposed a remarkable impact on the development of the NEVs. This chapter intends to discuss the government's role in promoting NEVs to fulfill the carbon peak and carbon neutralization goals in China. Hence provide the government with some useful suggestions to accelerate the process of developing the NEVs by using the case study in China.

Here we use case 1 to illustrate the effort the Chinese government has taken to promote the development of NEV in Hainan province.

Case: The Development of NEV in Hainan Province China

To speed up the construction of the national ecological civilization pilot zone and help achieve the goal of "carbon peak, carbon neutral", according to the Ministry of Industry and Information Technology on the organization of the pilot application of NEVs for electricity, the Department of Industry and Information Technology of Hainan Province, the People's Government of Hainan Province (2021) formulated the "Hainan Province New Energy Vehicles for Electricity Application Pilot Implementation Plan", which was officially implemented on November 5, 2021.

On February 17, 2022, the Department of Industry and Information Technology of Hainan Province (2022) was informed that in January this year, Hainan Province's additional NEVs increased 571% year-on-year. In the past year, the province's new energy vehicle promotion achievements, new energy vehicle holdings in the national "first square", holdings growth rate ranked first in China. Meanwhile,

Figure 6. Total number of public electric vehicle charging stations in China from 2015 to 2020 (in 1,000s)
Source: EVCIPA; Forward Intelligence (Qianzhan)

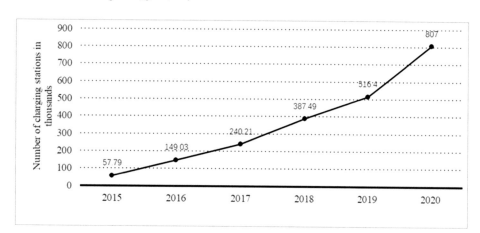

on March 31, 2022, according to the People's Government of Hainan Province (2022), the government issued "Several Measures to Encourage the Use of New Energy Vehicles in Hainan Province in 2022", which accelerates the promotion and application of NEVs in Hainan province, ensure that the proportion of NEVs in new vehicles in our province exceeds 30% in 2022, and promote the realization of the goal of "carbon peak" and "carbon neutral" in the field of transportation.

The Measures encourage the use of new energy trucks, and from January 1 to December 31, 2022, for new vehicles purchased in the province (including postal express and urban logistics distribution vehicles) and registered in the province, after the mileage reaches 30,000 km within one year from the registration of the vehicle, each vehicle can apply for operating subsidies of 30,000 yuan, 20,000 yuan and 10,000 yuan for heavy-duty, medium-duty and light-duty vehicles respectively (https://www.hainan.gov.cn/).

Hainan is the first province in China to propose a 2030 timeline to ban the sale of fuel vehicles, and the first island economy in the world to propose a region-wide clean energy vehicle development strategy (Chen et al., 2022). In recent years, Hainan has continued to introduce policies and measures to accelerate infrastructure development and cultivate the new energy vehicle consumer market.

In 2021, a total of 58,700 NEVs will be promoted in Hainan Province, accounting for 7.2% of ownership, 4.6 percentage points higher than that of the whole country, and 25.8% of new vehicles, 13.1 percentage points higher than that of the whole country, with a growth rate of 91%, ranking first in the country (http://www.stats.gov.cn/).

According to the statistics of the Development and Reform Commission of Hainan Province (2021), more than 20,000 new charging piles were added in Hainan Province, more than 47,200 charging piles were built in total, and 32 exchange stations were built in total, which is the only province in China with 100% full coverage of charging piles built on highways, and the vehicle-pile ratio of NEVs is 2.4:1, which is better than the national average level of vehicle-pile ratio (figure 6).

In 2022, the work report of the Hainan Provincial Government put forward the target of "NEVs accounting for more than 30% of new vehicles". Xu Tao, director of the New Energy Vehicle Development and Supervision Department of the Department of Industry and Information Technology of Hainan Province, said that this year Hainan Province will take effective measures to accelerate the promotion and market consumption of NEVs, study the introduction of incentive policies to stabilize market expec-

Figure 7. Annual sales of new energy vehicles in China 2011-2020, by type
Source: CAAM

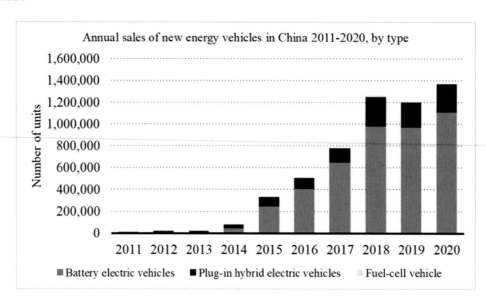

tations, and continue to carry out new energy vehicle activities in rural areas to promote the upgrading of consumption in those areas.

From the perspective of energy structure transformation, Hou (2022) analysis that, although Hainan is one of the lowest carbon emissions in the country, has a "carbon neutral" comparative advantage, if we want to take the lead in the country to achieve a "carbon neutral ", the realization of ecological civilization and high-tech industry linkage development, there are still energy structure adjustment innate deficiencies (Li & Nam, 2022). Meanwhile, the green development of high-tech industry has not yet come online, due to slow technological innovation, lack of industrial support and other issues (Dong et al., 2019). In this regard, he proposed to support Hainan to accelerate the construction of a clean energy island; support Hainan to accelerate the development of high-end manufacturing of NEVs; support the establishment of NEVs in the Hainan national laboratory and international technology innovation cooperation platform.

With the step-by-step promotion of government policies, Hainan Province has a greater incentive to develop and promote NEV and thereby obtain better environmental conditions and economic benefits. From the example of Hainan province, the effectiveness of the policy is greatly illustrated.

WITH THE GOAL OF CARBON PEAK AND CARBON NEUTRALIZATION, WHAT'S THE IMPACT OF NEV ON REDUCING CARBON EMISSIONS TO MITIGATE CLIMATE CHANGE

Today and in the future, carbon emissions have become a hot topic for economic development and automotive transformation (Pichler et al., 2021). In the face of the challenges of climate change, the Chinese government has proposed a "double carbon" target of achieving carbon peaking by 2030 and carbon neutrality by 2060, while major European countries, the US and Japan have also proposed carbon-neutral strategies to promote a green and low carbon transformation of their economies and societies.

In the field of transport carbon emissions, major countries around the world have put forward clearer targets for a zero-carbon transition in the automotive sector.

For example, Germany proposes to electrify all passenger cars by 2050, France proposes to stop using fossil fuels in all passenger cars and light commercial vehicles by 2040, Norway proposes to electrify all passenger cars and light vans sold by 2025, the UK proposes to stop selling fuel-fired passenger cars and vans by 2030, Japan proposes to stop selling new fuel-fired vehicles by 2035, and the US proposes that NEVs account for 50% of new vehicle sales by 2030.

President Xi Jinping, who attended the opening ceremony of the Second United Nations Global Conference on Sustainable Transport by video on 2021 November 14 and delivered a keynote address, pointed out that it is necessary to accelerate the formation of green and low-carbon transport modes, strengthen the construction of green infrastructure, promote new energy, intelligent, digital and lightweight transport equipment, encourage and guide green travel, to make transport more environmentally friendly and travel more low-carbon.

In the aspect of road transport, the number of new energy buses and new energy trucks is increasing, reaching 400,000 and 430,000 respectively by the end of 2019; in express delivery and logistics, the proportion of new energy and clean energy vehicles in the postal delivery vehicle fleet is steadily increasing.

As shown by previous studies, China's tremendous economic progress has been achieved at the expense of the environment. The use of coal-based fossil fuels emits large amounts of greenhouse gases (mainly carbon dioxide) and polluting gases (Tan et al., 2018). The promotion of NEVs can reduce the concentration of greenhouse gases in urban air. At the same time, the development of NEVs is the best option to solve the problem of carbon emissions from vehicles, as exhaust gas purification and treatment technologies are not effective in dealing with the following problems (Bonsu, 2020).

The electric sector is a typical example. It plays an important role in energy conservation and emission reduction in China, and to reduce environmental pollution it needs to consider the impact of electricity consumption on carbon emissions (Wang & Li, 2016). As is demonstrated in the component bar chart (figure 7) below, battery electric vehicles, which mainly used electricity, are a driving force that accounts for a large part of the Chinese market. Therefore, NEVs and especially electric cars are of great significance to solve the problems.

Benefits have been confirmed through many empirical and theoretical studies of NEVs. Therefore, accelerating the promotion of NEVs is an important step to optimizing the energy structure, promoting energy-saving and emission reduction, and developing economic sustainability (Tian, Zhang, Chi, & Cheng, 2021).

To further demonstrate the issue, this chapter will involve two typical cases in China (BYD & LEAD, the former is the company that occupied the largest share of the market in China, the latter is the leading enterprise in the field of power battery).

THE ADDED VALUE OF THE INNOVATIVE OUTCOMES OF NEW POLICIES AND TECHNOLOGY CONGLOMERATION TO NEV

The environmental issues of NEV are worthwhile to lucubrate, which will give guidance for the authorities to take action to deal with the related problems that exist. This chapter demonstrates China's determination as a responsible power to address climate change and pursue a green and low-carbon development path but also invigorates the transition to green and sustainable development in manufacturing sectors.

Simultaneously, policies about the NEVs and Lithium-ion batteries (LIBs) are discussed below, it shows the different methods government used to the problems of over-reliance on subsidy policies.

Case: BYD - The Company Which Occupied the Largest Share of the Market in China

Because reducing carbon emissions is the driving force behind the popularity of electric vehicles, commercial vehicles should rightly be the primary target. Large vehicles with serious carbon emissions, if they can accelerate their electrification, will certainly be of great help in improving atmospheric pollution, which will not underestimate BYD's influence around the world (Masiero, Ogasavara, Jussani, & Risso, 2016).

As a leader in NEVs in China (figure 8), BYD is actively at the forefront of the road to energy efficiency and emission reduction. Since 2021, BYD commercial vehicles have continued to win back-to-back orders in overseas markets (figure 9), achieving zero carbon targets through sustainable development and advanced technology.

BYD has always adhered to the development concept of "technology is king, innovation is the foundation" (https://www.bydauto.com.cn/), using technological innovation to help the transportation industry to save energy and reduce emissions, leading the change of electrification in the automotive industry while opening up the second half of intelligent development.

BYD is the only company in the world that has mastered the core technologies of new energy vehicle chips, battery, motor, electric control and charging support, vehicle manufacturing, etc (Liu & Meng, 2017). By 2020, BYD commercial vehicles will have achieved a comprehensive innovation from the inside out, with a new electric commercial vehicle platform based on the core technology of "26111", and a high level of integration and standardization in the manufacturing of new energy commercial vehicles (https://www.bydauto.com.cn/). Increased production costs caused by constrained technological options will then negatively affect local industry performance. However, this long-standing position positing a substitutive economy-environment relationship has been increasingly challenged, since upward convergence in global environmental regulations may create room for a synergistic one (Li & Nam, 2022).

Driven by the new round of technological revolution, the automotive industry has entered a whole new era from the traditional era (Li & Nam, 2022). 5G, big data, autonomous driving, artificial intelligence and other cutting-edge technologies are changing the way people travel. Autonomous driving is also gradually catching fire in the commercial vehicle sector. In the field of intelligence, commercial vehicles can take faster steps than passenger cars.

On 12 November, the 26th United Nations Climate Change Conference (COP26) concluded successfully in Glasgow, Scotland, UK. During the conference, BYD was invited to participate in several related activities and give panel discussions and keynote speeches (Wyns & Beagley, 2021). Under the initiative of COP26, BYD signed three agreements with various organizations, including governments, enterprises and environmental protection agencies, to accelerate the global electrification process and the construction of supporting infrastructure. At the same time, BYD's purely electric buses were also used as official shuttle vehicles to provide green travel services for delegates from more than 190 countries and regions (Hepburn et al., 2021).

Figure 8. Electric vehicle sales across China in 1st half of 2020, by leading original equipment manufacturer (OEM)
Source: EV-Volumes.com

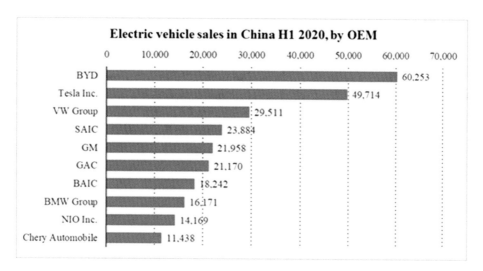

As is stated on BYD's official website, BYD sold 87,473 NEVs in February 2022, up 764.1% year-on-year (https://www.bydauto.com.cn/). With a share of 19.07%, BYD's sales volume is the first in the country.

However, as the leading enterprise in China, Ji et al. (2020) stated that BYD's development still has some problems. The immaturity of core technology has greatly affected driving safety. The NEVs

Figure 9. BYD's export trade volume and value of complete vehicle products, January 2017 - September 2021
Source: http://www.customs.gov.cn/

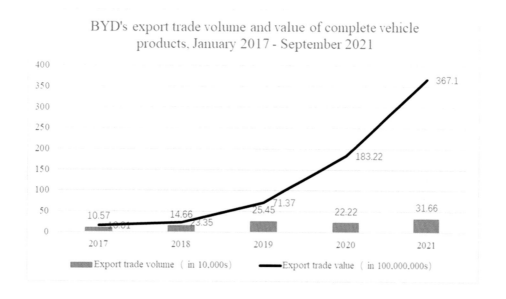

produced by BYD have their problems such as inaccurate power display and fragile and easily damaged parts. Compared to other brands, BYD's NEVs still have much room for improvement in terms of performance and durability.

Meanwhile, Lu et al. (2021) stated that over-reliance on subsidy policies is a severe problem either. In recent years, BYD's development has relied heavily on national and local government subsidies for new energy vehicle enterprises. With the reduction of subsidies, BYD's business situation has also suffered a certain impact. As a matter of urgency, BYD needs to upgrade its technology to reduce production costs and stabilize the impact of the subsidy withdrawal.

In conclusion, under the general trend of electrification of buses, trucks and special vehicles, new energy logistics vehicles still have a huge market space and volume advantage (Deng et al., 2022). By promoting the low-carbonization of public transportation and the comprehensive electrification of urban trucks and special-purpose vehicles, we are helping to realize the low-carbonization of engineering and logistics. To date, BYD has sold more than one million NEVs (including passenger cars and commercial vehicles) and reduced CO_2 emissions by more than 5.48 million tons, equivalent to planting 460 million trees, demonstrating the responsibility of Chinese enterprises (https://www.byd.com/cn/NewsAndEvents/News.html).

As the world's determination to combat climate change grows, the global electric vehicle market is booming like never before. On the sustainable development path of carbon emission reduction, big companies with the social responsibility and commitment of an international enterprise and the strong technical force of sustainable development should continue to take practical actions to fulfill the initiative of "cooling the earth by 1°C".

Case: LEAD - The Leading Enterprise in the Field of Power Batteries in China

LEAD is a professional supplier of equipment and solutions for manufacturers of energy-saving and new energy products such as film capacitors, lithium batteries and photovoltaic cells/modules. After more than a decade of entrepreneurial development, it has become an up-and-coming leading company covering its major product areas (https://www.leadchina.cn/).

Before 2008, there have been several domestic lithium battery equipment manufacturing enterprises, but most enterprises do relatively simple equipment, many are semi-automatic products (Jiang & Chen, 2022), coupled with the manufacturers are not large-scale, relatively inexperienced technology, in competition with Japan and South Korea and other advanced equipment companies are at a disadvantage (Guo et al., 2022). In addition, the manufacturers are not large-scale, and their technical experience is relatively lacking, so they are at a disadvantage in the competition with Japan and South Korea and other advanced equipment companies (Ko & Yoon, 2022). Therefore, before 2008 China's high-end lithium equipment still mainly relied on imports. After years of technology transformation and R&D investment, the company has mastered the core technology of the lithium battery equipment industry, including winding technology, high-speed slitting technology, automatic welding technology, automatic adhesive technology and vacuum injection technology. The core technologies of the lithium battery equipment industry, mainly winding technology, high-speed slitting technology, automatic welding technology, automatic adhesive technology and vacuum liquid injection technology, have all reached an international advanced level.

Previous studies have found a reason for its achievement. Lu (2021) stated that the company, as a leading domestic lithium equipment enterprise, will enjoy the big opportunity in the industry brought by

Figure 10. Analysis of the development dynamics of new energy vehicles in China
Source: https://www.miit.gov.cn/

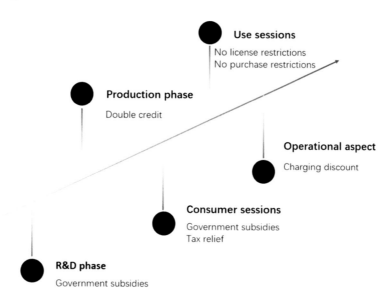

the new energy market outbreak (Liu, Ma, & Liu, 2022). The country vigorously promotes the development of the new energy vehicle industry and has released the "Energy-saving and New Energy Vehicle Industry Development Plan (2012~2020)".

NEVs are one of China's strategic emerging industries. The government attaches great importance to the development of the new energy vehicle industry and has introduced a full range of incentive policies (table 2), from government subsidies for research and development, double points for production, financial subsidies and tax breaks for consumption, unlimited licensing and no purchase restrictions for use, and charging incentives for operation, covering almost the entire life cycle of NEVs (figure 10).

Secondly, Since the State Council put forward the strategy of "innovation-driven development" and the strategy of scientific & technological innovation, the Chinese investment in technological R&D has been continuously increased, and the output of technological achievements in colleges and universities has also been increasingly rich, and the quantity and quality of the achievements have been continuously improved (Xin-gang & Wei, 2020). However, at present, there are several major trends in the lithium-related industries. First and foremost, lithium battery market share is becoming more and more concentrated, with small producers slowly being eliminated due to the capital Technology gap. After the lithium safety problems highlighted these issues, market share is increasingly concentrated in high-end lithium manufacturers, and upstream high-end lithium equipment manufacturers first benefit (Pu et al., 2022). Furthermore, the lithium equipment domestic replacement accelerates the process, mainly due to the price of domestic equipment, service advantages and technology (Sun et al., 2021). However, lastly, safety issues highlighted, based on safety considerations, high-quality lithium battery demand is expected to further increase (Jiang et al., 2022c).

Lithium-ion batteries (LIBs) have been playing a leading role in the energy storage modules of NEVs (Sun et al., 2021), in turn, it will greatly promote the NEV industry. Now that the entire world attaches great importance to environmental pollution, LEAD has cooperated with major automobile brands to replace fossil fuel with electricity, reducing pollution caused by the increasing number of vehicles, and replacing fossil fuel vehicles with NEVs. It is just a matter of time (Yu, Jiao, et al., 2022).

CONCLUSION AND RECOMMENDATIONS

NEVs are one of China's strategic emerging industries. The government attaches great importance to the development of the new energy vehicle industry and has introduced a full range of incentive policies, from government subsidies for research and development, double points for production, financial subsidies and tax breaks for consumption, unlimited licensing and no purchase restrictions for use, and charging incentives for operation, covering almost the entire life cycle of NEVs.

Energy-saving and emission reduction are of practical use from the user side. However, in the short term, it does not do much from an environmental point of view and amounts to the centralization of pollution consumption.

Under the pressure of energy and environmental protection, NEVs will undoubtedly become the future direction of automotive development. If NEVs are developed rapidly, in 2020 China will have 140 million vehicles, saving 32.29 million tons of oil and replacing 31.1 million tons of oil, saving and replacing a total of 63.39 million tons of oil, which is equivalent to a 22.7% reduction in the demand for oil for automobiles. By 2030, the development of NEVs will save 73.06 million tons of oil and replace 91 million tons of oil, saving and replacing a total of 164.06 million tons of oil, equivalent to a 41% reduction in automotive oil demand. By then, biofuels and fuel cells will play an important role in the replacement of oil in vehicles (Ma et al., 2019).

In the long run, whereas, it is more conducive to the transformation of the global energy mix, i.e. to getting rid of traditional fossil energy sources, which is conducive to curbing global warming.

The joint response to climate change and the continued promotion of green development is the only way for mankind to achieve sustainable development. The policy constraints and industrial changes that accompany the vision of achieving carbon neutrality by 2060 will have a profound impact on all sectors and will also bring opportunities and challenges to the green development of the transport sector. After years of efforts, China's green and low-carbon development policy system has been increasingly improved, and energy-saving and emission-reducing technologies for transport vehicles have been continuously advanced, so the transport sector has a high potential for carbon emission reduction. However, the high level of carbon emissions in the transport sector has become a common dilemma for all countries in the world, and road transport, with automobiles at its core, has become the hardest hit by carbon emissions in the transport sector. Carbon emission reduction in the transport sector involves a wide range of different groups. From the improvement of engine technology in the production of vehicles, to the increase in the proportion of NEVs in the purchase and use of vehicles, to the optimization of the transport structure in the freight transport sector, to the fundamental change in the way people travel in the daily travel sector, all are closely related to energy saving and emission reduction and green development in the transport sector. To achieve the goal of carbon emission reduction in transport, it is necessary to take technical feasibility, economic reasonableness and social acceptability as the criteria, focusing on the new energy of transport, the optimization of transport structure, and the transformation of the daily travel pattern of the residents, etc., and to coordinate the interests of multiple parties to jointly promote.

ACKNOWLEDGMENT

The authors extend sincere gratitude to:

- Our colleagues from Soochow University, the Australian Studies Centre of Shanghai University and Krirk University as well as the independent research colleagues who provided insight and expertise that greatly assisted the research, although they may not agree with all of the interpretations/conclusions of this chapter.
- China Knowledge for supporting our research.
- The Editor and the International Editorial Advisory Board (IEAB) of this book who initially desk reviewed, arranged a rigorous double/triple blind review process and conducted a thorough, minute and critical final review before accepting the chapter for publication.
- All anonymous reviewers who provided very constructive feedbacks for thorough revision, improvement, extension, and fine tuning of the chapter.

REFERENCES

Ahmad, N. N. N., & Hossain, D. M. (2015). Climate Change and Global Warming Discourses and Disclosures in the Corporate Annual Reports: A Study on the Malaysian Companies. *Procedia: Social and Behavioral Sciences*, *172*, 246–253. doi:10.1016/j.sbspro.2015.01.361

Bonsu, N. O. (2020). Towards a circular and low-carbon economy: Insights from the transitioning to electric vehicles and net zero economy. *Journal of Cleaner Production*, *256*, 120659. Advance online publication. doi:10.1016/j.jclepro.2020.120659

Chen, X., Huang, L., Liu, J., Song, D., & Yang, S. (2022). Peak shaving benefit assessment considering the joint operation of nuclear and battery energy storage power stations: Hainan case study. *Energy*, *239*, 121897. Advance online publication. doi:10.1016/j.energy.2021.121897

Deng, Y., Mu, Y., Dong, X., Wang, X., Zhang, T., & Jia, H. (2022). Coordinated operational planning for electric vehicles considering battery swapping and real road networks in logistics delivery service. *Energy Reports*, *8*, 1019–1027. doi:10.1016/j.egyr.2021.11.185

Ding, R., Liu, Z., Li, X., Hou, Y., Sun, W., Zhai, H., & Wei, X. (2022). Joint charging scheduling of electric vehicles with battery to grid technology in battery swapping station. *Energy Reports*, *8*, 872–882. doi:10.1016/j.egyr.2022.02.029

Dong, H., Li, P., Feng, Z., Yang, Y., You, Z., & Li, Q. (2019). Natural capital utilization on an international tourism island based on a three-dimensional ecological footprint model: A case study of Hainan Province, China. *Ecological Indicators*, *104*, 479–488. doi:10.1016/j.ecolind.2019.04.031

Ge, J., & Lin, B. (2021). Impact of public support and government's policy on climate change in China. *Journal of Environmental Management*, *294*, 112983. doi:10.1016/j.jenvman.2021.112983 PMID:34119988

Gu, T. J. (2021). *CATL breaks through the game by positioning itself as an upstream equipment vendor to LEAD*. Electronic Publishing House. https://kns.cnki.net/kcms/detail/detail.aspx?dbcode=CJFD&dbname=CJFDLAST2021&filename=YCYW202101014&uniplatform=NZKPT&v=iWDo6ZOTDFRkpBf0YWWr8P1mrVPhEcFfy_VKUBMkB9tIrmynrXq8zWzxLFuYU3My

Guo, J., Zhang, Y.-J., & Zhang, K.-B. (2018). The key sectors for energy conservation and carbon emissions reduction in China: Evidence from the input-output method. *Journal of Cleaner Production*, *179*, 180–190. doi:10.1016/j.jclepro.2018.01.080

Guo, Y., Wu, S., He, Y.-B., Kang, F., Chen, L., Li, H., & Yang, Q.-H. (2022). Solid-state lithium batteries: Safety and prospects. *eScience*. doi:10.1016/j.esci.2022.02.008

He, J., Li, J., Zhao, D., & Chen, X. (2022). Does oil price affect corporate innovation? Evidence from new energy vehicle enterprises in China. *Renewable & Sustainable Energy Reviews*, *156*, 111964. Advance online publication. doi:10.1016/j.rser.2021.111964

Hepburn, C., Qi, Y., Stern, N., Ward, B., Xie, C., & Zenghelis, D. (2021). Towards carbon neutrality and China's 14th Five-Year Plan: Clean energy transition, sustainable urban development, and investment priorities. *Environmental Science and Ecotechnology*, *8*, 100130. Advance online publication. doi:10.1016/j.ese.2021.100130 PMID:36156997

Hu, J.-W., Javaid, A., & Creutzig, F. (2021). Leverage points for accelerating adoption of shared electric cars: Perceived benefits and environmental impact of NEVs. *Energy Policy*, *155*, 112349. Advance online publication. doi:10.1016/j.enpol.2021.112349

Ji, X. Y., Xu, W. H., Shen, P. Y., Ma, X. R., & Lu, R. (2020). *Analysis of the prospect of China's automobile industry under the adjustment of new energy subsidy policy - Taking BYD as an example*. Anhui University of Finance and Economics. https://kns.cnki.net/kcms/detail/detail.aspx?dbcode=CJFD&dbname=CJFDLAST2020&filename=YXJI202034009&uniplatform=NZKPT&v=7N3vuKiq-sacH2kA3cJtwPQMdOATxh1q0ch6TwSP-XYfyho7wOy0ChF-GwDpaNIV

Jiang, D., Li, W., Shen, Y., & Yu, S. (2022a). Does air pollution affect earnings management? Evidence from China. *Pacific-Basin Finance Journal*, *72*, 101737. Advance online publication. doi:10.1016/j.pacfin.2022.101737

Jiang, T., Yu, Y., Jahanger, A., & Balsalobre-Lorente, D. (2022b). Structural emissions reduction of China's power and heating industry under the goal of "double carbon": A perspective from input-output analysis. *Sustainable Production and Consumption*, *31*, 346–356. doi:10.1016/j.spc.2022.03.003

Jiang, W., & Chen, Y. (2022). The time-frequency connectedness among carbon, traditional/new energy and material markets of China in pre- and post-COVID-19 outbreak periods. *Energy*, *246*, 123320. Advance online publication. doi:10.1016/j.energy.2022.123320

Jiang, X., Chen, Y., Meng, X., Cao, W., Liu, C., Huang, Q., Naik, N., Murugadoss, V., Huang, M., & Guo, Z. (2022c). The impact of electrode with carbon materials on safety performance of lithium-ion batteries: A review. *Carbon*, *191*, 448–470. doi:10.1016/j.carbon.2022.02.011

Jin, B. (2021). Research on performance evaluation of green supply chain of automobile enterprises under the background of carbon peak and carbon neutralization. *Energy Reports*, *7*, 594–604. doi:10.1016/j.egyr.2021.10.002

Ko, J., & Yoon, Y. S. (2022). Lithium phosphorus oxynitride thin films for rechargeable lithium batteries: Applications from thin-film batteries as micro batteries to surface modification for large-scale batteries. *Ceramics International*, *48*(8), 10372–10390. doi:10.1016/j.ceramint.2022.02.173

Li, G., Luo, T., & Song, Y. (2021). Climate change mitigation efficiency of electric vehicle charging infrastructure in China: From the perspective of energy transition and circular economy. *Resources, Conservation and Recycling*. Advance online publication. doi:10.1016/j.resconrec.2021.106048

Li, J., & Sun, C. (2018). Towards a low carbon economy by removing fossil fuel subsidies? *China Economic Review*, *50*, 17–33. doi:10.1016/j.chieco.2018.03.006

Li, J.-F., Ma, Z.-Y., Zhang, Y.-X., & Wen, Z.-C. (2018). Analysis on energy demand and CO2 emissions in China following the Energy Production and Consumption Revolution Strategy and China Dream target. *Advances in Climate Change Research*, *9*(1), 16–26. doi:10.1016/j.accre.2018.01.001

Li, X., & Nam, K.-M. (2022). Environmental regulations as industrial policy: Vehicle emission standards and automotive industry performance. *Environmental Science & Policy*, *131*, 68–83. doi:10.1016/j.envsci.2022.01.015

Liu, J., & Meng, Z. (2017). Innovation Model Analysis of New Energy Vehicles: Taking Toyota, Tesla and BYD as an Example. *Procedia Engineering*, *174*, 965–972. doi:10.1016/j.proeng.2017.01.248

Liu, W., Ma, Q., & Liu, X. (2022). Research on the dynamic evolution and its influencing factors of stock correlation network in the Chinese new energy market. *Finance Research Letters*, *45*, 102138. Advance online publication. doi:10.1016/j.frl.2021.102138

Lu, S. T., & Liu, C. (2021). *BYD under the background of "One Belt One Road" Analysis of Competitive Advantages of New Energy Vehicle Exports*. Jilin International Studies University. https://kns.cnki.net/kcms/detail/detail.aspx?dbcode=CJFD&dbname=CJFDAUTO&filename=SCXH202202029&uniplatform=NZKPT&v=K0y7w0XLaFQ50_5cSpT1L5dyjnfS5EVUNqit6jJYWm1WUfwZf3NhVQbJeNCWw5dh

Ma, Y., Shi, T., Zhang, W., Hao, Y., Huang, J., & Lin, Y. (2019). Comprehensive policy evaluation of NEV development in China, Japan, the United States, and Germany based on the AHP-EW model. *Journal of Cleaner Production*, *214*, 389–402. doi:10.1016/j.jclepro.2018.12.119

Masiero, G., Ogasavara, M. H., Jussani, A. C., & Risso, M. L. (2016). Electric vehicles in China: BYD strategies and government subsidies. *RAI Revista de Administração e Inovação*, *13*(1), 3–11. doi:10.1016/j.rai.2016.01.001

Pichler, M., Krenmayr, N., Schneider, E., & Brand, U. (2021). EU industrial policy: Between modernization and transformation of the automotive industry. *Environmental Innovation and Societal Transitions*, *38*, 140–152. doi:10.1016/j.eist.2020.12.002

Pu, G., Zhu, X., Dai, J., & Chen, X. (2022). Understand technological innovation investment performance: Evolution of industry-university-research cooperation for technological innovation of lithium-ion storage battery in China. *Journal of Energy Storage*, *46*, 103607. Advance online publication. doi:10.1016/j.est.2021.103607

Shi, H., Chai, J., Lu, Q., Zheng, J., & Wang, S. (2022). The impact of China's low-carbon transition on economy, society and energy in 2030 based on CO2 emissions drivers. *Energy*, *239*, 122336. Advance online publication. doi:10.1016/j.energy.2021.122336

Sun, S., Jin, C., He, W., Li, G., Zhu, H., & Huang, J. (2021). Management status of waste lithium-ion batteries in China and a complete closed-circuit recycling process. *The Science of the Total Environment*, *776*, 145913. doi:10.1016/j.scitotenv.2021.145913 PMID:33639457

Tan, R., Tang, D., & Lin, B. (2018). Policy impact of new energy vehicles promotion on air quality in Chinese cities. *Energy Policy*, *118*, 33–40. doi:10.1016/j.enpol.2018.03.018

The People's Government of Hainan Province. (2021). *Hainan Province New Energy Vehicles for Electricity Application Pilot Implementation Plan*. Retrieved from https://www.hainan.gov.cn/hainan/tingju/202103/2c7e2327668348e2ac8b5126c471ed93.shtml

The People's Government of Hainan Province. (2022). *More than 20,000 new charging posts in Hainan province by 2021*. Retrieved from https://www.hainan.gov.cn/hainan/tingju/202201/0f3a34589322481ca2cfbddc3fc90ecb.shtml

The People's Government of Hainan Province. (2022). *Several Measures to Encourage the Use of New Energy Vehicles in Hainan Province in 2022*. Retrieved from https://www.hainan.gov.cn/hainan/tingju/202203/7119c88fcac64961ace02720f3229b6f.shtml

The World Bank. (2021). World Development Indicators: *Trends in greenhouse gas emissions. The World Bank*. Retrieved from http://wdi.worldbank.org/table/3.9

Tian, X., Zhang, Q., Chi, Y., & Cheng, Y. (2021). Purchase willingness of new energy vehicles: A case study in Jinan City of China. *Regional Sustainability*, *2*(1), 12–22. doi:10.1016/j.regsus.2020.12.003

Wang, J., & Li, L. (2016). Sustainable energy development scenario forecasting and energy saving policy analysis of China. *Renewable & Sustainable Energy Reviews*, *58*, 718–724. doi:10.1016/j.rser.2015.12.340

Wang, M., & Feng, C. (2020). The impacts of technological gap and scale economy on the low-carbon development of China's industries: An extended decomposition analysis. *Technological Forecasting and Social Change*, *157*, 120050. Advance online publication. doi:10.1016/j.techfore.2020.120050

Wang, X., Huang, L., Daim, T., Li, X., & Li, Z. (2021). Evaluation of China's new energy vehicle policy texts with quantitative and qualitative analysis. *Technology in Society*, *67*, 101770. Advance online publication. doi:10.1016/j.techsoc.2021.101770

Wang, Y.-J., Chen, Y., Hewitt, C., Ding, W.-H., Song, L.-C., Ai, W.-X., Han, Z.-Y., Li, X.-C., & Huang, Z.-L. (2021). Climate services for addressing climate change: Indication of a climate livable city in China. *Advances in Climate Change Research*, *12*(5), 744–751. doi:10.1016/j.accre.2021.07.006

Wiryadinata, S., Morejohn, J., & Kornbluth, K. (2019). Pathways to carbon neutral energy systems at the University of California, Davis. *Renewable Energy*, *130*, 853–866. doi:10.1016/j.renene.2018.06.100

Wu, Z., Shao, Q., Su, Y., & Zhang, D. (2021). A socio-technical transition path for new energy vehicles in China: A multi-level perspective. *Technological Forecasting and Social Change*, *172*, 121007. Advance online publication. doi:10.1016/j.techfore.2021.121007

Wyns, A., & Beagley, J. (2021). COP26 and beyond: Long-term climate strategies are key to safeguard health and equity. *The Lancet. Planetary Health*, *5*(11), e752–e754. doi:10.1016/S2542-5196(21)00294-1 PMID:34774113

Xin-gang, Z., & Wei, W. (2020). Driving force for China's photovoltaic industry output growth: Factor-driven or technological innovation-driven? *Journal of Cleaner Production*, *274*, 122848. Advance online publication. doi:10.1016/j.jclepro.2020.122848

Yang, X., Zhao, C., Xu, H., Liu, K., & Zha, J. (2021). Changing the industrial structure of tourism to achieve a low-carbon economy in China: An industrial linkage perspective. *Journal of Hospitality and Tourism Management*, *48*, 374–389. doi:10.1016/j.jhtm.2021.07.006

Yu, P., Jiao, A., & Sampat, M. (2022). The Effect of Chinese Green Transformation on Competitiveness and the Environment. In *Handbook of Research on Green* (pp. 257–279). Circular, and Digital Economies as Tools for Recovery and Sustainability. doi:10.4018/978-1-7998-9664-7.ch014

Yu, P., Weng, Y., & Ahuja, A. (2022). Carbon Financing and the Sustainable Development Mechanism. In Handbook of Research on Energy and Environmental Finance 4.0 (pp. 301-332). Academic Press.

Zhang, S., Bai, X., Zhao, C., Tan, Q., Luo, G., Wu, L., Xi, H., Li, C., Chen, F., Ran, C., Liu, M., Gong, S., & Song, F. (2022). China's carbon budget inventory from 1997 to 2017 and its challenges to achieving carbon neutral strategies. *Journal of Cleaner Production*, *347*, 130966. Advance online publication. doi:10.1016/j.jclepro.2022.130966

Chapter 7
Global Citizen Science Programs and Their Contribution to the Sustainable Development Goals

Alexandros Skondras
https://orcid.org/0000-0002-6869-3312
Aristotle University of Thessaloniki, Greece

Eleni Karachaliou
Aristotle University of Thessaloniki, Greece

Ioannis Tavantzis
Aristotle University of Thessaloniki, Greece

Nikolaos Tokas
Aristotle University of Thessaloniki, Greece

Margarita Angelidou
Aristotle University of Thessaloniki, Greece

Efstratios Stylianidis
Aristotle University of Thessaloniki, Greece

ABSTRACT

The aim of this chapter is to review how citizen science programs can contribute to each SDG, supported also by successful cases. Prerequisites have been the inclusion of at least one case from all five global areas and the field of interest is well spread through various study areas while the cases are only RFO or RPO projects. The design and implementation of this research is structured as follows. Each SDG is analyzed in a separate chapter. To begin with, the main goals of the relevant SDG are mentioned. The chapter continues with the relevance of citizen science projects to the respective goal. Ultimately, an example on how a citizen science project can make the goal effective is chosen. The case studies are selected based on their location, the timeframe of the project, and are structured beginning with the people or institutions involved, continuing with the project description and concluding with the project's contribution to the relevant SDG. The chapter then proceeds to a comparative case study analysis, with results and discussion in order to extract conclusions.

INTRODUCTION

In September 2015 the General Assembly of the United Nations stipulated a global agenda, namely

DOI: 10.4018/978-1-6684-4829-8.ch007

Global Citizen Science Programs and Their Contribution to the Sustainable Development Goals

Transforming Our World: the 2030 Agenda for Sustainable Development, which includes 17 Sustainable Development Goals (SDGs). It also includes global targets that should be accomplished by 2030, in order to achieve a sustainable function of society, economy and the environment for the long run. The targets address five fields, namely the well-being of people (1, 2, 3, 4, 5, 6, 11), the well-being of the planet (6, 7, 11, 13, 14, 15), the prosperity of communities (5, 8, 9, 11, 12, 14), the achievement of peace and social cohesion (10, 11, 16) and the creation of partnerships (11,17) (ICOMOS, 2021).

In parallel, during recent years, new technologies for data collection and collaboration have radically transformed the ways though which researchers can work together with citizens, as well as the ways through which they can collect and process data in research. The approach of citizen science, more particularly, is based on citizen participation in research, with a particular focus on implementing research in collaboration with scientists and research institutions (McKinley et al., 2015). Striking examples of applications of this nexus can be found in recent projects related to the observation of land and marine ecosystems, as well as astronomy and conservation. These represent areas in which citizen participation has vastly improved the pace and quality of data collection, analysis and evaluation (Schleicher & Schmidt, 2020). In overall, citizen science has created new potentials for synergies between academia and citizens, enabling learning and knowledge and raising awareness about important topics of interest for researchers, institutions and communities (Vohland et al., 2021). As a natural consequence of all these exciting developments, citizen science has also been increasingly gaining traction within the context of formal Research and Innovation (R&I) performed by Research Funding Organizations (RFOs) or Research Performing Organizations (RPOs), namely universities and research organizations, while several citizen science labs and hubs are being funded and developed therein.

Citizen science can have a positive impact on progress towards the achievement of many SDGs, with positive effects also at national and regional level (Dorler et al., 2021). Notable examples include culture integration in sustainable development approaches (UCLG, 2018), heritage, creativity and diversity (ICOMOS, 2021), equality and social inclusion, and urban planning and public spaces (UCLG, 2018). Following this realization, during the last few years, discussions from various institutes have shown how citizen science can promote the SDGs, and what new metrics and targets for their implementation and their monitoring could be used (West et al., 2021). Discussions have also been focusing on the potential to contribute in accelerating the achievement of SDG targets (Fritz et al., 2019).

The aim of this paper is to review how citizen science projects and initiatives can contribute and be relevant to each SDG, supported also by successful cases. Essential prerequisites have been the inclusion of at least one case from all five global areas and the field of interest is very well spread through various study areas, while the cases are only Research Funding Organizations (RFOs) or Research Performing Organizations (RPOs) projects.

The design and implementation of this research is structured as follows. Each Sustainable Development Goal is analyzed in a separate chapter. To begin with, the main goals of the relevant SDG are mentioned. The chapter continues with the relevance of citizen science projects to the respective goal. Ultimately, an example on how a citizen science project can make the goal effective is chosen. The case studies are selected based on their location, the timeframe of the project (every case study was founded in the last twenty years and is still functional) and are structured beginning with the people or institutions involved, continuing with the project description and concluding with the project's contribution to the relevant SDG. The selection and analysis of data sources are based on previously published peer-reviewed journals and/or manuscripts published by SDG related organizations. The paper then proceeds to a comparative case study analysis, with results and discussion in order to extract conclusions.

SUSTAINABLE DEVELOPMENT GOALS

The Sustainable Development Goals (SDGs) act as a universal reference for the transition of sustainable development between the years 2015-2030, and provide a methodological approach through which governments can assess actions and policies, in order to achieve these targets by 2030. These goals are universal and are applicable to all countries, either developing or developed. The proposed targets and goals serve as a network, through which all countries can set a common sustainable future (Delli Paoli & Addeo, 2019). The SDGs cover a plethora of areas including energy, food security, health and well-being, poverty, inequality, production and consumption, gender, urbanization and environmental issues affecting climate change and marine and land ecosystems (Fritz et al., 2019).

Every SDG can be formulated as a series of goals and targets, which success can be monitored through 232 indicators (Ballerini & Bergh, 2021). This global indicator framework was developed by international and regional agencies and member states of the United Nations. The data acquisition is sourced primarily from global databases harvested by governments, statistical offices and international organizations. The process is reported to the High-level Political Forum of Sustainable Development (HLPF) which reviews the agenda and monitors the progress at global level (Fritz et al., 2019). The SDGs take into account capacities, development levels and policies from different national realities and while the targets are set as global, each nation can decide, guided by their level of global ambition, their own targets and how these goal shall be incorporated in strategies, policies and national planning processes (Delli Paoli & Addeo, 2019).

SDG 1: End Poverty in all its Forms Everywhere

This goal aims at ending poverty in all its forms by 2030. It also aims to build protection for vulnerable and poor societies, improve availability to basic services and provide support to women and men that have been exposed to events that are climate-related and other social, environmental and economic disasters and vulnerable situations (Global indicator, n.d.).

In the direction of achieving this SDG, citizen science can improve processes and methods for the development of high quality resilience and economic strategies that engage indigenous communities, improving, for example, culture and heritage management that preserves and nurtures sustainability related capacities and resources (UCLG, 2018). Equal access can and should be guaranteed, regardless of gender and economic status. The sound preservation of natural habitats and conservation sites can promote sustainable and inclusive development to services such as infrastructure, sanitation systems and access to water and food. Furthermore, citizen science initiatives can reduce vulnerability and exposures to extreme events related to climate and foster entrepreneurship, job creation, innovation, and creativity (ICOMOS, 2021).

Case Study: Cape Citizen Science (Africa)

Cape Citizen Science is a program co-hosted by two universities, the University of Pretoria, and Stellenbosch University (Hulbert et al., 2019). It has been selected as a case due to its potential to contribute the elimination of food poverty.

More particularly, Cape Citizen Science is a project that monitors plant diseases caused by groups of microorganisms and fungi. Citizens can observe and contribute data with respect to perishing plants through online tools and report sampling locations, through which researchers can respond faster to new diseases and pest invasions, with the goal of contributing to the conservation of biodiversity in South African regions (Hulbert, 2016). That said, the program relies heavily on the participation of citizens to promote education and research (Hulbert et al., 2019).

The contribution of the project to SDG 1 is the access to food of better quality, to be achieved through the discovery and monitoring of plant diseases and the reduction of the vulnerability of the plants and the seeds to extreme events.

SDG 2: End Hunger, Achieve Food Security, and Improved Nutrition and Promote Sustainable Agriculture

This goal aims at ending hunger and improves nutrition and sustainable agriculture by 2030. It also ensures universal accessibility to sufficient, nutritious, and safe food during the entire year. The goal tries to enhance sustainable production of high-quality food and promotes sustainable agricultural practices and equitable land access, boosts the technology required to food production, and enhances agricultural productivity with international economic support in technology and infrastructure (Global indicator, n.d.).

Citizen science initiatives can use traditions, knowledge and practices of local people and communities and apply them for the successful building of strategies on sustainability, both economic and environmental (natural resources), such as the diversification of seeds or the fauna welfare conservation (UCLG, 2018). The fair and equal distribution of food and the security of these good practices have been enhanced by many citizen science projects, particularly in cultural and agricultural landscapes, while the diversity of forests, fisheries and agricultural projects regarding agroeconomics, indigenous fishing systems and researches of indigenous medicines and herbs can help support biodiversity (ICOMOS, 2021).

Case Study: Neighborhood Nestwatch (North America)

Neighborhood Nestwatch is a citizen science project established by the Smithsonian's National Zoo & Conservation Biology Institute with funding provided by the National Science Foundation (Evans et al., 2005).

This project uses interaction between citizens and researchers in order to improve citizens' knowledge about the wild birds that live in their backyards. Participants are taught on how to collect valid data that can be used to examine the impact of urban policies on wildlife and the project helps them understand ecology, the nesting patterns and the population of birds in the area. The collection of data can promote science literacy among the participants, as well as opportunities for local communities to act upon their local environment (Evans et al., 2005).

The main contribution to the SDG 2 is the improvement of the genetic diversity of wild species of birds through research, the utilization of the scientific data for associated knowledge and the improvement of sustainable nutrition.

SDG 3: Ensure Healthy Lives and Promote Well-being for all at all Ages

SDG 3 aims to promote well-being and ensure healthy lives for all, during all stages of life. The goal aims to improve health-related issues, such as environmental diseases, child and maternal well-being, provision of adequate health services for all and universal access to affordable vaccines and medicines.

Furthermore, it aims to improve relevant research practices, health risk management and boosts international financing in the fields of health and well-being (Global indicator, n.d.).

Citizen science labs can contribute to the integration of traditional health systems and inform local health and well-being policies, for example by enhancing and analyzing active health and welfare practices at local level. These labs can promote participation in local capacity-building programs that foster accessibility to high quality health settings (UCLG, 2018). Furthermore, citizen science labs can promote health and well-being for all, by implementing projects that affect urban quality of life, such as pedestrian mobility in historic areas, or take advantage of non-medical health and well-being practices to restore connections among nature, culture, health and well-being (ICOMOS, 2021).

Case Study: Floating Forests (North America)

Floating Forests is a citizen science project established by the University of Massachusetts in Boston and funded by the National Aeronautics and Space Administration (NASA) (Rosenthal et al., 2018).

The project aims to map the giant kelp forests in the Atlantic and Pacific Ocean through a satellite system that collects photos from coastline ecosystems and then relies on citizens' help for data processing and consensus classification. The information they provide is used by researchers to trace kelp patches in order to detect where species grow and to improve kelp health and distribution globally. Sound kelp forest management has the potential to improve therapeutic practices in health and well-being (Rosenthal et al., 2018).

The contribution of the project to the SDGs is of paramount importance as it promotes the connection between nature and health. Giant kelp forests can provide the primary materials for the development of alternative medicines, thus the conservation of these marine ecosystems and the scope of the project is in line with SDG 3.

SDG 4: Ensure Inclusive and Equitable Quality Education and Promote Lifelong Learning Opportunities for all

This goal aims to promote equitable access to qualitative primary, secondary education and early development of knowledge for girls and boys, and the same for tertiary education and subsequent employment for women and men. It also seeks to reduce gender inequalities in education and improve the level of literacy and numeracy among all youth and adult people, especially their knowledge and skills to contribute to sustainable development (Global indicator, n.d.).

Citizen science labs can advance this goal by enhancing cultural and educational initiatives in a direction that increases employability of individuals. Individuals can then engage in meaningful lifelong learning, acquiring new knowledge and skills, as well as an appreciation of intercultural dialogue and diversity. Moreover, citizen science projects can provide the training, education and other safe learning contexts in which professionals of all ages and genders will enhance their appreciation of cultural, history and heritage, while recognizing the paramount role of equal education (UCLG, 2018). Furthermore, the recognition of locally relevant languages and abilities can help the design of universally accepted and equitable curricula that widen the range of employment opportunities and educational prospects (McGhie, 2019).

Case Study: Project Budburst (N. America)

Project Budburst is run by the Chicago Botanic Garden and the National Ecological Observatory Network and is sponsored by the National Science Foundation (Henderson et al., 2011).

The project main goal is to engage volunteers in monitoring plants as seasons change. Volunteers collect data about flowering, leafing, leaf falling and fruiting and report their observations on an online database. Scientists can then use the data in order to study the responsiveness of plant species to local changes in climate. By comparing historical data, scientists can also detect long-term impacts of climate change and increase awareness (Henderson et al., 2011).

The contribution of this citizen science project to SDG4 is outstanding, as the project provides volunteers and other citizens with the opportunity to educate themselves about plant changes in response to seasons, climate change and specific environmental conditions. In overall, this increases the quality of their knowledge and literacy (Henderson et al., 2011).

SDG 5: Achieve Gender Equality and Empower all Women and Girls

This goal aims at eliminating violence and discrimination against girls and women in all forms, in both in the public and the private domains. The goal also tries to secure equal opportunities for women in decision-making in public, economic and political life, including their representation in leadership positions. Moreover, it seeks to ensure access to reproductive rights and sexual health for all (Global indicator, n.d.).

Citizen science hubs can provide for the gender dimension in policy making, reducing gender discrimination and boosting gender equality, as well as by suggesting measures to eradicate violence, harmful practices and any type of discrimination. The hubs can offer a space from which their voices can be heard on media. They can also enhance recognition and visibility of citizen science initiatives carried out by girls and women (UNESCO, 2014). Moreover, citizen science hubs can ensure the effective and full participation of all genders and opportunities for equitable gender leadership at all levels and stages of decision-making (ICOMOS, 2021).

Case Study: The Technion - Mada Tech Citizen Science Lab (Asia)

The Lab is co-established by the collaboration of MadaTech - Israel National Museum of Science, Technology and Space and The Faculty of Education in Science and Technology, Technion (Mada Tech, n.d.).

The faculty's teaching and Research and Development (R&D) activities are varied and focus on learning science, educational technologies, science education and public engagement. Education, biochemistry and gender-related humanities are the main topics of the citizen science projects that are implemented in the Citizen Science Lab (About Madatech, n.d.). An example of such a project is "Gender Engagement with posts authored by female scientists on Facebook". The purpose of this project is to emphasize how the representation of female scientists in popular media emphasizes their feminine qualities, rather than their professional achievements (Rozenblum et al., 2018).

The SDG5 (Gender Equality) is the main goal that is targeted by this project, as it highlights gender discriminations against women and empowers women's in terms of gender equality.

SDG 6: Ensure Availability and Sustainable Management of Water and Sanitation for all

This goal aims to provide equitable and universal access to affordable and clean drinking water, hygiene and sanitation, particularly for girls and women, especially those that are in a vulnerable situation. The goal also aims to reduce illegal dumping and pollution from hazardous materials and chemicals in order to improve the quality of water and freshwater supply, as well as to reduce water scarcity in the world (Global indicator, n.d.).

Citizen science initiatives can put into practice indigenous policies for the sustainable use of ecosystems that are water-related and can promote activities and events for raising awareness about water scarcity and pollution (UCLG, 2018). Citizen science projects can also be developed for water-related heritage management (for example historical water infrastructures), incentivizing the adoption of sustainable practices for water management and sanitation for all (ICOMOS, 2021).

Case Study: Que Pasa Riachuelo? (South America)

Que Pasa Riachuelo? is a citizen science initiative established by the Foundation for Environment and National Resources (FARN), the Center for Legal and Social Studies (CELS), Fundation Metropolitana and the Citizens' Association for Human Rights and Fundacion Ciudad (Fressoli et al., 2016).

The initiative's purpose is to help local communities oversee the execution of the clean-up of the river Riachuelo, a small and contaminated river which is at the southern border between the city of Buenos Aires and the province. The tool used for the citizen initiative is a free access online platform that includes a georeferenced interactive map. The map shows available information on the area of the Matanza-Riachuelo Basin Authority (ACUMAR) and citizens' reports of the state of pollution in this area. The main goal is to report breaches and generic problems that affect the general well-being and the environment (Fressoli et al., 2016) and to promote the community's ability to influence decision making and public policies in the area.

The main contribution to SDG 6 is the improvement of the quality and access of drinkable water, especially for underage people, regarding the high rates of infant mortality due to hygiene, sanitation, and inadequate water supply.

SDG 7: Ensure Access to Affordable, Reliable, Sustainable and Modern Energy for all

This goal aims to provide access to reliable, modern and affordable energy services to all, increase substantially the renewable energy share in the percentage of global energy, and double the rate of global improvement in the efficiency of energy production. The goal also ensures the international cooperation for universal access to clean energy and the expansion of infrastructure for the supply of sustainable and modern energy services (Global indicator, n.d.).

Citizen science projects can explore creative processes and potentials in order to promote state of the art approaches in energy consumption and production, as well as collaborating with organizations to carry out activities that raise awareness on sustainability and create environmental impact (UCLG, 2018). Citizen science initiatives can also promote energy efficiency both in new buildings with the reduction of energy consumption and in traditional buildings with the research of modern materials and performance and waste management strategies. Moreover, these projects can research local lifestyle and climate choices related to indigenous building methods and materials and techniques. This would lead to an increased usage of local materials and reduction of energy consumption, applicable both in urban and rural settlements (ICOMOS, 2021).

Case Study: Birds and Windows Project (North America)

Birds and Windows Project is a citizen science project established in 2013 by the University of Alberta and funded by the University of Alberta, Environment Canada and the Alberta Cooperative Conservation Research Unit (ACCRU) (Kummer et al., 2016).

The project aims to reduce the number collisions that happen between birds and windows - which are a reason of a significant number of deaths of North American birds. It further aims to identify the relevance of the typology of windows and houses, in order to aid the design of future mitigation strategies for the eradication of bird-window collisions. The main goal of the project is accomplished by means of data collection of accidents in the area by citizens, the identification of the species (feeder birds or non-feeder birds) and subsequent data analysis (Kummer et al., 2016).

The contribution of the project to the SDG is important, as it aims to enhance the energy efficiency of windows on buildings, with regards to which typologies of windows are more prone to bird crash and the subsequent expansion and upgrade of sustainable infrastructure and technology.

SDG 8: Promote Sustained, Inclusive, and Sustainable Economic Growth, Full and Productive Employment and Decent Work for all

This goal aims for a successful economic growth for each person, with a particular aim to guide the development of policies that are well positioned to promote economic productivity and development. Such policies will lead to the creation of decent jobs for all, including persons with disabilities and young people. The goal also aims to ensure the reduction of unemployed youth, the protection of labor rights, such as equal remuneration for work that has equitable value, and the elimination of forced labor (Global indicator, n.d.).

Citizen science can research the potentials of the economic, cultural, and social sectors to foster sustainable, inclusive and fair employment and sustainable tourism models, especially with regards to local communities and ecosystems. In sequence, these initiatives can support international mobility, boost employment with proper labor conditions, while also helping to preserve local cultural identities (Balta Portoles and Dragicevic Sesic, 2017). Furthermore, they can reduce the exploitation of workers for political ends, and they can mitigate the risk of applying economic business models that are potentially harmful for local beneficiaries (ICOMOS, 2021).

Case Study: Lookit (North America)

Lookit is a citizen science project established by MIT (Massachusetts Institute of Technology) and more specifically by the MIT Early Childhood Cognition Lab. It is funded by the National Science Foundation (NSF) (Scott & Schulz, 2017).

Lookit studies the cognitive development of children through an online platform that translates efficiently children's abilities and learning mechanisms on a computer-based testing environment. Parents can access the platform remotely and upload the data collected from their webcam for analysis (Scott & Schulz, 2017).

The project promotes inclusion, without affecting families from lower social status, and it ensures the diversity of education levels, incomes, language background and races among the participants (Scott & Schulz, 2017), and thus has an important contribution in SDG 8.

SDG 9: Build Resilient Infrastructure, Promote Inclusive and Sustainable Industrialization and Foster Innovation

This goal aims at enhancing the development of sustainable, reliable and resilient infrastructure in order to ensure equitable and affordable access, sustainable and inclusive industrialization and increase knowledge-driven human capital. Furthermore, the goal aims to foster the inclusion of small-scale enterprises, particularly from developing countries, increasing their prospects to penetrate high value markets and access affordable credit (Global indicator, n.d.).

Citizen science projects can provide access to a range of facilities and spaces dedicated to creation, training and production of science and they can enhance social innovation and inclusivity in projects. Examples include sponsorship and microcredit projects implementing alternative financing mechanisms. These initiatives can increase understanding of connections that exist between social innovation and grassroots initiatives, as well as the development of resilient infrastructure that may be available for all (UCLG, 2018). Furthermore, they can promote tangible and intangible assets (ranging from traditional architecture and natural wonders, to social support systems and traditional knowledge) that can enhance the resilience and adaptability of local communities (ICOMOS, 2021).

Case Study: Aurorasaurus (North America)

The citizen science project Aurorasaurus began in 2014 and is based on a strong coalition between the government funded National Aeronautics and Space Administration (NASA) and the National Science Foundation (NSF) (Hall, 2015).

Aurorasaurus is a citizen science project that gathers sightings and real-time tracking of Northern and Southern Lights in order to understand the Auroras (Hall, 2015). The public can enter their observations through the project website, as well as through a mobile application platform, with the goal of better analyse their frequency, visual characteristics and location (Tapia et al., 2014).

The main output of this project is the creation of a scientific tool that allows a community of users to better predict sightings of the auroras (Tapia et al., 2014), and is in line with SDG 9, which aims to foster innovation.

SDG 10: Reduce Inequalities within and Among Countries

This goal aims to reduce income inequalities as well as those related to sex, age, race, disability, origin, ethnicity, economic or religion within a community. This can be done by eradicating legislations, practices and policies that are discriminatory, and promoting and adopting others that enhance greater equality. The goal also seeks to reduce inequalities among countries, related for example to migration and development (Global indicator, n.d.).

Citizen science projects can become integrated within policies that are related to the ability of people to produce, create, and disseminate their own expressions, as well as celebrate and recognize cultural and social diversity, especially among the most vulnerable individuals. These projects can promote policies that have to do with inequalities, such as internal displacement, migration and people with disabilities, in order to give voice to voiceless people, irrespective of sex, race, origin, ethnicity, age, religion or other status (UCLG, 2018). Furthermore, they can help and have a huge role in preserving cultural diversity and support social cohesion and dignity within communities with the main goal of alleviating social inequalities (ICOMOS, 2021).

Case Study: Journey North (North America)

The citizen science initiative Journey North was established by the Annenberg Learner and is now funded by the University of Wisconsin-Madison Arboretum (About Journey, n.d.).

Journey North gathers sightings of Monarch Butterflies waves migrations. These are mapped on a real-time basis across the North American continent. Sightings are reported by volunteers during spring and fall, by taking pictures and reporting of first eggs, larvae and early adult monarchs (in the spring), migration events, breeding and roosting monarchs (in the fall), while data are used for the development of real-time maps (Davis & Howard, 2005).

The main goal is the creation of a scientific pattern about the rate of recolonization of the monarchs, based on the rate of movement of the wave migration and the overall area that the wave front occupies (Davis & Howard, 2005), and is in line with SDG 10, as the project aims in safeguarding natural diversities and it enables various American states to adopt legislations and preservation policies that align their environmental policies.

SDG 11: Make Cities and Human Settlements Inclusive, Safe, Resilient and Sustainable

This goal aims to provide global access to safe, adequate and affordable housing, basic services and sustainable and accessible transport systems. It also seeks to promote sustainable and inclusive urbanization, enhance the resilience of cities in the face of disasters and reduce the impact of cities on the environment by promoting sound municipal waste management and air pollution control (Global indicator, n.d.).

Citizen science hubs in this domain can act as promoters of social inclusion and liveability, walkability and well-being, putting forward the use of good practices for urban planning and design at human scale, more open and connected spaces, as well as optimized air and noise levels. They can also promote actions for disaster risk management and sociotechnical adaptation of local communities in the face of climate change (ICOMOS, 2021). Citizen science hubs can moreover guide the design and implementation of policies for the preservation of cultural heritage in all its forms, tangible and intangible, while building the capacities of people and organization to act for sustainable local development. They can also elevate cultural heritage in the regeneration of historic neighborhoods, city centers, green and public spaces, informing also plans for local and regional development (UNESCO, 2016).

Case Study: Brooklyn Atlantis

Brooklyn Atlantis is a citizen science project established by the Polytechnic Institute of NYU and funded by the National Science Foundation (Laut et al., 2014).

The project aims to aid the environmental control and monitoring of the Gowanus Canal in New York City and its polluted waters. The water is analysed by an aquatic surface vehicle, which can gather data and take pictures using a plethora of sensors. It provides a web-based interface for social participation of citizen scientists where they can help with the analysis of data collected and the identification and classification of wildlife discovered in the data gathered. The project investigates human-machine interaction, and the effectiveness of the community involvement and field testing (Laut et al., 2014).

The contribution of the project to the SDG is important as it aims to reduce the impact of human activity on the urban environment by raising awareness about urban water pollution. It also enhances local communities' well-being and overall contributes to making the city of New York safer and more sustainable by means of sound pollution management.

SDG 12: Ensure Sustainable Consumption and Production Patterns

This goal aims to enhance sustainable production and consumption, a more efficient use and sustainable management of natural resources, and the uptake of environmentally friendly practices regarding the management of chemicals. Other aims also include the reduction of waste generation by means of raising awareness on sustainable practices and lifestyles, in compliance also with national priorities (Global indicator, n.d.).

Citizen science projects can enhance management of cultural and historical assets by way of designing and implementing projects that promote the consumption and production of local products. They can also facilitate citizen initiatives for the sustainable management of public and open spaces, especially those that have to do with gardening and urban crops (UCLG, 2018). These projects instigate the behavioral change needed in order to reduce current wasteful practices and increase awareness about the limited availability and compatibility of some traditionally used raw materials with modern standards in animal welfare and tourism (ICOMOS, 2021).

Case Study: Local Environment Observer Network (LEON) (N. America)

The Local Environmental Observer Network (LEON) was established by the Center for Climate and Health at the Alaska Native Tribal Health Consortium and is funded by the U.S. Environmental Protection Agency (Brubaker et al., 2013).

This project combines native knowledge with western technology and science for the documentation of climate change. Events observed include those related with extreme weather, flooding, erosion, droughts, unusual flora and fauna, and events in general that can threaten water and food security and community health. The main goal of the project is to improve the understanding of how indigenous communities are changing; identifying the threats and connecting local community members with experts that can provide scientific support on these topics (Brubaker et al., 2013).

The contribution of the project to the SDG is evident, as it promotes the inclusion of local culture and heritage for the sustainable management of natural resources and to ensure sustainable production and consumption of goods.

SDG 13: Take Urgent Action to Combat Climate Change and its Impacts

This goal aims to enhance resilience against hazards related to climate change and global natural disasters. It also aims to integrate climate change mitigation measures in any kind of strategy and policy and improve awareness, education and institutional and human capacity on climate change impact mitigation and adaptation (Global indicator, n.d.).

Citizen science labs can act as proponents of environmental sustainability, by establishing projects that work in line with local governments and environmental professionals, with the goal of fostering strategies to devise adaptation for climate change (UCLG, 2018). These labs can also design strategies to combat desertification, droughts, increased temperature and extreme weather events, avoiding landscape destruction, coastal erosion and structural damages. Furthermore, they can inform the development of responsive strategies for agricultural and architectural adaptations, especially in historic centers and peri-urban areas (ICOMOS, 2021).

Case Study: Citizen Science Lab (Europe)

The Citizen Science Lab was established by the University of Leiden and supports researchers, citizens, and organisations to generate new knowledge together.

The Lab is coordinated by researchers from various fields, such as environmental science and society. The Citizen Science Lab is participating in various projects, with the goal to engage stakeholders in community events and provide examples of community-based research. It is also supporting the development of citizen science across Europe thorough its central platform for sharing knowledge, initiating action and supporting mutual learning ("Plastic spotter: Spot plastic in the canals of Leiden", 2019). A relevant project hosted by the lab is "Plastic Spotter". In this project volunteers study, collect and depict floating litter and macro plastics from Leiden's canals. The main goal of the project is to promote cleaning-up efforts, reduce plastic pollution and prevent litter from re-entering the water (Tasseron et al., 2020).

The contribution to the SDG 13 is important, as the project ensures the reduction of riverine litter and raises awareness that help climate change mitigation in the city, in line with local strategies.

SDG 14: Conserve and Sustainably use the Oceans, Seas and Marine Resources for Sustainable Development

SDG14 aims to prevent and significantly reduce marine pollution, as well as to provide ways to protect and manage sustainably coastal and marine ecosystems and minimize ocean acidification. It also emphasizes the need to end overfishing, to preserve marine and coastal areas, and to foster the economic benefits of underdeveloped countries and Small Island developing States by the use of marine and coastal resources in a sustainable way (Global indicator, n.d.).

Citizen science projects can contribute to the protection and sustainable management of seas, oceans and marine resources through the establishment of initiatives that tackle with the preservation and reinforcement of coastal and marine ecosystems (UCLG, 2018). These projects can also ensure the identification and preservation of underwater archeological assets and develop systems and knowledge bases for resource management, fishing and aquaculture and indigenous practices. Furthermore, they can ensure clean water and food security for all while reducing hazards, such as those emerging as a result of climate change (ICOMOS, 2021).

Case Study: Australian Museum Center of Citizen Science (Australia)

The Australian Museum Centre for Citizen Science (AMCCS) was established in 2015 in Sydney by the Australian Museum Research Institute (The Australian Museum, n.d.).

AMCCS is a national and international leader in citizen science, giving the opportunity to anyone around the world to be engaged in projects and contribute to Australian scientific research. The main scientific projects of research are related to the monitoring of climate change impacts on biodiversity, the detection of pest species and the understanding about the drivers of effective biodiversity conservation (Get Involved, n.d.). Such a project is "FrogID", the main purpose of which is to help understand the dramatic decline of frog populations in Australia, by means of developing a biodiversity database of frogs that relies on acoustic data gathered by citizens and validated from the researchers (Rowley et al., 2019).

The SDG targeted by this project is SDG 14, due to the conservation efforts that it prioritizes towards amphibian ecosystems and the biodiversity in Australia.

SDG 15: Protect, Restore, and Promote Sustainable use of Terrestrial Ecosystems, Sustainably Manage Forests, Combat Desertification, and Halt and Reverse Land Degradation and Halt Biodiversity Loss

This goal aims to promote the sustainable use, restoration, and conservation of inland and terrestrial ecosystems of freshwater, including forests, mountains, wetlands and drylands. It also aims to put a halt to desertification and reduce natural habitats degradation. Other targets stipulate the necessity to end trafficking and poaching of protected species, and the integration of development processes and biodiversity values into local and national planning (Global indicator, n.d.).

Citizen science networks can develop projects that ensure the preservation of traditional knowledge and the sustainable management of biodiversity and terrestrial ecosystems (UCLG, 2018). The conservation, safeguarding, enhancement and management of cultural landscapes and tangible and intangible heritage should be embraced to achieve long-term inclusion of indigenous people and promote sustainable development. Furthermore, these projects can mitigate intensive farming, desertification, land degradation, wildlife illegal hunting and the proliferation of invasive species (ICOMOS, 2021).

Case Study: Snapshot Safari (N. America/Africa)

Snapshot Safari began in 2010, as a citizen science initiative from the University of Minnesota and has since become an independent network based on collaboration between the Nelson Mandela University, the Department of Science and Technology of South Africa and the South African National Biodiversity Institute (SANBI) (Pardo et al., 2021).

The Snapshot Safari Network is a large-scale international network of camera traps to monitor and study the ecological dynamics and the diversity of eastern and southern African mammals. The project is applied into a wide variety of African habitats, protected area sizes and wildlife communities. This multidisciplinary citizen science network combines advanced machine learning techniques (Willi et al., 2019) in order to analyse millions of animal photographs (Pardo et al., 2021), and in sequence to process rapidly large and reliable data volumes (Norouzzadeh et al., 2018).

The main contribution to SDG 15 is the identification of fauna through camera, in order to better understand the biodiversity values, the dynamics and the conservation of wildlife populations across Africa (Pardo et al., 2021).

SDG 16: Promote Peaceful and Inclusive Societies for Sustainable Development, Provide Access to Justice for all and Build Effective, Accountable and Inclusive Institutions at all Levels

SDG16 aims at the global reduction of violence and related death rates, the eradication of abuse, trafficking, exploitation and all forms of torture and violence against children, the reduction of illegal flow of money and arms and the equal and inclusive access to justice for all, by promoting the rule of law at international levels. It also focuses on substantially reducing corruption in all forms and to build effective, inclusive, and accountable institutions at all levels (Global indicator, n.d.).

Citizen science projects can promote the prevention and resolution of local conflicts and the sound implementation and evaluation of policies. Relevant initiatives can provide access to open and free information on local media, ensuring freedom of speech and transparency of local institutions. The activities can take place both on the international and the local level. These projects can ensure that strategies for the reduction of violence and promotion of cultural policies and peace can be established and cultural assets that have been stolen due to wars and conflicts can be returned to their indigenous communities (Zarate, 2015). Furthermore, they can ensure that increased international mobility can foster mutual tolerance, understanding and peace between people (ICOMOS, 2021).

Case Study: Portland Urban Coyote Project (N. America)

The Portland Urban Coyote Project is established by the Portland State University and the Oregon State University in collaboration with Portland Audubon (Rasmussen, 2015).

The project's main goals are to provide information about the interactions between humans and coyotes in the city, map and collect sightings in the metropolitan area of Portland and produce in-depth analyses of data and reports through interactive maps of coyote sightings in the city. The benefit of the project is its contribution to the interaction between the community and stakeholders, the circulation of information through strong communication channels and the improvement of the effectiveness of individuals' knowledge, as well as community action (Rasmussen, 2015).

The contribution of the project to the SDG is the reduction of violence and torture towards coyotes through strong policies and the promotion of the inclusion and participation of the community in decision-making about laws safeguarding the safety of humans and wildlife.

SDG 17: Strengthen the Means of Implementation and Revitalize the Global Partnership for Sustainable Development

The last SDG goal enhances international support and domestic resource mobilization, particularly in developing countries. It also contributes in providing developing countries with the necessary support to manage sustainably their long-term debts, adopting investment friendly regimes and promoting the uptake of state of the art technologies on favorable terms (Global indicator, n.d.).

Citizen science labs can foster the participation of local stakeholders and cultural associations in local and national strategies to promote the SDGs, promoting international cooperation on these topics. These labs can improve the production and distribution of services and goods, particularly those that are needed to understand the importance of global partnerships for sustainable development (UCLG, 2018). Furthermore, these labs can enable a broader participatory process, building the capacities of the stakeholders more effectively and promoting good practices for inter-sectoral, international, and intergenerational partnerships. Such partnerships may well aim at the implementation and development of sustainability-oriented practices (ICOMOS, 2021).

Case Study: Citizen Cyberlab (Europe)

Citizen Cyberlab is co-established by the European Particle Physics Laboratory (CERN), the United Nations Institute for Training and Research (UNITAR) and the University of Geneva (About, n.d.).

The aim of Cyberlab is to level up citizens' engagement as they develop projects that encourage citizens and scientists to collaborate in new ways, to solve challenges in the fields of research and public participation and for sustainable development (About, n.d.).. Such a project is "Rethinking Science and Public Participation" which contextualizes the citizen science field within the topics of public participation in research, education and lifelong learning and helps evaluate the future of the field and the partnerships between citizens and researchers (Strasser et al., 2018).

The contribution of the project to SDG 17 is relevant and of huge importance, as it focuses on the participatory research and improves the revitalization of global partnerships through coproduction between society and science.

RESULTS AND DISCUSSION

The data analysis indicates that citizen science data is produced mainly for community well-being, public governance and scientific research. Most of the projects are primarily used to engage, empower and educate citizens, and the majority of the data produced by the case studies surveyed (ex. *Local Environment Observer Network* and *Neighboorhood Nestwatch*), appears to have a positive impact on communities. Especially for the projects based in developing countries data collection is meant to activate the participation of citizens in scientific research and education (ex. *Que Pasa Riachuelo?* in South America).

The use of citizen science data in labs from developed countries (mainly Europe and North America) is shown to foster accountable and transparent public governance and policy making and help institutions raise awareness of certain topics of interest (ex. *Mada Technion Lab* and gender equality initiatives or public participation and science in *Citizen Cyberlab*). Several initiatives also promote the advance of scientific research through citizen science projects (ex. *Project Budburst*). Furthermore, various projects (especially those who have to do with data collection and the environment) place emphasis on the data collection itself (ex. *Snapshot Safari* in Africa).

The majority of the case studies analyzed, have responded efficiently to the relevant SDG addressed and the main priorities of the goal that are trying to achieve. A fair amount of them, can even assist other SGDs (ex. *Cape Citizen Science* in Africa can also fulfill SDG2 requirements). It generally appears that SDGs that have to do with well-being of people (ex. SDG 1,2) or well-being of the planet (ex. SDG 7,11,13) can be of service to plural SDGs simultaneously.

With regards to the case studies analyzed, citizen science projects have shown a tendency to generate a significant impact, with initiatives that focus both on local areas and communities (ex. *Brooklyn Atlantis*) but also globally (ex. *Aurorasaurus*). Nevertheless, it has to be said that in various citizen science projects their existence and association to the SDG addressed is not directly correlated, so it can be difficult at times, to establish the connection between a case study and the SDG that it belongs to.

CONCLUSIONS

In the analysis for the contribution of citizen science projects to the Sustainable Development Goals (SDGs), it is found that all continents have ensured the implementation of citizen science initiatives that can actively contribute to the SDGs. It is remarkable that all SDGs have at least one citizen science project that contributes to the respective SDG and is relevant to the Sustainable Development Goal mentioned.

It needs to be pointed out that the projects that have been analyzed cover the majority of the fields of interest of research programs (ranging from biodiversity, conservation, nature and indigenous communities to culture, heritage, education, society and gender equality). Furthermore, the range of the project scales is vast, as citizen science projects can be developed by small-scale grassroots initiatives up to well-known universities that establish citizen science hubs and labs for the purpose. From the latter statement, the conclusion is that Research Funding Organizations (RFO) and Research Performing Organizations (RPO) make a significant contribution to every SDG addressed, as all the projects that have been explored, are included in these categories. All things considered, the postulation is that citizen science projects can evolve and progress in correlation to the Sustainable Development Goals and their goals can be relevant and help to the sustainable development cause.

Citizen science projects, initiatives and hubs play a paramount role in the achievement of the 17 SDGs. Ultimately, effective implementation at international, local and regional level must happen simultaneously with the involvement of communities, institutions, research entities and governments, both in the design of the projects but also in the monitoring and evaluation of policies and strategies of sustainable development programs. The empowerment of local actors, citizens and communities by grassroots initiatives is bringing knowledge and fosters education in a plethora of fields and it seems a safe assumption - that the exponential curve regarding the foundation of projects, hubs, and initiatives - can only bring bigger contribution and relevancy between SDGs and citizen science.

REFERENCES

About Citizen Cyberlab. (n.d.). Retrieved February 8, 2022, from https://www.citizencyberlab.org/about

About Journey North. (n.d.). Retrieved February 8, 2022, from https://journeynorth.org/about-journey-north

About Mada Tech. (n.d.). Retrieved February 8, 2022, from https://www.madatech.org.il/ en/about

Ballerini, L., & Bergh, S. I. (2021). Using citizen science data to monitor the Sustainable Development Goals: A bottom-up analysis. *Sustainability Science, 16*(6), 1945–1962. doi:10.100711625-021-01001-1 PMID:34316319

Balta Portoles, J., & Dragicevic Sesic, M. (2017). Cultural rights and their contribution to sustainable development: Implications for cultural policy. *International Journal of Cultural Policy, 23*(2), 159–173. doi:10.1080/10286632.2017.1280787

Brubaker, M., Berner, J., & Tcheripanoff, M. (2013). LEO, the Local Environmental Observer Network: A community-based system for surveillance of climate, environment, and health events. *International Journal of Circumpolar Health, 2013*(72), 22447. doi:10.3402/ich.v72i0.22447

Davis, A., & Howard, E. (2005). Spring recolonization rate of monarch butterflies in eastern North America, new estimates from citizen-science data. *Journal of the Lepidopterists Society, 59*(1), 1–5.

Delli Paoli, A., & Addeo, F. (2019). Assessing SDGs: A methodology to measure sustainability. *Athens Journal of Social Sciences, 6*(3), 229–250. doi:10.30958/ajss.6-3-4

Dorler, D., Fritz, S., Voigt-Heucke, S., & Heigl, F. (2021). Citizen science and the role in Sustainable Development. *Sustainability, 13*(10), 5676. doi:10.3390u13105676

Evans, C., Abrams, E., Reitsma, R., Roux, K., Salmonsen, L., & Marra, P. P. (2005). The neighborhood nestwatch program: Participant outcomes of a citizen-science ecological research project. *Conservation Biology, 19*(3), 589–594. doi:10.1111/j.1523-1739.2005.00s01.x

Fressoli, M., Arza, V., & Castillo, M. D. (2016). *Entina Argentina. The impact of citizen-generated data initiatives in Argentina.* Civicus Datashift.

Fritz, S., See, L., Carlson, T., Haklay, M. M., Oliver, J. L., Fraisl, D., Mondarini, R., Brocklehurst, M., Shanley, L., Schade, S., When, U., Abrate, T., Anstee, J., Arnold, S., Billot, M., Campbell, J., Espey, J., Gold, M., Hager, G., & West, S. (2019). Citizen science and the United Nations Sustainable Development Goals. *Nature Sustainability, 2*(10), 922–930. doi:10.103841893-019-0390-3

Get Involved. (n.d.). *The Australian Museum.* Retrieved February 8, 2022, from https://australian.museum/get-involved/

Hall, L. (2015, June 16). Aurorasaurus. *NASA.* Retrieved February 8, 2022, from https://www.nasa.gov/feature/aurorasaurus

Henderson, S., Newman, S., Ward, D., Havens-Young, K., Alaback, P., & Meymaris, K. (2011). Project BudBurst: Continental-scale citizen science for all seasons. *AGU Fall Meeting Abstracts.*

Hulbert, J. M. (2016). Citizen science tools available for Ecological Research in South Africa. *South African Journal of Science, 112*(5/6), 2. Advance online publication. doi:10.17159ajs.2016/a0152

Hulbert, J. M., Turner, S. C., & Scott, S. L. (2019). Challenges and solutions to establishing and sustaining citizen science projects in South Africa. *South African Journal of Science, 115*(7/8). Advance online publication. doi:10.17159ajs.2019/5844

International Council of Monuments and Sites. (2021). *Heritage and the sustainable development goals: policy guidance for heritage and development actors.* International Council on Monuments and Sites – ICOMOS.

Kummer, J. A., Bayne, E. M., & Machtans, C. S. (2016). Use of citizen science to identify factors affecting bird–window collision risk at houses. *The Condor, 118*(3), 624–639. doi:10.1650/CONDOR-16-26.1

Laut, J., Porfiri, M., & Nov, O. (2014). Brooklyn Atlantis: a robotic platform for environmental monitoring with community participation. *Proceedings of the International Conference of Control, Dynamics Systems, and Robotics.*

McGhie, H. A. (2019). *Museums and the Sustainable Development Goals: a How-to Guide for Museums, Galleries, the Cultural Sector and their Partners.* Curating Tomorrow.

McKinley, D. C., Miller-Rushing, A. J., Ballard, H. L., Bonney, R., Brown, H., Evans, D. M., French, R. A., Parrish, J., Phillips, T., Ryan, S., Shankley, L., Shirk, J., Stepenuck, K., Weltzin, J., Wiggins, A., Boyle, O., Briggs, R., Chapin, S., Hewitt, D., & Soukup, M. (2015). *Investing in citizen science can improve natural resources management and environmental protection. In Issues in Ecology.* Ecological Society of America.

Norouzzadeh, M. S., Nguyen, A., Kosmala, M., Swanson, A., Palmer, M. S., Packer, C., & Clune, J. (2018). Automatically identifying, counting, and describing wild animals in camera-trap images with Deep Learning. *Proceedings of the National Academy of Sciences of the United States of America, 115*(25). Advance online publication. doi:10.1073/pnas.1719367115 PMID:29871948

Pardo, L. E., Bombaci, S. P., Huebner, S., Somers, M. J., Fritz, H., Downs, C., Guthmann, A., Hetem, R. S., Keith, M., le Roux, A., Mgqatsa, N., Packer, C., Palmer, M. S., Parker, D. M., Peel, M., Slotow, R., Strauss, W. M., Swanepoel, L., Tambling, C., ... Venter, J. A. (2021). Snapshot safari: A large-scale collaborative to monitor Africa's remarkable biodiversity. *South African Journal of Science, 117*(1/2). Advance online publication. doi:10.17159ajs.2021/8134

Plastic spotter: Spot plastic in the canals of Leiden. (2019, November 27). Leiden University. Retrieved from https://www.universiteitleiden.nl/en/news/2019/11/plastic-spotters-launch

Rasmussen, Z. A. (2015). Coyotes on the Web: Understanding human-coyote interaction and online education using citizen science. *Dissertations and Theses, Paper 2643.* . doi:10.15760/etd.2639

Rosenthal, I., Byrnes, J., Cavanaugh, K., Bell, T., Harder, B., Haupt, A., Rassweiler, A., Perez-Matus, A., Assis, J., Swanson, A., Boyer, A., McMaster, A., Trouille, L. (2018). *Floating forests: Quantitative validation of citizen science data generated from consensus classifications.* Academic Press.

Rowley, J., Callaghan, C., Cutajar, T., Portway, C., Potter, K., Mahony, S., Trembath, D., Flemons, P., & Woods, A. (2019). FrogID: Citizen scientists provide validated biodiversity data on frogs of Australia. *Herpetological Conservation and Biology, 14*(1), 155–170.

Rozenblum, Y., Dalyot, K., Lachman, E., & Baram-Tsabari, A. (2018). Gendered engagement with posts authored by female scientists on Facebook. *Proceedings of the 15th Chais Conference for the Study of Innovation and Learning Technologies: Learning in the Digital Era.*

Schleicher, K., & Schmidt, C. (2020). Citizen science in Germany as research and Sustainability Education: Analysis of the main forms and foci and its relation to the Sustainable Development Goals. *Sustainability, 12*(15), 6044. doi:10.3390u12156044

Scott, K., & Schulz, L. (2017). Lookit (part 1): A new online platform for Developmental Research. *Open Mind: Discoveries in Cognitive Science*, *1*(1), 4–14. doi:10.1162/OPMI_a_00002

Strasser, B. J., Baudry, J., Mahr, D., Sanchez, G., & Tancoigne, E. (2018). "Citizen science"? Rethinking science and public participation. *Science & Technology Studies*, 52–76. doi:10.23987ts.60425

Tapia, A., Lalone, N., MacDonald, E., Hall, M., Case, N., & Heavner, M. (2014). *AURORASAURUS: Citizen science, early warning systems and space weather*. Proceedings of the AAAI Conference on Human Computation and Crowdsourcing.

Tasseron, P., Zinsmeister, H., Rambonnet, L., Hiemstra, A.-F., Siepman, D., & van Emmerik, T. (2020). Plastic hotspot mapping in Urban Water Systems. *Geosciences*, *10*(9), 342. doi:10.3390/geosciences10090342

United Cities and Local Governments. (2018). *Culture in the sustainable development goals: a guide for local action*. UCLG Committee on Culture.

United Nations Educational Scientific and Cultural Organization. (2014). *Gender equality. Heritage and creativity*. UNESCO.

United Nations Educational Scientific and Cultural Organization. (2016). *Cultural urban future: Global report on culture for sustainable urban development*. UNESCO.

United Nations. (n.d.). *Global indicator framework for the Sustainable Development Goals and targets of the 2030 Agenda for Sustainable Development*. Retrieved February 8, 2022, from https://unstats.un.org/sdgs/indicators/Global%20Indicator %20Framework%20after%20refinement_Eng.pdf

Vohland, K., Land-Zandstra, A., Ceccaroni, L., Lemmens, R., Perello, J., Ponti, M., Samson, R., & Wagenknecht, K. (2021). *The science of citizen science*. Springer International Publishing. doi:10.1007/978-3-030-58278-4

West, S., & Paterman, R. (2017). *How could citizen science support the sustainable development goals?* Stockholm Environment Institute.

Willi, M., Pitman, R. T., Cardoso, A. W., Locke, C., Swanson, A., Boyer, A., Veldthuis, M., & Fortson, L. (2018). Identifying animal species in camera trap images using Deep Learning and Citizen Science. *Methods in Ecology and Evolution*, *10*(1), 80–91. doi:10.1111/2041-210X.13099

Zarate L. (2015). Right to the city for all: a manifesto for social justice in an urban century. *The Nature of Cities*.

KEY TERMS AND DEFINITIONS

Biodiversity: The biological variety of life on the planet. It is a measure of dissimilarities at the ecosystem and genetic level. It surrounds the ecological, cultural, and evolutionary procedures that support life on Earth.

Citizen Science: Scientific research managed, in part or as a whole, by nonprofessional scientists and/or citizens. Citizens help scientific research by public participation, collection and analysis of data, research and monitoring actions which frequently offer advancements in improving public understanding of science and increasing awareness to the community.

Climate Change Mitigation: Consists of steps to curb global warming and its related consequences. It involves limitations of human-made greenhouse gases and the reduction of their accumulation in the atmosphere.

High-Level Political Forum of Sustainable Development (HLPF): A division of both the United Nations Economic and Social Council and the United Nations General Assembly, in charge of the organization's strategy on sustainable development. It reviews the progress and commitments of the 17 Sustainable Development Goals.

National Science Foundation (NSF): An independent organization of the government of the United States that assists cardinal research in the fields of engineering and science (with the exclusion of health-related projects).

Research Funding Organizations (RFO): Organizations that fund projects and research programs at their capability level. These organizations can range from research institutes and agencies to higher education institutions, and they set regulations and standards on research and science at a political level.

Research Performing Organizations (RPO): Organizations that enable and empower researchers to take action responsibly and reduce the threats of violations to research ethics. They also make sure that proper policies and procedures, facilities and government arrangements are established.

Sustainable Development Goals (SDG): A collection of 17 interconnected goals drafted in 2015 by the United Nations General Assembly and created to be a map for a global sustainable future. Each goal has specific targets to be achieved until 2030 and indicators that are used to evaluate the progress towards respective targets.

Chapter 8

Global Warming on Business Planning in Mexico and the Impact of Best Practices on Quality of Life

José G. Vargas-Hernández
https://orcid.org/0000-0003-0938-4197
Posgraduate and Research Department, Tecnológico Mario Molina Unidad Zapopan, Mexico

Pedro Antonio López de Haro
Universidad Autónoma Indígena de México, Mexico

ABSTRACT

The aim of this chapter is to identify and analyze some of the best practices in relation to the impact of global warming on business planning in Mexico. The phenomenon of global warming as changes in the business environment has impacted on the strategic planning processes of Mexican companies in the creation of objectives and methodologies in various fields and sectors, highlighting the importance of adapting to the needs and the limitations that may arise with climate change in the various markets we are focused.

INTRODUCTION

The following chapter is an exploratory research on the opinions of climate change on firms in Mexico. 145 firms were surveyed using a website called surveymonkey.com, and the question referred to their stance in regards to climate change, quality of life and how to introduce it into the business practices of the firm. We found that only 33% of the firms surveyed said they took the threat of climate change seriously enough to factor it in their strategic planning. Some of the problems related to climate change that Mexican companies have to deal with are: Pollution and deterioration in the environment, shortage of

DOI: 10.4018/978-1-6684-4829-8.ch008

raw materials, an increase in health problems for workers and costumers, water pollution, brand image due to environmental policies, etc.

Environmental management in business tends to evaluate the performance of business units with indicators of profitability and return on investment, in addition to their contributions to reduce carbon dioxide emissions, which has been identified as a very important greenhouse gas. Global warming has been a major issue in business administration and public policy in recent years. Internationally, climate change has received attention in various economic and social forums.

Although science provides clear and conclusive evidence of the impacts of climate change on the quality of life, both the private and public sectors are reacting slowly, privileging their interests. In sum, collective efforts are insufficient to design and implement policies and strategies aimed at mitigating the effects of climate change on quality of life. Until now, the same market forces and the economic rationality of business have maximized profits with lower costs, which drive investment in the use of resources and technologies that produce greenhouse gas emissions, accelerating global warming with all its consequences for the earth.

In Al Gore's documentary (Pérez, 2007), An Inconvenient Truth, the US former vice president describes the relationship between global warming and CO_2 emissions and predicts that the phenomenon of global warming will get increasingly worse if greenhouse emissions aren't reduced significantly. However, Gore also warns how certain interests' agents have spread doubt about global warming, and its relationship to human activity. Guhl (2008) talks about the role of mass media for the public's misunderstanding of the problem.

Global warming also operates within the context of the recent global change, associated with increased concentrations of certain gases, such as greenhouse gases (GHG) and in the atmosphere. The last IPCC report (2007) warns that the biggest increases in GHG emissions between 1970 and 2004 come from power generation, transportation, and industry, followed by construction, forestry and agriculture.

There is incontrovertible evidence on how the dependence on fossil fuels is the most important factor contributing to global warming (Guhl, 2008; Sussman and Freed, 2008).

The aim of this research is to understand the impacts of climate change in the planning processes of companies and businesses in Mexico.

STATE OF THE ART

As a result of the evaluation of the impact of climate change, several reports have been produced. This problem has been addressed in the past by the oil companies (Van den Hove, Le Menestrel and Bettignies, 2002; Levy and Kolk, 2002), and three strategies were developed: (1) prioritizing trade implications while weakening perception that anthropogenic greenhouse gas emissions are causing climate change, (2) avoiding responsibility, and (3) prioritizing the need for a change in the business process while limiting the negative effect in terms of commercial consequences. Boiral (2006) mentions that managers, especially in large industries, should try to develop an inventory of alternatives and sources of GHG, in order to determine which options are more efficient for replacement.

The implementation of a proactive strategy can help improve employee involvement and the image of a corporation. Many factors such as the sector, technological innovation, the market price of CO_2 emissions, social pressures and public policy, can significantly change the assessment of the opportunities and threats arising from the Kyoto Protocol. This depends on the vision, values, and skills of managers.

Jorqueta and Orrego (2010) mention that addressing the impact of possible climate change will allow the industry to define some strategies that could be useful for future planning, including sustainability strategies.

According to Sussman and Fredd (2008), the projected effects of climate change are:

1. increase in temperature, with fewer cold days and warmer nights, and the duration, intensity, and duration of heat.
2. Rising sea level, between 18 and 59 centimeters.
3. Increased precipitation and humidity, including at high latitudes, and winter precipitation.
4. Extreme winds, tropical storms, and other events, including the activity of hurricanes, floods, and fires in arid and semi-arid areas; and
5. Other related effects, including reduction in the duration of snow seasons, loss of glaciers and permafrost. All of them are labeled as elements of climate risk.

Meanwhile, Proverbs (2010) states that global warming increases the risk of flooding, provided that the construction industry must consider as an element in its planning risk and consider their impacts to the sustainability of locations in the real estate developments, such as: reduction in price of the land, increased costs of formalities for the construction, increasing construction costs, reduced salability, and difficulty in securing long-term viability of skills construction.

According to Garcia (2006), when applying the methodology of cost-benefit analysis (CBA) to the phenomenon of global warming, there are several complications:

1. Proper data collection. Because there is still uncertainty about what will be the environmental impact that cause greenhouse gases and about the social, economic, and probable environmental impacts, along with a lack of agreement on the economic costs of those impacts.
2. The use of gross national product (GNP). The greenhouse effect damage existing production flows and welfare, measuring the cost and loss of total annual production, is detrimental to the results of past production and existing stocks of natural resources.
3. The use of the discount rate. The use of CBA does not lead to a policy of sustainability. The cost-benefit analysis is related to total quantities (net costs or benefits), while sustainability is related to equity in the distribution over time; and
4. The uncertainty. Scientists still do not understand some of the key mechanisms that determine climate change.

De la Vega (2011) mentions that raising the average temperature of the planet's surface does not occur uniformly, and, given its geographical position, Mexico can be very vulnerable. The global average temperature has increased by more than 1 °C over the last thirty years, with some regions having increased 3 °C. And while this can be explained by natural processes, such as atmospheric circulation of the planet and the relationship of ocean currents, the phenomenon is aggravated by overexploitation of natural resources and damage to ecosystems by deforestation. According to De la Vega (2011), Mexico is having significant impacts due to global warming: higher maximum temperatures, duration, frequency, and intensity of extreme events such as warmer heat waves ... causing decrease in the availability of water.

Currently, for many organizations, climate change is a priority issue on their agenda. The threat of climate change is, for Ernst & Young Business Risk Report, the fourth place in business risks (Ernst &

Young, 2010). This danger was exceeded only by the risks of the credit crisis, regulation and compliance and recession. This issue concerns global leaders and high-growth companies. In a recent study of global organizations with more than $25 billion of market capitalization, performed by Ernst & Young, 73% of the companies had been committed to reducing its greenhouse gas (Ernst & Young, 2010).

So far, most of the efforts to combat climate change are disjoined and uncoordinated -actors meet frequently, but they don't seem to agree on much and, because they are negatively affected by the same policies they are deciding on, they fail to propose meaningful solutions, in fact, actors have formed strong coalitions against carbon tax and the General Law on Climate Change (Ortega Díaz & Casamadrid Gutiérrez, 2018).

RESEARCH METHODS

The methodological design of the study is a non-experimental, quantitative, field research, transversal and exploratory-descriptive. Quantitative and exploratory research was selected because of the scope of the investigation: to determine the impact on business planning processes of global warming.

Business planning is a process usually written in a document that determines and describes, in detail, the goals, objectives, strategies, policies, and actions of the whole firm, including human resources, financial, production, operations, and marketing activities and how they are going to be achieved. Business planning is a fundamental tool that provides a road map to promote the business development and growth, focusing on dealing with the business risks to make profits and ensuring business prosperity and success.

The research is considered transversal because no historical comparisons are to be made over a period. In addition, the survey technique was chosen because the source of information is the executive and managerial personnel of firms.

Survey Design

For the survey design, we developed a model that relates the variables to be measured (figure 1).

The applied model (Figure 1) was developed from the contributions of Ernst & Young (2010), Boiral (2006) and Sussman and Freed (2008), organizing the factors according to the components of the external environment of a company According to Thompson, Strickland, and Gamble (2008). The variables identified as impact of global warming are grouped into two dimensions:

1. Identification of risks and
2. Identification of opportunities; both elements in the environment analysis as input for strategic planning and business.

In this case, the variables in the models were measured using Likert scales.

Data Collection

The target population of this research are the companies located in Mexico, of any size, business orientation and location, this geographic coverage being chosen by the exploratory nature of the study. The

Figure 1. Environmental factors of climate change in the Mexican firm.
Source: Own construction

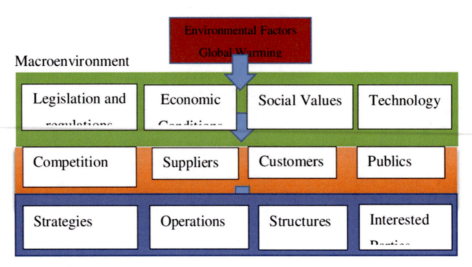

Source: Authors

universe of companies was determined as infinite population, given that they are more than 500,000 units of study, according to the Mexican Business System (Secretaría de Economía, 2012).

Survey technique using online information systems (www.surveymonkey.com) was used for data collection. The strategy for contacting these companies was a combination of personal visits, telephone contact and email. The general intention was that managers or business owners were the ones who provided the information.

In total, 145 surveys were conducted, gathering information from 15 states. Applying the criteria of the Ministry of Economy (Secretaría de Economía, 2009), of these 145, 25% were micro enterprises, 13% small enterprises, 30% were medium and 33% were large. On the other hand, 31% mentioned having an international geographical coverage, 36% national, 26% regional and only 7% local. Most of those surveyed belong to the service sector (70%), while 20% to industry and 10% to trade. Of those surveyed, 78% were private and 22% public.

Data Analysis

Once the information was collected, we proceeded to process and analyze the data, using SurveyMonkey's tools themselves. The first activity was to build a description based on the responses, identifying profiles according to different respondents.

The second activity involved the construction of a relevance tree (Chevalier and Buckles, 2009; Godet, 2000).

RESULTS

Practice of Strategic Planning

61% of respondents regularly are performing strategic planning, while 17% said they engage in strategic planning but periodicity, 9% said they did once, and 13% said no strategic planning is done.

Of respondents who engage in strategic planning, 80% answered participating directors, 56% the board of directors, 48% managers, 5% and 2.5% Financial advisors' entities.

About Climate Change

Of respondents who perform strategic planning, 93% answered that they know about the issue of climate change. Moreover, when asked if they considered its effects on the strategic planning of the organization, 33% said yes, they have done analysis of risks and opportunities, 30% mentioned that have been touched on, and 38% mentioned that have not been considered.

On the risks that are considered in strategic analysis, cost increases are the elements that are considered, mainly the cost of energy (85%), supplies (50%), damage to facilities and infrastructure (35%), and the cost of public health factors (30%). In the second place, respondents cited increased uncertainty in planning elements, such as water supply (55%), energy (45%), supplies (40%) and food (20%). Third, respondents expressed increased compliance requirements of regulations (50%) and pressures on the public image of the organization (45%).

Among the less considered risks, investors said discounts (5%), tax compliance from climate change (10%), and changes in the behavior of stakeholders (10%), product and service competition substitutes (10%), and changes in expectations of provider (15%), dissatisfaction with climate change (15%), brand awareness (20%), economic sanctions (20%) and changes in the segment of customers (25%).

On identified opportunities, 75% of respondents mentioned the development of new products and services, 70% implementation of environmental education programs, 60% for training, 55% improved corporate image as well as the application of technologies Green and ecological values of employees, 45% implementation of green strategies, 40% by improvements in energy and operational efficiency for power generation, as well as greater participation and responsibility of employees, 35% new customer segments.

Of those surveyed, 33% said that, derived from the analysis of climate risks and opportunities, they have invested in several strategic projects; 5% said that they had invested in a single project, 33% replied that they had not invested but have analyzed various initiatives; and 29% replied that they had not invested.

DISCUSSION

The problem tree (Figure 2) identifies the main causes and effects found in this research. As the main problem, we considered the effects of climate change of the firm.

In this way, the scheme was developed grouping the effects the global warming such as external public, social and political pressures. Compliance requirements of global, regional, local regulations and other sectors, uncertainty factors in the effect on public health and uncertainty in the supply of raw materials. And the causes that have led directly the effects mentioned above are: There has been a

Figure 2. Problem tree for climate change on the Mexican firm
Source: Own creation

shortage of resources (raw materials) caused by climate change, social groups can begin to put pressure on companies that pollute, and it has had increased health effects from climate change and pollution in vulnerable areas where the companies are established.

As we can see, there is a clear relationship between the effects and causes which are based on conditions of social, health and scarcity of resources (raw materials) for organizations. These indicators have permeated the effects on business planning which has led to strategies in various sectors, such as the development of new products and services, green brand strategy, enhance corporate image, improvements in energy efficiency are created and operative to generate clean energy, ecological values on employees, and environmental education programs.

These strategies are part of the effects that global warming has generated directly in planning organizations, as indicated by the surveys. Another important element in organizations is the creation of brands focused on specific segments of green consumers, as well as diversification in green bonds and green technologies management.

BEST PRACTICES ON QUALITY OF LIFE REGARDING THE IMPACT OF GLOBAL WARMING ON BUSINESS PLANNING IN MÉXICO

All business units in Mexico should adopt some of the best practices already tried and tested and learn advanced technology innovations that can add greater efficiency to resources and thereby increase the quality of life. The efficiency of resource use to reduce global warming in the business units is crucial to improving the quality of life and economic development.

Among the best practices for climate protection is a priority to reduce emissions from buildings operations and economic units in Mexico. Rational use of other energy sources such as geothermal waters, which do not contribute to global warming, are some of the many options to increase the environmental performance of business units that also brings benefits to improving environmental performance, quality life and public health. The prevention of climate change by establishing protective measures have an impact on local economic and environmental benefits. The strategic and direct confrontation threat of climate change, by governments and communities make them involved in various prevention activities.

Among other recommended measures considered as the best practices in action plan for counteract the global warming phenomenon, are the development and use of biofuels, promoting the use of alternative and renewable energy, the development and promotion of clean development mechanisms and projects, design and implementation of a watershed development strategies, development policy for reservoirs, waters and irrigation capacities, promotion of sustainable forests and green urban areas, etc.

Among the best practices to mitigate global warming of economic units in Mexico is the creation of green spaces and recreational green parks and gardens around the most polluting manufacturing plants. These areas and green gardens will help promote the quality of life by providing opportunities for further relaxation, sports, and health. Rational use of other energy sources such as geothermal waters, which do not contribute to global warming, are some of the many options to increase the environmental performance of business units that also brings benefits to improving environmental performance, quality life and public health.

To reduce the contribution of economic units in Mexico to climate change, it is evident that they require reducing energy consumption based on fossil and increase the use of green energy through processes and innovative measures and cost effectiveness, contributing to improve air quality and ultimately a better quality of life. There is not an inclusive list of best practices that local governments and communities have in terms of protection activities that demonstrate their potential for prevention activities of global climate through local action. Each case is different (O'Neill and Rudden, 2013).

Some of these best practices to improve the standard of quality of life are: The establishment of standards of sustainable energy through mechanisms of energy codes for use in buildings and processes in businesses as mechanisms supervised by local governments. A procurement program of environmentally friendly products is important to be implemented such as purchasing green energy for operations. Reducing greenhouse gas emissions in operations and buildings of economic units is also critical by increasing standards of energy efficiency of the processes using machinery and heavy equipment.

Some more practices require the development and implementation of programs of energy conservation, the establishment of audit, quotas, and incentives for improvements in energy efficiency and the use of other green energy such as solar, wind, etc. Cooperative programs can start purchasing green power for more efficient use of machinery and equipment. Programs for planting trees and green areas around the buildings of the plant to improve climate and has other benefits. Firms must reduce the use of vehicles transporting goods and people by promoting improvements in pedestrian areas and transit bicycles and other vehicles, as well as to establish incentive programs for green transit-oriented development. Land development programs through policy areas and discouragement of activities and permit fees that invade the occupation of floor space.

Climate changes severely affect companies due to the potential of environmental conditions associated with the risks of adverse conditions, such as natural disasters, which are not always identified as vulnerable or considered in the planning of business operations situations. Small and medium businesses that have limited resources and are located only in few places are very vulnerable to the ravages of natural disasters caused by climate change. With limited access to resources and capital, small businesses may suffer from severe economic damage. The impacts of climate change on small and medium businesses have a long reach to break the chains of supply consumer markets.

Among other recommended measures to improve the environment and mitigate the effects of climate change on small and medium-sized business practices, we recommend working in partnership with communities and local governments to design and implement a joint plan and programs to share data and information, promote actions that build resilience of the value chain, so that vulnerabilities are reduced, to be sustained with less economic damage and recover quickly from disasters and calamities. These plans and programs should include strategies and policies for cost effectiveness of measures to reduce emissions and bring great benefits to the quality of life of people and the economy in general.

CONCLUSION

Climate change is a complex phenomenon, which has been very polemic, and has presented various effects on societies and businesses.

The implementation of complex environmental innovations as a distinct organizational strategy, encourage entrepreneurship to mitigate the effects of global warming and increase the quality-of-life standards of all workers and other stakeholders, while preserving environmental conditions in favor of long-term life.

The literature review showed how various organizations have considered the issue at international and national level. This research shows that the issue has been considered by Mexican firms in their planning processes and strategic projects are being generated from this consideration.

Climate change has had an impact on the strategic planning process of Mexican companies as a factor that is changing the economic, political, and legal environment in which it operates, and has included three key elements: the increase in costs and the increase in the uncertainty of key factors such as the development of new products and markets.

This exploratory study opens the door for further descriptive and explanatory analysis studies highlighting success stories of strategies in Mexican and foreign companies.

REFERENCES

Boiral, O. (2006). Global Warming: Should Companies Adopt a Proactive Strategy? *Long Rabge Planning. International Journal of Strategic Management, 39*(2), 315–330.

Chevalier, J. M., & Buckles, D. J. (2009). *SAS2 Guide to Collaborative Research and Social Mobilization* [SAS2 Guía para la Investigación Colaborativa y la Movilización Social]. Plaza y Valdés.

De la Vega-Salazar, M. Y. (2011). Possible risk factors for Mexico's epicontinental fish in the face of global warming [Posibles factores de riesgo de los peces epicontinentales de México ante el calentamiento global]. *Avances en Investigación Agropecuaria, México, 15*(2), 65–78.

Ernst & Young. (2010). *The response of companies to climate change* [La respuesta de las empresas al cambio climático. Elegir el camino correcto]. Recovered from: https://www.ey.com/MX/es/Services/Specialty-Services/Climate-Change-and-Sustainability-Services

García Fernández, C. (2006). Cost-benefit analysis and the difficulty of its application to climate change [El análisis coste-beneficio y la dificultad de su aplicación al cambio climático]. *Estudios de Economía Aplicada, España, 24*(2), 751–762.

Godet, M. (2000). The Toolbox of Strategic Foresigh [La caja de herramientas de la prospectiva estratégica] (4th ed.). España: Laboratoired'Investigation Prospectiva et Stratégique y Prospektiker – Instituto Europeo de Prospectiva y Estrategia.

Guhl Corpas, A. (2008). Ethical aspects of global warming [Aspectos éticos del calentamiento climático global]. *Revista Latinoamericana de Bioética, Colombia, 8*(2), 20–29.

Jorqueta-Fontena, E., & Orrego.Verdugo, R. (2010). Impact of global warming on the phonology of a vine variety grown in southern Chile [Impacto del calentamiento global en la fonología de una variedad de vid cultivada en el sur de Chile]. *Agrociencia, México, 44*(4), 427–436.

Levy, D. L., & Kolk, A. (2002). Strategic Responses to Global Climate Change: Conflicting Pressures on Multinatrionals in the Oil Industry. *Business and Politics, 4*(3), 275–300. doi:10.2202/1469-3569.1042

O'Neill; K. & Rudden, PJ. (2013). *Environmental Benchmarking & Best Practice Report RPS Group, Ireland*. European Green Capital Award 2012 & 2013.

Ortega Díaz, A., & Casamadrid Gutiérrez, E. (2018). Competing actors in the climate change arena in Mexico: A network analysis. *Journal of Environmental Management, 215*, 239–247. doi:10.1016/j.jenvman.2018.03.056 PMID:29573674

Pérez Salazar, B. (2007). Review of "An Inconvenient Truth. The Planetary Crisis of Global Warming and How to Deal with It" by Al Gore [Reseña de "Una verdad incómoda. La crisis planetaria del calentamiento global y cómo afrontarla" de Al Gore]. *Revista de Economía Institucional, Colombia, 9*(17), 385–395.

Proverbs, D. (2010). The Impact of Climate Change on Local Scale Flood Risk for Indivual developments. School of Engineering and the Built Enviroment, University of Wolverhampton, United Kingdom. *Systematic Reviews*, 89.

Secretaría de Economía. (2009, June 30). Agreement establishing the stratification of micro, small and medium-sized enterprises [Acuerdo por el que se establece la estratificación de las micro, pequeñas y medianas empresas]. *Diario Oficial de la Federación*.

Secretaría de Economía. (2012). *Mexican Business Information System* [Sistema de Información Empresarial Mexicano]. Recovered from http://www.siem.gob.mx/siem/

Sussman, F. G., & Freed, J. R. (2008). *Adapting to climate change: A business approach. Pew Center on Global Climate Change*. Recovered from https://www.c2es.org/business-adaptation

Thompson, A. A., Jr., Strickland, A. J., III, & Gamble, J. E. (2008). Strategic Management [Administración Estratégica]. In Teoría y casos (15th ed.). McGraw Hill.

Van den Hove, S., Le Menestrel, M., & de Bettignies, H.-C. (2002). The oil industry and climate change: Strategies and ethical dilemmas. *Climate Policy, 2*(1), 3–18. doi:10.3763/cpol.2002.0202

KEY TERMS AND DEFINITIONS

Business Planning: It is a process, usually written in a document, that determines and describes the goals, objectives, strategies, policies, and actions of the whole firm, including the human resources, financial, production, operations, and marketing activities and how they are going to be achieved. Business planning is a fundamental tool that provides a road map to promote the business development and growth, focusing on dealing with the business risks to make profits and ensuring business prosperity and success.

Climate Change: A gradual earth's increase of average global temperature in the overall atmospheric and oceanic warming, generally predicted to occur and attributed to the gradual rise in greenhouse effect resulted by pollution and caused by high levels of gases such as carbon dioxide and chlorofluorocarbons. These gases are collected in the air, which traps heat in the atmosphere, preventing it from going into space.

Quality of Life: The quality of life is a subjective concept that refers to the conditions of well-being in which people live in all areas and levels of the environment, society, and the community, etc. These conditions of well-being respond to the satisfaction of the ecological, physiological, material, social, psychological, emotional, and human development needs, in such a way that they have a pleasant existence and a life worth living.

Chapter 9
Green Intellectual Capital as a Catalyst for the Sustainable Development Goals:
Evidence From the Spanish Wine Industry

Javier Martínez Falcó
https://orcid.org/0000-0001-9004-5816
University of Alicante, Spain

Bartolomé M. Marco-Lajara
https://orcid.org/0000-0001-8811-9118
University of Alicante, Spain

Patrocinio Zaragoza-Sáez
University of Alicante, Spain

Lorena Ruiz-Fernández
https://orcid.org/0000-0001-8894-9701
University of Alicante, Spain

ABSTRACT

The purpose of this research is to analyze the different effects of the green intellectual capital (GIC) of wine companies on the fulfillment of the Sustainable Development Goals (SDGs), contributing to academic literature in a remarkable way, since, to the authors' knowledge, there is no previous research that has addressed this relationship. In order to achieve the proposed objective, the research follows a qualitative approach, since the single case study was used. The research results demonstrate how the three dimensions of the GIC (green human capital, green structural capital, green relational capital) act as catalysts for the fulfillment of SDGs 3, 5, 6, 7, 8, 9, 11, 12, 13, 15, and 17.

DOI: 10.4018/978-1-6684-4829-8.ch009

INTRODUCTION

Growing customer awareness of environmental challenges and strict environmental regulations mean that sustainability has become a business paradigm present in various economic sectors (Annunziata et al., 2018). There is some consensus in considering sustainability as a necessary challenge given the negative externalities generated by organizations to the environment. Thus, although the intellectual background of this debate dates back to the 1950s of the last centuries, today it is still necessary to make further progress in this direction.

In the wine industry, sustainability is an extremely important issue for two main reasons. On the one hand, the industry faces serious threats as a consequence of climate change, as well as water and energy scarcity (Aigrain et al., 2016; Gilinsky et al., 2016). On the other hand, proper environmental management of wineries can mean obtaining a competitive advantage, by allowing to increase market share and organizational innovation processes (Fiore et al., 2017). In this sense, previous work has shown that customers tend to select wines that have been developed following sustainable practices, despite not knowing what this means in practice (Schäufele & Hamm, 2017). From the producers' point of view, it has been shown that the implementation of environmental practices improves the quality and economic efficiency of employees (Szolnoki, 2013).

The concept of sustainability in the wine industry is supported by official documents of the International Organisation of Vine and Wine (OIV), defining what sustainable viticulture is (OIV, 2004), what its guidelines are (OIV, 2008) and its general principles (OIV, 2016). Similarly, the main wine regions have developed their own national programs in order to implement the principles of sustainability in their territories (Marco-Lajara et al., 2022a; Marco-Lajara et al., 2022b; Marco-Lajara et al., 2022c; Marco-Lajara et al., 2022d; Marco-Lajara et al., 2022f; Marco-Lajara et al., 2022g; Marco-Lajara et al., 2022h; Marco-Lajara et al., 2022i; Marco-Lajara et al., 2022j; Marco-Lajara et al., 2022k; Marco-Lajara et al., 2022l). For the case of Spain, the Spanish Wine Federation (FEV, by its acronym in Spanish) developed the Strategic Plan 2019-2024 (FEV, 2019) with the aim of guaranteeing future wine production and its legitimacy in society through a comprehensive sustainability strategy. Said document establishes the set of policies that can be developed by the wine industry value chain to achieve the Sustainable Development Goals (SDGs) established by the United Nations (UN). According to this strategic plan, the 11 SDGs related to the wine industry and, therefore, those in which the sector can have a relevant impact, are the following: SDG 3 -health and well-being-, SDG 5 -gender equality-, SDG 6 -clean water and sanitation-, SDG 7 -affordable and clean energy-, SDG 8 -decent work and economic growth-, SDG 9 -industry, innovation and infrastructure-, SDG 11 -sustainable cities and communities-, SDG 12 -responsible production and consumption-, SDG 13 -climate action-, SDG 15 -living terrestrial ecosystems-, SDG 17 -partnerships to achieve the goals-.

Compliance with the SDGs by organizations allows them to balance their economic, social and environmental objectives. In order to meet these objectives, companies must generate new knowledge that allows them to develop their activities through a sustainable approach (Boons et al., 2013), and they can adopt to increase their endowment of intangibles in its three aspects: human capital, structural capital and relational capital (Davenport & Prusak, 1998). In this sense, Intellectual Capital (IC) that incorporates environmental aspects, i.e., Green Intellectual Capital (GIC), takes on special relevance as it represents an essential element for achieving corporate sustainability.

The motivation for this research stems from two premises. First, a higher GIC enables companies to better meet their environmental challenges. Secondly, the GIC acts as a transforming element in the organization through the knowledge achieved. Based on these arguments and using the case study applied to a winery, the present study aims to answer the following research question: does GIC have a positive effect on the fulfilment of the SDGs promoted by the UN?

To answer the question posed, the research is structured as follows. First, after this brief introduction, section 2 presents the theoretical framework, providing a theoretical review of the GIC and the SDGs. Section 3 explains the methodology followed in the empirical part to address the stated objective. Section 4 presents the main results obtained in the research. Finally, section 5 presents the conclusions obtained from the work, highlighting the main contributions, its limitations and future lines of research.

BACKGROUND

IC refers to the set of intangible assets of the organization that can derive a source of competitive advantage for the organization (Edvinsson & Malone, 1997), including individual and collective knowledge, intellectual property, experience and the organization's relationships with its different stakeholders (Bayraktaroglu et al., 2019). IC has traditionally been classified into three blocks: human, structural and relational capital (Mouritsen, 2003). These three groupings represent the set of assets that can exist in an organization, since they include intangibles related to people (human capital), to the company (structural capital) and to the organization's relationships with its stakeholders (relational capital).

Despite the vast amount of scientific production focused on IC, few studies have attempted to link IC with environmental management of organizations. In fact, until Chen's (2008) seminal article, which introduced the pioneering construct of GIC, no one had previously explored IC under the premises of environmental management. Chen's (2008) study divided GIC into Green Human Capital (GHC), Green Structural Capital (GSC) and Green Relational Capital (GRC), marking the beginning of a new stream within the study of IC, as it empirically demonstrated the positive relationship between the three dimensions that make up GIC and competitive advantage. Subsequently, Huang & Kung (2011) revealed that GIC represented a fundamental element for environmental compliance, organizational value creation and customer satisfaction of environmental demands. In the last lustrum, there has been an intensification of the study of GIC, with several scholars devoting their efforts to empirically demonstrate the linkage of GIC with other environmental variables (Thiagarajan & Sekkizhar, 2017; Jirawuttinunt, 2018; Yong et al., 2019; Yusliza et al., 2020; Mansoor et al., 2021). However, to our knowledge, the link between the GIC and SDGs has not been previously addressed. These goals were adopted in September 2015 with the aim of promoting sustainable development under the context of the 2030 Agenda. There are 17 SDGs and 169 targets that address different challenges related to climate change, as well as environmental and social inequality issues. In this way, the SDGs represent a call for global action among businesses, institutions and governments to achieve sustainable prosperity over time.

Since the creation of the 2030 Agenda in 2015, organizations began to analyse how they could adhere to these goals, given that they are essential in the implementation of the SDGs. Thus, although the goals correspond to a global agenda among all types of institutions to address environmental, social and economic challenges, companies are playing a notorious role in their achievement (van der Waal & Thijssens, 2020), thus satisfying the requirements of the UN. In particular, the UN stated, "We call on all businesses to apply their creativity and innovation to solving sustainable development challenges" (United Nations General Assembly, 2015, p. 25). The increasing visibility of the SDGs in the organizational context has raised the interest of several scholars specialized in the sustainable management of organizations. Thus, while several studies analyse the factors that lead to the adoption of the SDGs (Rosati & Faria, 2019; Van der Waal & Thijssens, 2020), others, on the other hand, explore the potential that the activities carried out by the organization possess to contribute to their fulfilment (Boiral et al., 2019; Pineda-Escobar, 2019). Our research contributes to the first typology of studies, since the GIC is considered as a catalytic variable for the fulfilment of the SDGs.

GHC increases as workers acquire new knowledge, skills and experiences related to environmental protection, improving the efficiency and sustainable performance of the organization. Such a block of intangibles, moreover, enables companies to cope with stringent international environmental regulations and the increasing environmental awareness of customers (Chang & Chen, 2012). Therefore, it is important to explore the effect of GHC on organizations' achievement of the SDGs, given that the stock of green knowledge held by workers is an essential element in achieving corporate sustainability (Jirakraisiri et al., 2021). Academic literature has addressed the positive effect of GHC on other variables such as economic performance, environmental performance, green innovation, green human resource management or green supply chain, among others (Delgado-Verde et al., 2014; Yadiati et al., 2019; Yong et al., 2019; Yusliza et al., 2020; Mansoor et al., 2021). However, the link between GHC and the SDGs has not yet been addressed through empirical research. By virtue of the above arguments, the following proposition is formulated:

Proposition 1: GHC is positively related to the achievement of the SDGs.

Several previous researches have recognized the role of the GSC in the improvement of corporate sustainability, being a necessary piece to achieve sustainable development. These intangible resources possessed by the organization allow obtaining sustainable competitive advantages over time (Zameer et al., 2020; Chen, 2008a), reducing the negative externalities generated by the activity carried out by the organization. Similar to what happened for the case of GHC, the relationship of GSC with green product innovation, economic performance, green culture of the organization or green human resources management has been empirically demonstrated (Liu, 2010; Chang and Chen, 2012; Delgado-Verde et al., 2014; Malik et al., 2020; Nisar et al., 2021). However, the relationship between GSC and the SDGs has not yet been addressed in the academic literature. For this reason, we put forward the following proposition:

Proposition 2: GSC is positively related to the achievement of the SDGs.

Organizations establish links with their suppliers, to exploit their expertise and knowledge and build achieve synergies that allow them to achieve economic, social and environmental results (Dal Mas & Paoloni, 2019). In fact, several researchers point out that more sustainable societies emerge as a consequence of linkages between companies and public and private institutions (Zhang et al., 2020). In this sense, green partnerships enable the exchange of green knowledge between different stakeholders of organizations, which allows increasing the sustainable performance of organizations (Yusoff et al., 2020). In addition, knowledge sharing and business cooperation have been considered decisive elements in implementing a sustainable approach. Such is the case that there is a vast scientific production that has addressed the study of the positive relationship of GRC on organizational performance, sustainable competitive advantage, environmental practices and green human resource management, among other aspects (Chang & Chen, 2012; Thiagarajan & Sekkizhar, 2017; Febrianti et al., 2020; Asiaei et al., 2021;

Ulla et al., 2021). The link between GRC and the SDGs, on the other hand, has not been previously analyzed. Therefore, we put forward the following proposition:

Proposition 3: GRC is positively related to the achievement of the SDGs.

METHODOLOGY

The research follows a qualitative approach, since the single case study has been used. This is conceived as a way of approaching a particular fact, phenomenon, event or situation in depth and in context. The objective with the use of this methodology is to deepen the understanding of the case under study and not to generalize the results obtained to the population (Yin, 2012).

There were three main reasons for conducting a single case study as opposed to a multiple case study. Firstly, the single case study selected is a method conducive to test the proposition put forward, being able to contribute significantly to the process of knowledge and theory building around the link between GIC and the SDGs. Secondly, a revealing case has been selected given the high degree of environmental management of the organization under study. Third, the case study provides a deep and broad understanding of the phenomenon under analysis.

Sample

The main objective of case study sampling is to collect specific cases, events or actions that can improve the phenomenon under study, i.e., the link between GIC and the SDGs. As Ishak et al. (2014, p. 2) note, "it is about finding cases or units of analysis that enhance what other researchers have learned about a particular phenomenon." For this reason, the case method involves the use of non-probability sampling.

However, despite being a non-probability sampling, a series of criteria were followed to select the sample. Firstly, it had to be a winery with Wineries for Climate Protection (WfCP) certification as a reflection of its environmental proactivity (there are currently only 40 wineries in Spain with this distinction). Secondly, they had to have integrated environmental issues before other companies in the wine industry. Thirdly, they had to be a winery with wide national and international recognition in the environmental field. After an initial screening, it was decided to select Bodegas Juan Gil for meeting the three established criteria.

Data Collection and Analysis

The case study conducted consisted of three phases: (1) literature review, (2) case study analysis, and (3) data collection. First, the first phase consisted of exploring the literature related to the GIC and the SDGs established by the UN. Second, a thorough description of the Bodegas Juan Gil case was carried out to identify its suitability for the present research. Third, given the qualitative nature of the research, triangulation was used for data collection to increase the validity and reliability of the conclusions drawn from the study. To carry out the triangulation, three sources of data were used: (1) in-depth interview with the winery's quality manager, Miguel Ángel Abad; (2) direct observation (visit to the winery and contact with employees); (3) access to documentation both internal (environmental declarations, bulletins, recognitions, etc.) and external (corporate videos, website, the Iberian Balance Sheet Analysis

System database, etc.). The interview was structured in three blocks (see Appendix 1). The first block of questions was of a general nature and aimed to learn about the history of the winery and its environmental commitment at a general level. In the second block of questions, the degree of specificity was increased, since questions were asked about the dimensions of the GIC and its relationship with the winery's environmental management. Finally, the third block of questions addressed those issues related to the company's degree of compliance with the SDGs that had not been asked in the other blocks.

The interview was conducted on December 3, 2021, lasted one hour and was recorded in its entirety for later transcription. The use of interview transcripts to provide theoretical frameworks has been widely used in the management literature (Murray, 1996; Eisenhardt & Graebner, 2007; Lawrence, 2010). This process involves identifying key words and phrases from the interviews to determine areas of convergence and possible gaps that remain to be filled. In this sense, Patton (2002), content analysis in the case study refers to "any qualitative data reduction and sense-making effort that takes a volume of qualitative material and attempts to identify consistencies and core meanings" (p. 453).

During the site visit, it was possible to speak with other winery staff members who emphasized the organization's commitment to the environment. Data collection was interrupted when additional data provided a minimal understanding of the phenomenon under study. Once the transcription was completed, the interviewees' opinions on the first draft were analyzed to check its validity and the interviewee, Miguel Angel Abad, reviewed and accepted the transcriptions of his interview.

RESULTS

History of the Winery Group

The origins of Bodegas Juan Gil date back to 1916, the year in which Juan Gil Giménez, great-grandfather of the current generation, abandoned his trade to enter the wine industry by building a small winery located in the center of the town of Jumilla. Juan Gil Guerrero, son of the founder, dedicated his whole life to the winery. However, it was the third generation who consolidated and laid the foundations of what we know today as Gil Family Estates. In this way, Juan Gil González and Francisco Gil González, grandsons of the founder, began to develop the image of quality, seriousness and efficiency for which their wines are recognized today.

For more than a century, the passion for wine has been passed down from father to son and the philosophy of the great-grandfather Juan Gil Jiménez is still alive: to do a good job and proudly carry the name of his brand. Today, the fourth generation, with Miguel and Ángel Gil Vera at the helm, has driven its development to consolidate the company as a large winemaking group. This expansion began at the end of 2001, the year in which the great-grandchildren made their first great strategic leap. Miguel Gil Vera, with the support of his brother Ángel, undertook the relaunching of the first family winery, involving the remaining nine siblings as partners. At that time, together with the strategic reconversion of Bodegas Juan Gil, Bodegas El Nido was founded in parallel, key in the great strategic leap of the group due to the appearance of its two great wines Clio and El Nido. Both wines contributed to the prestige of a wine region, Jumilla, which until then was just one more on the national scene. Thus, El Nido 2004 achieved the unimaginable recognition of 99 Parker points, something that very few Spanish wines had achieved until then. In fact, El Nido was the first Jumilla wine to be considered one of the ten best wines

Green Intellectual Capital as a Catalyst for the Sustainable Development Goals

Table 1. Wineries belonging to the group

Wineries	Year	Designation of Origin
Bodegas Juan Gil	1916	Jumilla
El Nido	2004	Jumilla
Can Blau	2004	Montsant
Ateca	2005	Calatayud
Atalaya	2007	Almansa
Tridente	2008	Zamora
Shaya	2008	Rueda
Lagar de Condesa	2013	Rías Baixas
Morca	2014	Campo de Borja
Rosario Vera	2016	Rioja

Source: own elaboration

in Spain, being the result of a collaboration between the Gil family and the Australian winemaker Chris Ringland, who continues to be linked to the winery.

Between 2004 and 2008, a strategy of geographic expansion was carried out, creating new wineries in other appellations. Thus, after the creation of El Nido 2004, in the same year Can Blau was founded in the Montsant appellation, followed by Ateca in Calatayud in 2005, Atalaya in Almansa in 2007, Tridente in Zamora in 2008, and Shaya in Rueda. This major expansion effort in new wineries between 2001 and 2008 involved an overall investment of 18.5 million euros. In addition, the efforts to expand throughout the different Spanish DOs did not cease, as they continued their growth in 2013 with Lagar da Condesa, in Rías Baixas, and in 2014 Morca, in Campo de Borja, was established, investing between 2009 and 2015 a total of 45 million euros. Finally, in 2016 Rosario Vera was founded, being the most recent winery of Gil Family Estates, located in the Rioja DO (see Table 1).

In the last five years, the group's export turnover has doubled, with 75% of its annual production going to foreign markets. As Miguel Angel explains "what allows us to be present in more than 40 countries around the world is our powerful portfolio and a successful logistic base, having a great penetration in the US and Canadian markets". In addition to these two foreign markets, the winery exports to other countries such as Sweden, Switzerland, Germany, Japan, United Kingdom, Russia, Belgium, China, Denmark, South Korea, Ukraine, Mexico, Dominican Republic and Poland. Also to France, Australia, Brazil, Czech Republic, Serbia, Kazakhstan, Singapore, Taiwan, Guatemala, Cuba, Peru, among others. As a result of the expansion and internationalization strategy, both the group's operating income and total assets have gradually increased to 40 million and 130 million, respectively (see Graph 1).

The group's wineries are designed to operate as sustainable facilities, according to a philosophy of energy efficiency and use of resources, which mark the activity throughout the production cycle and that responds to the need to always seek excellence in processes and products. By way of example, the Can Blau winery is 100% energy autonomous, being a benchmark for energy autonomy and environmental sustainability as it is self-sufficient with a solar energy system supported by generators, which avoids 66 tons of CO_2 emissions per year. In addition, the values of protection and respect for the environment have been transmitted to all the companies of the group, since the company wanted to expand geographically, not only to start operating in new Spanish wine regions, but also to transfer the environmental values

Figure 1. Operating revenues and total assets of the Gil Family Estates group
Source: Prepared by the authors based on the Iberian Balance Sheet Analysis System database (SABI, for its acronym in Spanish)

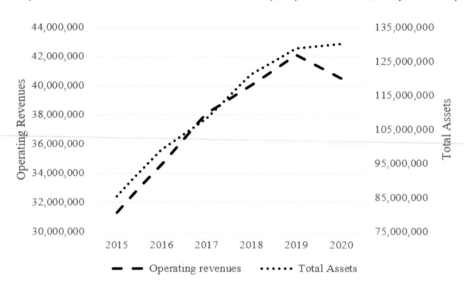

so deeply rooted in Bodegas Juan Gil to the rest of the companies. In the words of Miguel Ángel, "the expansion of Bodegas Juan Gil has served to transfer the environmental values of our organization to the rest of the wineries." Likewise, the wineries are located next to the vineyards to facilitate the monitoring of the evolution of the grapes, on the one hand, and to increase the value of the companies' assets, on the other. As Miguel Ángel points out, "the organization's slogan, we are land, is not random because from the organization we conceive the land as the main strategic asset, being able to guarantee future vintages, allow us to demonstrate our commitment to the territory in which we operate and guarantee the quality of our wines."

It is currently in the process of expanding in terms of hectares of production and winery equipment, having successfully overcome the difficulties arising from COVID-19. According to Miguel Ángel, "the coronavirus has been a difficult period with a lot of tension, but diversifying distribution channels and working hand in hand with customers has helped us overcome the period of uncertainty." Therefore, in line with the findings of Marco-Lajara et al. (2021a) and Marco-Lajara et al. (2021b), diversification of distribution channels in the wine industry has been a key part of overcoming the adversities caused by COVID-19.

Green Intellectual Capital in Bodegas Juan Gil

As previously explained, the GIC is made up of: the GHC, the GSC and the GRC. The factors influencing the formation of the three blocks of environmental intangibles for the case studied are set out below (see Table 2).

With regard to the GHC, it is worth highlighting the efforts made by the winery to raise awareness among its workers of the environmental challenges it faces. Bodegas Juan Gil invests in training courses for workers to acquire green knowledge that will subsequently result in cost savings (through the efficient use of resources) and greater differentiation (through its high level of specialization in sustainable practices). It attends trade fairs to detect new trends in the wine industry, such as natural, organic and biodynamic wines. The winery codifies its explicit knowledge through the code of conduct and good practices so that all workers can access this stock of knowledge, and promotes motivation, job satisfaction and the feeling of belonging to the group among its workers through the practices, given that workers are aware that environmental practices are increasing the economic profitability of the organization and, as a consequence, are helping to secure employment.

As for the GSC, there are different factors that influence its formation. The winery has recycling programs for cardboard, glass and wood, as well as a circular economy project that consists of making its own compost from grape stalks. It measures its carbon footprint and has the prestigious WfCP certification to endorse its environmental commitment. It has a brand with wide national and international recognition that is working to ensure that its organic wines are recognized by its customers. It has a flat organizational structure that facilitates the transmission of green knowledge, the values of its organizational culture are framed in the pillars of sustainability and it carries out R&D&I projects to improve the sustainability of the winery, which subsequently translate into tangible results for the organization, such as the manufacture of its own compost from the stalks.

Regarding the GRC, Juan Gil is part of a large number of associations that allow it to obtain and update the organization's environmental knowledge, such as its membership in the Designation of Origin (DO) Jumilla, the Jumilla Wine Business Association (ASEVIN, for its acronym in Spanish), the Jumilla Wine Route, the Murcian Institute for Agricultural and Food Research and Development (IMIDA, for its acronym in Spanish), the Integrated Center for Agricultural Training and Experiences (CIFEA, for its acronym in Spanish) in Jumilla or research agreements with several Spanish universities. The winery also requires green certificates from its suppliers in order to work with it, with the aim of ensuring the existence of a green supply chain.

Sustainable Development Goals in Bodegas Juan Gil

Bodegas Juan Gil contributes to the fulfillment of the Sustainable Development Goals related to the wine industry. The winery is adhered to the Wine in Moderation program (see Illustration 1) to promote actions aimed at promoting moderate wine consumption and combating the effects of alcohol abuse under 3 pillars: training/education to the sector and the consumer, promotion of moderate consumption and responsible communication, thus promoting compliance with SDG 3 linked to health and wellness.

Despite being a traditionally male sector in its origins, the winery has been making progress in gender equality in recent years, with the increasing incorporation of women in all areas of work related to the sector. As Miguel Ángel points out, "currently 30% of the people working in the winery are women. However, we are working to achieve parity between men and women." This underlines their commitment to improve their contribution to SDG 5 related to gender equality.

In addition, WfCP certification enables the winery to comply with SDGs 6, 7 and 13. On the one hand, water management is one of the requirements to achieve certification, including specific actions such as calculating the water footprint or implementing a plan to reduce consumption and improve discharges. On the other hand, using alternative energy sources to cover part of the winery's thermal needs and implementing energy auditing systems to save consumption in the production activity is also another of the essential pillars to obtain the certificate. Likewise, wineries with WfCP certification must follow the action plan drawn up by the certifying body to address climate change.

Table 2. Dimensions of GIC in Bodegas Juan Gil.

GHC	Training courses	"The company offers environmental awareness courses to its employees, as this is a key measure for saving resources."
	Trade fairs	"The company participates in important sector fairs, having our own national and international sales team."
	Best practice manuals	"We have a code of conduct and the code of good environmental practices. However, both documents are currently being improved thanks to the momentum of the Wineries for Climate Protection certification."
	Motivation, satisfaction and a sense of belonging to the organization.	"Workers are motivated and satisfied at the winery because we care about the environment and their working conditions. In addition, sustainability favors the winery's economic profitability which ensures employment."
GSC	Circular economy programs	"The winery has recycling programs for cardboard, glass and wood. We have also developed our own circular economy program based on generating our own compost from grape stalks."
	Technological systems to measure carbon footprint	"In the winery we measure our carbon footprint and audit it externally, it has a significant reduction in carbon footprint since 2018."
	Eco-efficient facilities	"We currently have solar panels that meet 20% of our energy demand. Similarly, the facilities were designed to save as much energy as possible."
	Brand	"Although we are a benchmark in terms of environmental management, the end customer still does not recognize us for our sustainable work. This is an aspect we need to continue working on."
	Environmental department	"It is the quality department that takes on environmental tasks. The winery dedicates great resources to carry out its environmental actions."
	Certifications	"We have Wineries for Climate Protection and organic wine certification."
	Organizational structure	"The hierarchy is quite flat. In fact, there are only three levels, management, area managers and operators."
	Organizational culture	"The core value of the organization is respect for the land. We have to give back to the earth everything it has given us."
	Investments in R&D&I	"We develop R&D&I projects that bring us value. Many of these projects have helped us launch new products to the market and improve our processes. Without going any further, the entire compost development was done with a research project with the University of Castilla la Mancha."
GRC	Relations with institutions	"We are in ASEVIN, the Jumilla Wine Route, the DO JUMILLA, IMIDA and CIFEA. In addition, we have developed several collaboration agreements with universities."
	Collaboration with suppliers and customers	"Of course we collaborate with our suppliers to meet common goals, in the case of the glassware we have developed a lightweight bottle for use in our products."
	Green supply chain	"They take eco-certificates into account when collaborating with their suppliers."

Source: own elaboration

Figure 2. Wine in Moderation label on the Juan Gil website.
Source: Juan Gil winery website (https://gilfamily.es/bodegas_juan_gil)

The organization is capable of generating 125 jobs, exerting a tractor effect on other activities in the auxiliary industry and related sectors. It also contributes to the growth, internationalization and international projection of the Spanish economy, through a consolidated trade surplus and export network, as well as the presence of Spanish wine products in more than 40 countries, thus contributing to SDG 8.

The wine sector is closely linked to the land and is therefore particularly aware that it can only continue to guarantee the quality of its products by respecting its environment and conserving the natural resources on which it depends so much. Bodegas Juan Gil is aware of the importance of the land and respect for the environment, which is reflected in its slogan, "we are land", and in the importance of organic vineyards, since 100% of the winery's vineyard area is organic, thus contributing to the fulfilment of SDG 12. Likewise, the wine tourism activities carried out by the winery contribute significantly to the protection of the cultural and natural heritage of Jumilla through its close historical link not only with gastronomy but also with the cultural heritage of the region, allowing the construction and development of sustainable cities and communities (meeting SDG 11).

In relation to the waste generated by the winery, great progress is being made in the eco-design of packaging to improve its recyclability, in the search for complementary systems such as reuse, as well as the recovery of field and winery by-products, accelerating its transition to a circular economy, enabling SDG 12 to be met.

Finally, the construction of public-private partnerships by the winery has been key in achieving the objectives to promote coherent policies in the different aspects that encompass sustainability, as well as the development of eco-innovations to improve its facilities, processes and equipment. These measures have favored the achievement of SDGs 9 and 17.

CONCLUSION

The purpose of this research is to analyze the different effects of the GIC of wine companies on the fulfillment of the SDGs, contributing to academic literature in a remarkable way, since, to our knowledge, there is no previous research that has addressed this relationship.

The origins of Bodegas Juan Gil date back to 1916, the year in which Juan Gil Giménez a small winery located in the center of the municipality of Jumilla. Subsequently, the second and the current third generation would consolidate the winery group Gil Family Estates, with ten wineries in nine wine regions of Spain. Currently, the group exports to more than 40 countries around the world, and is in the process of expanding in terms of hectares of production and winery equipment. These organizational milestones have made it possible to generate employment in the regions where they have been installed, as well as to exert a tractor effect on other activities in the auxiliary industry and related sectors (meeting SDG 8).

Regarding the GHC, the results indicate that Bodegas Juan Gil employees are in a constant process of acquiring environmental knowledge, since the members of the organization attend courses on environmental awareness and management organized by the company, participate in industry fairs to detect new trends in sustainability in the sector and codify their knowledge through codes of conduct and good environmental practices so that all members can access this stock of green knowledge. This acquired knowledge has allowed them to join pioneering projects in the industry, such as the Wine in Moderation program to promote actions aimed at promoting moderate wine consumption and combating the effects of alcohol abuse (meeting SDG 3), developing green process innovations, such as making its own compost from grape stalks (meeting SDG 9), as well as developing a sustainable production system based on a reduction of pesticides and fertilizers, soil conservation, rational use of water resources and proper waste management (meeting SDG 12). This allows us to confirm Proposition 1.

With regard to GSC, the winery has its own circular economy program, WfCP certification to endorse its environmental commitment, a brand linked to sustainable production, a flat organizational structure, a green organizational culture, as well as R&D&I projects to improve the winery's sustainability. This set of green organizational intangibles has enabled the organization to promote gender equality in the organization (meeting SDG 5), develop strategies for efficient water management and to reduce pollution and waste (meeting SDG 6), to make a clear commitment to self-consumption and clean energy (meeting SDG 7), to implement initiatives to adapt to and mitigate climate change (meeting SDG 13), to develop a model of wine tourism aimed at enhancing the value of the territory (meeting SDG 11) and to promote organic farming and local grape varieties: Monastrell (meeting SDG 15). This allows us to confirm Proposition 2.

As for the GRC, Juan Gil is part of numerous associations that allow it to acquire and transfer its environmental knowledge, such as the DO Jumilla, ASEVIN, the Jumilla Wine Route, IMIDA, CIFEA or research agreements with several Spanish universities. The winery also requires green certificates from its suppliers in order to work with it. This allows them to guarantee the construction of public-private alliances to promote policies consistent with the different aspects that affect the wine industry in terms of sustainability (complying with SDG17). This allows us to confirm Proposition 3.

FUTURE RESEARCH DIRECTIONS

Despite the relevant contributions made in this research, the study suffers from certain limitations. The main limitation is related to the impossibility of establishing comparisons in the single case study and to the fact that the winery analyzed stands out as an environmental benchmark within the wine industry. To overcome this limitation, as a future line of research we intend to carry out a multiple case study to analyze the GIC-SDG link both in wineries that stand out for their environmental work and those that are not guaranteed by this feature. In addition, after the multi-case study, we want to continue analyzing the GIC-SDG relationship by adopting a quantitative approach. To do so, we will use a Vector Autoregressive (VAR) model and consider more than one wine-producing country, in order to make the results generalizable and to establish differences between New and Old wine-producing countries.

ACKNOWLEDGMENT

This research has not received any specific subsidy from any public, commercial or non-profit sector funding agency. However, we would like to sincerely thank Bodegas Juan Gil for allowing us to enter their facilities, for allowing us to conduct an in-depth interview with their quality manager and for providing us with all the documentation necessary to carry out the research.

REFERENCES

Aigrain, P., Brugière, F., Duchêne, E., de Cortazar-Atauri, I. G., Gautier, J., Giraud-Héraud, E., & Touzard, J. M. (2016). Work of prospective on the adaptation of the viticulture to climate change: Which series of events could support various adaptation strategies? [Travaux de prospective sur l'adaptation de la viticulture au changement climatique: quelles séries d'événements pourraient favoriser différentes stratégies d'adaptation?]. In *BIO Web of Conferences* (Vol. 7). EDP Sciences.

Annunziata, E., Pucci, T., Frey, M., & Zanni, L. (2018). The role of organizational capabilities in attaining corporate sustainability practices and economic performance: Evidence from Italian wine industry. *Journal of Cleaner Production, 171*, 1300–1311. doi:10.1016/j.jclepro.2017.10.035

Asamblea General de las Naciones Unidas. (2015). Transforming our world: the 2030 Agenda for sustainable development. Resolution 15–16301, United Nations.

Asiaei, K., Bontis, N., Alizadeh, R., & Yaghoubi, M. (2021). Green intellectual capital and environmental management accounting: Natural resource orchestration in favor of environmental performance. *Business Strategy and the Environment, 31*(1), 76–93. doi:10.1002/bse.2875

Bayraktaroglu, A., Calisir, F., & Baskak, M. (2019). Intellectual capital and firm performance: An extended VAIC model. *Journal of Intellectual Capital, 20*(3), 406–425. doi:10.1108/JIC-12-2017-0184

Boiral, O., Heras-Saizarbitoria, I., & Brotherton, M. (2019). Corporate sustainability and indigenous community engagement in the extractive industry. *Journal of Cleaner Production, 235*, 701–711. doi:10.1016/j.jclepro.2019.06.311

Boons, F., Montalvo, C., Quist, J., & Wagner, M. (2013). Sustainable innovation, business models and economic performance: An overview. *Journal of Cleaner Production, 45*, 1–8. doi:10.1016/j.jclepro.2012.08.013

Chang, C. H., & Chen, Y. (2012). The determinants of green intellectual capital. *Management Decision, 50*(1), 74–94. doi:10.1108/00251741211194886

Chen, Y. (2008). The positive effect of green intellectual capital on competitive advantages of firms. *Journal of Business Ethics, 77*(3), 271–286. doi:10.100710551-006-9349-1

Chen, Y. (2008a). The driver of green innovation and green image–green core competence. *Journal of Business Ethics, 81*(3), 531–543. doi:10.100710551-007-9522-1

Dal Mas, F., & Paoloni, P. (2019). A relational capital perspective on social sustainability; the case of female entrepreneurship in Italy. *Measuring Business Excellence, 24*(1), 114–130. doi:10.1108/MBE-08-2019-0086

Davenport, T., & Prusak, L. (1998). *Working knowledge: How organizations manage what they know*. Harvard Business Press.

Delgado-Verde, M., Amores-Salvadó, J., Martín-de Castro, G., & Navas-López, J. (2014). Green intellectual capital and environmental product innovation: The mediating role of green social capital. *Knowledge Management Research and Practice, 12*(3), 261–275. doi:10.1057/kmrp.2014.1

Edvinsson, L., & Malone, M. (1997). *Intellectual Capital: Realizing your company's true value by finding its hidden brainpower.* Oxford University Press, New York.

Eisenhardt, K., & Graebner, M. (2007). Theory building from cases: Opportunities and challenges. *Academy of Management Journal, 50*(1), 25–32. doi:10.5465/amj.2007.24160888

Febrianti, F. D., Sugiyanto, S., & Fitria, J. R. (2020). Green Intellectual Capital Conservatism Earning Management, To Future Stock Return As Moderating Stock Return (Study Of Mining Companies In Indonesia Listed On Idx For The Period Of 2014-2019). *The Accounting Journal Of Binaniaga, 5*(2), 141–154. doi:10.33062/ajb.v5i2.407

FEV. (2019). *El sector del vino y el papel de la FEV en los Objetivos de Desarrollo Sostenible.* Disponible en: http://www.fev.es/fev/sostenibilidad-y-responsabilidad/objetivos-de-desarrollo-sostenible_122_1_ap.html

Fiore, M., Silvestri, R., Contò, F., & Pellegrini, G. (2017). Understanding the relationship between green approach and marketing innovations tools in the wine sector. *Journal of Cleaner Production, 142,* 4085–4091. doi:10.1016/j.jclepro.2016.10.026

Gilinsky, A. Jr, Newton, S., & Vega, R. (2016). Sustainability in the global wine industry: Concepts and cases. *Agriculture and Agricultural Science Procedia, 8,* 37–49. doi:10.1016/j.aaspro.2016.02.006

Huang, C., & Kung, F. (2011). Environmental consciousness and intellectual capital management: Evidence from Taiwan's manufacturing industry. *Management Decision, 49*(9), 1405–1425. doi:10.1108/00251741111173916

Ishak, N. M., Bakar, A., & Yazid, A. (2014). Developing Sampling Frame for Case Study: Challenges and Conditions. *World Journal of Education, 4*(3), 29–35.

Jirakraisiri, J., Badir, Y., & Frank, B. (2021). Translating green strategic intent into green process innovation performance: The role of green intellectual capital. *Journal of Intellectual Capital, 22*(7), 43–67. doi:10.1108/JIC-08-2020-0277

Jirawuttinunt, S. (2018). The Relationship between Green Human Resource Management and Green Intellectual Capital of Certified ISO 14000 Businesses in Thailand. *St. Theresa. Journal of the Humanities and Social Sciences, 4*(1), 20–37.

Lawrence, A. (2010). Managing disputes with nonmarket stakeholders: Wage a fight, withdraw, wait, or work it out? *California Management Review, 53*(1), 90–113. doi:10.1525/cmr.2010.53.1.90

Liu, C. (2010). Developing green intellectual capital in companies by AHP. In *2010 8th International Conference on Supply Chain Management and Information* (pp. 1-5). IEEE.

Malik, S., Cao, Y., Mughal, Y., Kundi, G., Mughal, M., & Ramayah, T. (2020). Pathways towards sustainability in organizations: Empirical evidence on the role of green human resource management practices and green intellectual capital. *Sustainability, 12*(8), 3228. doi:10.3390u12083228

Mansoor, A., Jahan, S., & Riaz, M. (2021). Does green intellectual capital spur corporate environmental performance through green workforce? *Journal of Intellectual Capital, 22*(5), 823–839. doi:10.1108/JIC-06-2020-0181

Marco-Lajara, B., Falcó, J. M., Fernández, L. R., & Larrosa, P. S. (2022a). Evolución del pensamiento en la disciplina de dirección estratégica: la visión de la empresa basada en las capacidades dinámicas y en el conocimiento. [Evolution of thinking in the discipline of strategic management: the dynamic capabilities and knowledge-based view of the enterprise] In *Investigación y transferencia de las ciencias sociales frente a un mundo en crisis* [Social science research and transfer in the face of a world in crisis]. (pp. 1801–1826). Dykinson.

Marco-Lajara, B., Sáez, P. D. C. Z., Falcó, J. M., & García, E. S. (2022b). El capital intelectual verde como hoja de ruta para la sostenibilidad: El caso de bodegas Luzón [Green intellectual capital as a roadmap for sustainability: the case of bodegas Luzón]. *GeoGraphos*, *13*(147), 137–146.

Marco-Lajara, B., Sáez, P. D. C. Z., Falcó, J. M., & García, E. S. (2022c). Las rutas del vino de España: el impacto económico derivado de las visitas a bodegas y museos. [Spain's wine routes: the economic impact of visits to wineries and museums] In *Investigación y transferencia de las ciencias sociales frente a un mundo en crisis* [Social science research and transfer in the face of a world in crisis]. (pp. 1774–1800). Dykinson.

Marco-Lajara, B., Seva-Larrosa, P., Martínez-Falcó, J., & García-Lillo, F. (2022d). Wine clusters and Protected Designations of Origin (PDOs) in Spain: An exploratory analysis. *Journal of Wine Research*, *33*(3), 146–167.

Marco-Lajara, B., Seva-Larrosa, P., Martínez-Falcó, J., & Sánchez-García, E. (2021b). How Has COVID-19 Affected The Spanish Wine Industry? An Exploratory Analysis. *NVEO-Natural Volatiles & Essential Oils Journal*.

Marco-Lajara, B., Seva-Larrosa, P., Ruiz-Fernández, L., & Martínez-Falcó, J. (2021a). The Effect of COVID-19 on the Spanish Wine Industry. In *Impact of Global Issues on International Trade* (pp. 211–232). IGI Global. doi:10.4018/978-1-7998-8314-2.ch012

Marco-Lajara, B., Zaragoza-Saez, P., Falcó, J. M., & Millan-Tudela, L. A. (2022h). Analysing the Relationship Between Green Intellectual Capital and the Achievement of the Sustainable Development Goals. In Handbook of Research on Building Inclusive Global Knowledge Societies for Sustainable Development (pp. 111-129). IGI Global.

Marco-Lajara, B., Zaragoza-Sáez, P., Falcó, J. M., & Millan-Tudela, L. A. (2022i). Corporate Social Responsibility: A Narrative Literature Review. Frameworks for Sustainable Development Goals to Manage Economic, Social, and Environmental Shocks and Disasters, 16-34.

Marco-Lajara, B., Zaragoza-Saez, P., Falcó, J. M., & Sánchez-García, E. (2022j). COVID-19 and Wine Tourism: A Story of Heartbreak. In Handbook of Research on SDGs for Economic Development, Social Development, and Environmental Protection (pp. 90-112). IGI Global.

Marco-Lajara, B., Zaragoza-Saez, P., & Martínez-Falcó, J. (2022g). Green Innovation: Balancing Economic Efficiency With Environmental Protection. In Frameworks for Sustainable Development Goals to Manage Economic, Social, and Environmental Shocks and Disasters (pp. 239-254). IGI Global.

Marco-Lajara, B., Zaragoza-Sáez, P., Martínez-Falcó, J., & Ruiz-Fernández, L. (2022k). The Effect of Green Intellectual Capital on Green Performance in the Spanish Wine Industry: A Structural Equation Modeling Approach. *Complexity*, ●●●, 2022.

Marco-Lajara, B., Zaragoza-Sáez, P., Martínez-Falcó, J., & Sánchez-García, E. (2022l). Green Intellectual Capital in the Spanish Wine Industry. In Innovative Economic, Social, and Environmental Practices for Progressing Future Sustainability (pp. 102-120). IGI Global.

Marco-Lajara, B., Zaragoza-Sáez, P. C., Martínez-Falcó, J., & Sánchez-García, E. (2022f). Does green intellectual capital affect green innovation performance? Evidence from the Spanish wine industry. British Food Journal, (ahead-of-print).

Marco-Lajara, B., Zaragoza Sáez, P. D. C., & Martínez-Falcó, J. (2022e). Does Green Intellectual Capital Affect Green Performance? The Mediation of Green Innovation. *Telematique*, *21*(1), 4594–4602.

Mouritsen, J. (2003). Intellectual capital and the capital market: The circulability of intellectual capital. *Accounting, Auditing & Accountability Journal*, *16*(1), 18–30. doi:10.1108/09513570310464246

Murray, G. (1996). A synthesis of six exploratory, European case studies of successfully-exited, venture capital-financed, new technology-based firms. *Entrepreneurship Theory and Practice*, *20*(4), 41–60. doi:10.1177/104225879602000404

Nisar, Q. A., Haider, S., Ali, F., Jamshed, S., Ryu, K., & Gill, S. (2021). Green human resource management practices and environmental performance in Malaysian green hotels: The role of green intellectual capital and pro-environmental behavior. *Journal of Cleaner Production*, *311*, 127504. doi:10.1016/j.jclepro.2021.127504

OIV. (2004). *Resolution CST 1/2004-Develpment of Sustainable Vitivinivulture*. Pareis.

OIV. (2008). *Resolution CST/2008-OIV Guidelines for Sustainable Vitiviniculture: Production, Processing and Packaging of Products*. Verone/it.

OIV. (2016). *Resolution CST 518/2016-OIV General Principles of Sustainable Vitiviniculture – Environmental - Social - Economic and Cultural Aspects*. Brento Gonçalves.

Patton, M. (2002). *Qualitative Research and Evaluation Methods* (3rd ed.). Sage.

Pineda-Escobar, M. (2019). Moving the 2030 agenda forward: SDG implementation in Colombia. *Corporate Governance: The International Journal of Business in Society*, *19*(1), 176-188.

Rosati, F., & Faria, L. (2019). Addressing the SDGs in sustainability reports: The relationship with institutional factors. *Journal of Cleaner Production*, *215*, 1312–1326. doi:10.1016/j.jclepro.2018.12.107

Schäufele, I., & Hamm, U. (2017). Consumers' perceptions, preferences and willingness-to-pay for wine with sustainability characteristics: A review. *Journal of Cleaner Production*, *147*, 379–394. doi:10.1016/j.jclepro.2017.01.118

Szolnoki, G. (2013). A cross-national comparison of sustainability in the wine industry. *Journal of Cleaner Production*, *53*, 243–251. doi:10.1016/j.jclepro.2013.03.045

Thiagarajan, A., & Sekkizhar, J. (2017). The Impact of Green Intellectual Capital on Integrated Sustainability Performance in the Indian Auto-component Industry. *Journal of Contemporary Research in Management*, *12*(4), 21–78.

Ullah, H., Wang, Z., Bashir, S., Khan, A., Riaz, M., & Syed, N. (2021). Nexus between IT capability and green intellectual capital on sustainable businesses: Evidence from emerging economies. *Environmental Science and Pollution Research International*, *28*(22), 27825–27843. doi:10.100711356-020-12245-2 PMID:33515153

Van der Waal, J., & Thijssens, T. (2020). Corporate involvement in sustainable development goals: Exploring the territory. *Journal of Cleaner Production*, *252*, 119625. doi:10.1016/j.jclepro.2019.119625

Yadiati, W., Nissa, N., Paulus, S., Suharman, H., & Meiryani, M. (2019). The role of green intellectual capital and organizational reputation in influencing environmental performance. *International Journal of Energy Economics and Policy*, *9*(3), 261–267. doi:10.32479/ijeep.7752

Yin, R. (2012). Case study methods. In H. Cooper, P. M. Camic, D. L. Long, A. T. Panter, D. Rindskopf, & K. J. Sher (Eds.), APA handbook of research methods in psychology, Vol. 2. Research designs: Quantitative, qualitative, neuropsychological, and biological (pp. 141–155). American Psychological Association. doi:10.1037/13620-009

Yong, J., Yusliza, M., Ramayah, T., & Fawehinmi, O. (2019). Nexus between green intellectual capital and green human resource management. *Journal of Cleaner Production*, *215*, 364–374. doi:10.1016/j.jclepro.2018.12.306

Yusliza, M., Yong, J., Tanveer, M., Ramayah, T., Faezah, J., & Muhammad, Z. (2020). A structural model of the impact of green intellectual capital on sustainable performance. *Journal of Cleaner Production*, *249*, 119334. doi:10.1016/j.jclepro.2019.119334

Yusoff, Y., Nejati, M., Kee, D., & Amran, A. (2020). Linking green human resource management practices to environmental performance in hotel industry. *Global Business Review*, *21*(3), 663–680. doi:10.1177/0972150918779294

Zameer, H., Wang, Y., & Yasmeen, H. (2020). Reinforcing green competitive advantage through green production, creativity and green brand image: Implications for cleaner production in China. *Journal of Cleaner Production*, *247*, 119119. doi:10.1016/j.jclepro.2019.119119

Zhang, Y., Sun, J., Yang, Z., & Wang, Y. (2020). Critical success factors of green innovation: Technology, organization and environment readiness. *Journal of Cleaner Production*, *264*, 121701. doi:10.1016/j.jclepro.2020.121701

ADDITIONAL READING

Corbo, C., Lamastra, L., & Capri, E. (2014). From environmental to sustainability programs: A review of sustainability initiatives in the Italian wine sector. *Sustainability*, *6*(4), 2133–2159. doi:10.3390u6042133

Maicas, S., & Mateo, J. J. (2020). Sustainability of wine production. *Sustainability*, *12*(2), 559. doi:10.3390u12020559

Merli, R., Preziosi, M., & Acampora, A. (2018). Sustainability experiences in the wine sector: Toward the development of an international indicators system. *Journal of Cleaner Production*, *172*, 3791–3805. doi:10.1016/j.jclepro.2017.06.129

Ouvrard, S., Jasimuddin, S. M., & Spiga, A. (2020). Does sustainability push to reshape business models? Evidence from the European wine industry. *Sustainability*, *12*(6), 2561. doi:10.3390u12062561

Pullman, M. E., Maloni, M. J., & Dillard, J. (2010). Sustainability practices in food supply chains: How is wine different? *Journal of Wine Research*, *21*(1), 35–56. doi:10.1080/09571264.2010.495853

KEY TERMS AND DEFINITIONS

Green Human Capital: Set of green intangibles held by employees.

Green Intellectual Capital: Set of intangible assets related to environmental protection that derive from the knowledge of employees, the codified knowledge held by the company and the organization's relationships with its various stakeholders.

Green Relational Capital: Set of green intangibles derived from the company's relationship with its stakeholders.

Green Structural Capital: Set of green intangibles held by the organization.

Sustainable Development Goals: Goals established by the UN in the context of the 2030 Agenda to improve the planet economically, socially, and environmentally.

Wine: Alcoholic beverage obtained by fermentation of grape juice.

Winery: Place where wine is aged and stored.

APPENDIX A

Block 1: History And Generic Data

1. When was the winery founded?
2. What do you consider to be the most important milestones of the organization?
3. Do you have other wineries in the national or international territory? How has the expansion/internationalization strategy been?
4. How many hectares of vineyards does the winery manage?
5. What is the surface area of organic vineyards and how long has there been organic vineyards?
6. What grape varieties do you work with?
7. How are the different distribution channels of the winery distributed? To which countries are they mainly exported?
8. Is organic, natural or biodynamic wine produced and to which countries is it mainly exported?
9. Is the wine distributed under the same brand or are there different brands within the group? Which ones?
10. What is the current situation of the winery?
11. How have you been able to combat the difficulties arising from COVID-19?

APPENDIX B: BLOCK 2: GREEN INTELLECTUAL CAPITAL

Green Human Capital

12. Do the employees attend seminars, workshops, events to improve their knowledge about sustainability in the industry?
13. Does the organization participate in industry fairs?
14. Does the organization have a manual of good practices, environmental statements, explicit commitments to improve its environmental performance?
15. How would you rate the motivation of employees in the winery to make it a benchmark for sustainability?
16. Do you consider that being a sustainable winery improves the employee's working situation?
17. Since when has the organization been firmly committed to the sustainability of the winery?

APPENDIX C: GREEN STRUCTURAL CAPITAL

18. Are there any recycling programs in the organization? Which ones?
19. Are there any emission control programs in the winery? Which ones?
20. Through which systems is the carbon footprint and water footprint measured?
21. Are there waste and energy reduction plans?
22. How is the waste generated in the winery reused? Are any products made from the by-products?
23. Do you have circular economy programs?

24. How do the winery facilities facilitate the efficient use of energy? Are there solar panels?
25. Do you consider that there is a link between the Bodegas Juan Gil brand and sustainability on the part of the customer?
26. Is there a department or person in charge of sustainability in the winery?
27. Does the winery have certificates that endorse its commitment to sustainability?
28. What is the organizational structure like? Is it hierarchical or flat?
29. What values of the organization do you consider that favor the sustainability of the winery?
30. Does the winery invest in R&D? Examples?

APPENDIX D: GREEN RELATIONAL CAPITAL

31. Does the winery participate with other associations to improve the environment?
32. Does the winery collaborate with its suppliers to improve its environmental objectives?
33. Do you take into account green certifications when collaborating with your suppliers?
34. Do you have collaboration agreements with universities?
35. How do you relate to your customers? Do you organize events to get to know their tastes and their preference for organic wine?

APPENDIX E: SUSTAINABLE DEVELOPMENT OBJECTIVES

36. How many employees does the winery currently have? Is there parity between men and women?
37. Do you develop wine tourism activities and to what extent do you consider that these activities contribute to the territorial development of Jumilla?
38. What is the link between the winery and the land where it was born (Jumilla)?
39. What are the winery's future challenges in terms of sustainability?

Chapter 10
Human Being and the Homeostasis Restoration of the Biosphere of the Earth and of Anthropocenosis:
Rethinking SDGs as for Dangerous Planetary Changes

Andrey I. Pilipenko
https://orcid.org/0000-0001-9446-345X
The Russian Presidential Academy of National Economy and Public Administration, Russia

Olga I. Pilipenko
https://orcid.org/0000-0001-5734-5673
Independent Researcher, Russia

Zoya A. Pilipenko
https://orcid.org/0000-0001-5734-5673
Bank of Russia, Russia

ABSTRACT

The Earth's biosphere could be interpreted as a meta-system that integrates anthropocenosis as its component. The anthropocenosis integrates numerous human-created systems in the economy, society, technologies that play the role of system elements of the biosphere. Nature's destruction is unequivocally connected with the activity of these elements. This is because the human-created systems have progressed and become more complicated due to the destruction of the Earth's biosphere. It means a violation of the dialectics of the interaction of the system integrity and its constituent elements. Homeostasis restoration of the Earth's biosphere and anthropocenosis makes the authors rethink the role of human being in the processes of system formation. Only an intellectually autonomous person with high social responsibility is able to integrate the goals of restoring the Earth's biosphere into his ethical values and realize them. So, the UN SDGs will be achieved only if individuals are formed with a conscious mission to restore the Earth's biosphere and its interaction with the anthropocenosis.

DOI: 10.4018/978-1-6684-4829-8.ch010

Human Being and the Homeostasis Restoration of the Biosphere of the Earth and of Anthropocenosis

INTRODUCTION

The COVID-19 pandemic has negatively affected almost all key aspects of human life, from the economy and commodity-money circulation, to society, relations with the state, organizational structures in companies, family ties, and natural disasters. The WEF experts commented on this as follows: "We are at a critical juncture for the future of human societies: we face an unprecedented global humanitarian and health crisis with the COVID-19 pandemic while the hour is late to stave off the worst of the climate and nature crises" (WEF, 2020). In other words, nature has demonstrated to the human community the result that its socio-economic progress has led to without taking into account harmony with the eco-natural environment. From a philosophical point of view, it is about the return of a dialectical approach to interaction with nature on the part of human-created systems. The coronavirus pandemic has provided humanity with a unique chance to rethink the universal laws of scientific, technological and socio-economic progress and form a person who is able to understand them and reasonably dispose of them both in organizing their systemic integrity and in relationships with nature.

COVID-19 has illustrated the marginal state of the economic system as it approaches the breakdown of the structural relations that have long held its integrity and provided progress. It is from these positions that the economic and social lockdown regimes introduced by the state in an effort to stop the spread of coronavirus infection should be regarded. And this fact is additional evidence of the exhaustion of the positive potential of the economic systems in statics. This remark is also true for states that ensured the structural integrity of such systems. In this regard all justifications for a Great Reset-oriented strategy (Sutcliffe, 2020; Billimoria and Bishop, 2020; Doumba, 2020; Schwab, et al., 2020) are meaningless. This is explained by the fact that the restoration of structural ties violated by the state in a system that has passed its optimum means counteracting the basic laws of self-organization and self-development, i.e. regress (Pilipenko, et al., 2022).

The specificity of the present moment is manifested in the accelerated divergence of socio-economic systems in the Globe for various reasons, ranging from environmental to humanitarian. This is fraught with fundamental consequences for all systems created by mankind over the past tens of thousands of years. Outwardly, it seems that the order that has taken root during this time has been replaced by a chaos of destruction of systemic relationships. However, if to recognize all human-created systems in the economy and society as integrities, then it should be stated that they have the ability to self-movement, which has been proven by centuries of their complication and progress. In other words, all allegedly chaotic changes in national socio-economic systems and of their interactions with each other and with nature are due to the dialectical laws of their self-movement. In this sense, the apparent chaos is a certain order that needs only to be understood (Pilipenko et al., 2021). The authors come to the statement of this position by consistently applying the theoretical tools of dialectical logic to rethinking unsolved puzzles in the eco-natural environment, in human-created systems, in humanitarian disasters, in the economic and societal crises of our time.

The goal of the authors is to search for the truth in this confusion of problems, which will make it possible to identify and understand the patterns that caused them, the principles that determine both order and chaos (Prigogine et al., 1984; Taleb, 2012), both in the Earth's biosphere (Gumilev, 2012a; 2012b), and in the anthropocenosis. It is about understanding the processes of complication of system integrity in the economy and society, and of the human being as a subjective component of all processes of system formation on the planet.

Human Being and the Homeostasis Restoration of the Biosphere of the Earth and of Anthropocenosis

The most acceptable research methods in this case are dialectical logic, analysis and synthesis, as well as a systematic approach to understanding self-movement in its various forms of self-organization and self-development. Moreover, at the heart of all these processes is a person and human activity. If it is necessary to return to the beginning of the process of system formation in human economic activity, then its separation from nature (Vernadsky, 1960a; 1960b) was associated with the desire to make labor more productive (efficient). It is in this context that the technology developed by man has become, on the one hand, a tool for increasing the effectiveness of human activity for the purpose of maximizing utility, and, on the other hand, a condition for creating artificial systems in the economy and society. As humanity moved forward in creating conditions to meet its needs, in accelerating scientific, technological and socio-economic progress, it realized its omnipotence. In the human mind, the systems artificially created by him in the economy flourished only as a result of human genius, outside and regardless of the Earth's biosphere, from which and thanks to which humanity emerged. As a result, a paradox arose from the point of view of systems theory, which is connected with the fact that the meta-system - the biosphere of the Earth began to be destroyed by socio-economic systems artificially created by mankind, being its elements. From the point of view of dialectical logic, these elements, even if organizationally complex, are designed to ensure the integrity of the meta-system, strengthening it with organizational ties that mediate its self-organization and self-development. This is explained by the fact that the Earth's biosphere as a meta-system and human socio-economic systems as its elements form a dialectical pair. And this means that they cannot exist without each other, and even their opposite vectors of complication should act only in the direction of strengthening the biosphere as a unity in diversity, i.e. her integrity.

In other words, humanity, which has managed to generate fantastic progress in all spheres of its activity, has missed one small but fundamentally significant truth - the Earth's biosphere and anthropocenosis cannot exist, much less develop without each other. And this position is an axiom. A break in their dialectical relationship is possible only in fantastic plots, such as the one in which humanity integrates into a different biosphere, or the biosphere finds other intelligent inhabitants. And since this is absolutely impossible today, therefore, natural disasters, violation of the ecological balance, i.e. the destruction of the homeostasis of the Earth's biosphere, is a consequence of human violation of the objective laws of self-movement of systemic integrity. So, the system elements in the composition of the anthropocenosis began to destroy structural ties with the Earth's biosphere, without which the human community is simply impossible (Bhattacharya and Stern, 2021; Dasgupta, 2004, 2019, February 2021).

The same the oretical approach allows the authors to build the logic of reasoning that will make it possible to understand the patterns of restoring the homeostasis of the Earth's biosphere and anthropocenosis. Thus, the "main culprit" of the violation of the systemic integrity of the Earth's biosphere as a meta-system is its structural elements, which are socio-economic systems artificially created by mankind. Then all modern natural disasters are a reaction of the meta-system to the destructive actions of its elements, which, instead of maintaining structural ties with it, constantly break them, trying to isolate themselves from it by all means of scientific and technological progress. As a result, all the chaos caused by climate change, the rapid spread of coronavirus infections, the shallowing of rivers and seas, the decrease in the ozone layer of the Earth's atmosphere, etc., becomes a logical consequence of the violation of the system laws of self-movement by the structure-forming elements in the composition of the anthropocenosis in relation to its meta- systems - the biosphere of the Earth. From the situation described above, the conclusion follows that the restoration of homeostasis means the revival of the dialectical relationships between elements and the system.

Human Being and the Homeostasis Restoration of the Biosphere of the Earth and of Anthropocenosis

If to abstract from the forms of organizing the professional activities of people within the sustainable companies, which play an important role in the processes of system formation, then its subjective component – a person – comes to the fore. It is in this context that the authors highlight the fundamental problem associated with the ignorance of the human phenomenon. Everything is explained by the fact that a person, as a subjective component of the processes of system formation in the economy and society, is called upon to play the role of a driver of scientific, technical and socio-economic progress. However, there are many "BUTs" along the way. The subject is also the subject because his behavior, reaction to events, desire or unwillingness to conscientiously perform his functions are predetermined by his personified ethical assessments of what is good or bad, what is fair or unfair, whether one should participate in this or no. And only with this in mind, a person either coordinates his actions with colleagues in professional activities or with fellow citizens in society, or not. In other words, only with a deep understanding by individuals, CEOs of self-sufficient companies, the political and intellectual elite of the fundamental significance of the problem of restoring the destroyed biosphere of the Earth, they will purposefully generate ways to restore it (BCG, 2021). And in this regard, there are problems of self-organization and self-development of the person himself, which are associated with his education and socialization, self-reflection, reflection and his transcending into the economy and society. The complex dialectic of the interrelations of these processes that mediate the self-movement of the subject actually predetermines the complexity of solving the problem of restoring the homeostasis of the Earth's biosphere and anthropocenosis.

In this context, the UN vision of the logic of restoring the fundamental harmony of the Earth's biosphere and anthropocenosis through the achievement of sustainable development goals (SDGs) is of great interest. Since the beginning of the 2000s, research and scientific developments have been intensified within the framework of the UN, which led to the definition of 17 SDGs in 2015 (Hansen, et al., January 2022). Their achievement, according to UN experts, will transform life on the planet in the interests of peace and prosperity. In fact, it was about the restoration of broken dialectical links in human-created systems in the economy, technology and society, as well as the homeostasis of the anthropocenosis and the Earth's biosphere (Koehring et al., 2021).

Taking into account all the above, the authors propose the following sequence of presentation of the material. First, the authors dwell on the interpretation of the paradox associated with the scientific and technological progress of mankind. Its results are fantastic given the opportunities that the modern technological revolution 4.0 brings. But the reverse side of this process is the destruction of the Earth's biosphere, the reduction of natural biodiversity, natural disasters, deadly pandemics, etc. In other words, instead of the full restoration of nature as it is used by mankind, scientific and technological progress has led to the prosperity of economic systems due to the irreparable destruction of the natural environment. This aggravated the problem of violation of the homeostasis of the Earth's biosphere and anthropocenosis. Second, the authors demonstrate the process of self-movement of human-created systems in the economy and society, which is accompanied by destruction of homeostasis of the Earth's biosphere and anthropocenosis. In accordance with dialectical logic, such a negative effect can be due solely to the violation of the laws of system formation and general organizational principles. Third, the subjective component always predetermines the processes of system formation. In this regard, it is human beings that are the key drivers of homeostasis restoration of the biosphere of the Earth and of anthropocenosis. However, in order for this human potential to give a real result, it is necessary to rethink the fundamental mechanisms of self-organization and self-development of the human being, to understand the dialectics of its education and socialization, the place of ethical norms of behavior in its motivation for action.

Human Being and the Homeostasis Restoration of the Biosphere of the Earth and of Anthropocenosis

Fourthly, the authors give their interpretation of the significance of 17 SDGs developed by the world community under the auspices of the UN. Only the structural hierarchy of these SDGs, taking into account the principle of dominance of the dialectics of education and human socialization, will make it possible to build an adequate subordination of these goals and a sequence of steps that will lead to the restoration of homeostasis of the Earth's biosphere and the anthropocenosis. The results of the study are summarized and the problem of organizing the activities of intellectually autonomous individuals with high ethical standards of behavior at the level of self-sufficient companies that are able to mediate the restoration of the dialectics of the economy and society, as well as the Earth's biosphere and anthropocenosis are discussed. In the context of empirical evidence, the authors propose models that describe the dialectic of the interaction between a person's education and his socialization. And to understand the specifics of human transcending into any system integrity, whether it be an economic system or the Earth's biosphere, dynamic sets proposed by Russian mathematicians (Mirimanoff, 1917; Chechulin, 2012) are used. In the conclusion, in a concentrated form, the authors' findings on the chapter are stated.

BACKGROUND

The problems of restoring the dialectics of interaction between the natural environment and human-created systems always worried great philosophers and thinkers. Many authors' findings were inspired by the geniuses of the past and present who anticipated the distant future: Hegel (1892), Kant (1781), von Schelling (1993), Marx (1995), von Mises (1998), Vernadsky (1960), Weber, (2009), Durkheim (1895), de Chardin (1987), Berger and Luckmann (1966) and many other brilliant researchers.They substantiated that any form of phenomenon has its own content, and any content has a form of its manifestation. Moreover, any system integrity, whether it is the Earth's biosphere (Vernadsky, 1960a; 1960b; 2018; WEF, 2020) or anthropocenosis, economy or society should be understood dialectically. It is about their capability of endless self-movement, including the stages of self-organization and self-development, as well as transition from statics to dynamics. It should be especially noted that all the elementary system components of the mega-system integrity must objectively coordinate their self-movement in statics and dynamics with the corresponding dialectical changes of the integrity that includes them at the mega-level. As a result an organically complex structure of the system, capable of self-enforcing, and of self-rebuilding arises. All these self-moving systemic structures are subject to the action of dialectical laws of unity and struggle of opposites, of the transition of quantitative changes into qualitative ones and of the negation of negation.

The authors applied these approaches to the interpretation of dialectical relationships and the homeostasis of the Earth's biosphere (Vernadsky, 1960a) and anthropocenosis (Gumilev, 2012a; 2012b). In this case the objectively existing biosphere of the Earth is interpreted as a mega-system with the quality of structural integrity, capable of self-movement and subject to the action of dialectical laws. As for the anthropocenosis as part of the Earth's biosphere, it is a set of structural integrities of artificially human-created systems in the economy, society, technology, etc., which in general act as its elements. Moreover, from the very beginning of the formation of the anthropocenosis, all human-created systems mediated the interaction of the humans with the nature. From the point of view of the authors, the dialectical approach to understanding these relationships between the mega-system and its human-created systemic elements makes it possible to conclude that their mutual self-movement can be possible only if their stages (mechanisms) are harmonized. It is about a coordinated change of structural relationships at all levels of systemic organization. Such a dialectical interaction becomes a necessary condition for the progress of the mega-system as well as its systemic elements (Arnol'd, 2003; Arthur, 2014).

In this context, the phenomenon of human being acquires great significance (Chardin, 1987). It was a human who formed artificial systems in the economy, technology and society without taking into account the dialectics of the interaction of the Earth's biosphere and anthropocenosis. As a result, the gradual rupture of the dialectical interaction of the whole (the Earth's biosphere) and its system elemental components (anthropocenosis) caused serious natural disasters. From a theoretical point of view, all of them are the result of the prevailing (rooted) human misunderstanding of the dialectics of the interaction of artificially human-created systems and the objectively existing biosphere of the Earth, from which mankind emerged many thousands of years ago. In other words, the authors prove that modern human-driven disasters and potential ones in future, in fact, are the result of a systematic disregard for dialectical laws of self-movement (self-organization and self-development) of any system integrity, including of the mega-system (Earth's biosphere) and its system elements (united within the anthropocenosis).

The COVID-19 pandemic has become clear evidence that humanity's ignorant attitude to dialectical laws for a long time leads to such disasters as a rapidly spreading life-threatening infection that humanity cannot cope with more than two years even in conditions of reaching the required level of collective immunity. Moreover, the uniqueness of the COVID-19 shock is manifested in the fact that it served as a detonator for the numerous crises development. From a theoretical point of view, they are forms of manifestation of the achieved margins of the systems created by mankind in the economy, technology, society, formal institutions, etc. This is due to the fact that the main acting figure of the anthropocenosis – human has ceased to be self-sufficient, i.e., unprotected from the negative impact of planetary threats (WEF, 2020; United Nations Framework Convention on Climate Change, 2021; Uitto and Batra 2022; Dasgupta, 2021).

So, due to pandemic humanity has not only destroyed the integrity of its own artificially created systems in the economy and society, but also succeeded in violating the dialectics of their interaction, as well as in exacerbating homeostasis destruction of the Earth's biosphere and anthropocenosis (United Nations, 2012). As a result the societal crisis caused by the COVID-19 pandemic becomes the evidence of inadequate socialization of people, of the destruction of the social system integrity. Forms of their manifestation are connected with many anti-social phenomena, such as the anti-vaccination movement in many countries of the world, the opponents' speeches of the regimes of economic lockdown and social distancing, an increase in the proportion of NEET social stratum among young people, etc. (OECD, 2021). But the most dangerous consequence of the societal crisis was the phenomenon of the population's loss of confidence in their states. In fact, the societal crisis was a logical consequence (in fact, it matured and was inevitable) of the nation states violation of the dialectics of interaction between the economy and society as systemic integrities. At one time, Marx (1995) wrote that the true progress of mankind was connected with its transition from the "realm of necessity" to the "realm of freedom". In fact, it was about replacing economic priorities with non-economic preferences in motivating people to work for the benefit of society. And this means that the economic system determining the progress of mankind within a long time, must objectively give way to the social priorities of human societies as an engine for the prosperity of humanity (Becker, 1976). Today, however, another fundamental problem has become apparent, without the solution of which it is impossible to ensure a bright future for mankind (Arrow, 1963). It is about adequate socialization of citizens of the whole world, which will allow them to agree on solutions to all the grandiose problems of mankind and understand the need to put on the agenda as an indisputable priority – homeostasis restoration of the biosphere of the earth and of anthropocenosis.

In this context, it becomes understandable such active efforts of the UN to convey awareness of the significance of this problem to all representatives of national elites and societies (United Nations, 2012). The deep meaning of 17 SDGs needs a systemic rethinking of the subordination of steps on the way to their implementation in the context of providing the multiplier effect as for homeostasis restoration of the biosphere of the Earth and of anthropocenosis.

The Paradox of the Scientific and Technological Progress of Mankind: Violation of the Homeostasis of the Earth's Biosphere and Anthropocenosis

Scientific and technological progress is undoubtedly a positive phenomenon in the processes of system formation with human participation. And, nevertheless, the current conditions of exacerbation of environmental problems on the planet make it necessary to think about the negative consequences of the technological revolutions implementation by humanity in the process of complicating economic systems. In other words, abstraction from the problem of the dialectics of interaction between the biosphere of the Earth and of anthropocenosis turns technological revolutions from the most important factors of socio-economic progress into a mechanism for the destruction of the biosphere as the natural capital of mankind. And this is a logical consequence of the violation of fundamental dialectical laws. Systems artificially created by man in the economy, technology, society act as complex system elements of the biosphere as a meta-system. If the system integrity at the meta level is reduced, then the only reason can be the absence of structure-forming actions on the part of its system elements.

Evidence of the destruction of the biosphere as a result of the functioning of human-created systems is the following data given by Partha Dasgupta (2021). Global GDP has risen enormously since the 1950s (Figure 1), and world economic output is 15 times higher. And these results are largely predetermined by the opportunities provided by nature to mankind. But over the same time, the biosphere has shrunk. Current

Figure 1. Expansion of the global economic activity of mankind (global economic output) and reduction of the Earth's biosphere (species population sizes) for the period 1750 - 2019
Source: Dasgupta, Partha. (2021b). Economics Nature's Way. Good economics demands that we manage Nature better. Finance & Development. A Quarterly Publication of the International Monetary Fund. September. Volume 58. Number 3

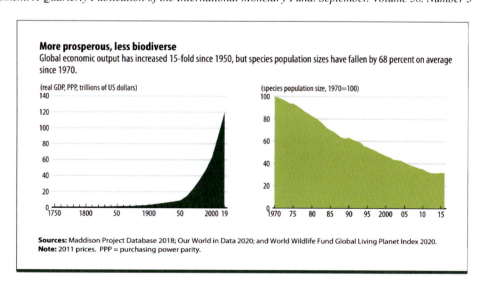

extinction rates are about 100 to 1,000 times higher than the background rate—the normal process of species loss—over the past several million years. And they are accelerating. Between 1970 and 2016, the population of species fell globally by 68 percent on average. A recent report by the Intergovernmental Science-Policy Platform on Biodiversity and Ecosystem Services showed that 14 of the 18 global ecosystem services assessed were in decline (Dasgupta, 2021).

The beginning of these transformations lies in human labor as a way of interaction between the human and the nature. Initially, the human himself opposed nature and interacted with it. But with the development of labor activity and the growth of its technological equipment, it was labor and technology that began to mediate the relationship between the man and the nature. It is no coincidence that philosophers interpret labor as the most important type of human social activity, which acts as its genetically starting point. In fact, human labor is a material process of exchange of objects, energy and information between a person and the natural environment. As a result, these relationships were divided into two components: (1) "Human – Labor (+ technology)" and (2) "Labor (+ technology) – Nature". The technological equipment of labor changed the quality of this interaction. While man was creating the most primitive tools of labor, he strove to adapt to natural conditions (Marx, 1995; Bogdanov, 1934). However, gradually, as its technological equipment grew, labor as a process of interaction between the man and the nature acquired structural qualities: (a) it became typical for a person, (в) it became a recurring phenomenon, (c) and it began to be distinguished by stability and sustainability. Thanks to these changes, the man has become increasingly persistent in adapting the conditions of nature to his needs. The fundamental reason for such "aggressive" behavior is the human desire to expand the sphere of his life (to settle the natural environment), aiming at maximizing his utility (meeting his needs).

This is due to the fact that in conditions of limited human energy (human resources, labor), the goal of maximizing utility can be achieved exclusively by increasing the productivity (efficiency) of human labor. According to Lem (1964, p. 66), "... a person must build between himself and the nature the whole chain of links, in which each subsequent link will be an amplifier of his capabilities more powerful than the previous". Technological support of human labor (economic) activity serves as the basis for the growth of its efficiency. This determines the subordination of progress in the field of technology and economics. The higher the rate of technological progress the more, all other things being equal, the effectiveness of human economic activity. Moreover, technological progress is interpreted in the context of the growing capabilities of humanity to achieve its goals.

Technological progress is realized in the process of complication of tools and means of labor, as well as technologies for organizing operations. At the first stage of technological evolution, the formation of new technologies was carried out purely empirically, by the method of numerous trial and error by obsessed experimenters who, as a rule, could not explain the nature and possibilities of using their engineering developments (Mokyr, 2005, 2009). In fact, until the 18th century, many outstanding inventions were made without a scientific basis to prove them: engineering was not supported by mechanics, iron production developed without metallurgy, agriculture without soil science, mining without geology, hydropower without hydraulics, dye production without organic chemistry, and medicine without microbiology and immunology. The second stage of technological progress is associated with the emergence and development of theoretical developments, which became the starting point for experimenters in the field of technology development. Theoretical knowledge as discoveries or progress in the material world is integrated into the technological process, during which the deliberately dead-end directions of experiments are cut off and a significant leap forward is provided because of time saving and the achievement of an effect in all spheres of human life. According to Lem (1964, p. 11), knowledge of a human brought him to the stage of exiting the empirical state of technology development in conditions when, in essence, the possibilities of technological progress predetermined exclusively by patience and perseverance, insights and glimpses of intuition of practitioners, had been exhausted. Moreover, theoretical knowledge, as a rule, significantly outstripped the technological developments of practical engineers. Lem (1964, P. 115) noted that if the growth of the theory stopped, then the "theoretical reserves" already accumulated by it would be enough to further improve the technology for a century ahead. In the middle of the 19th century, the growth of technology on the globe was about 6% per year, but the needs of a significant part of humanity were not met.

In the twentieth century, this gap deepened even more as the quality of technological progress changed. It follows that any slowdown in technological progress predetermines not just socioeconomic stagnation, but the beginning of regression. At the same time, this gigantic technological leap in the progress of the human community was accompanied by a build-up of huge problems, in which one should look for the origins of the current global economic slowdown, global geo-natural disasters, multiplying anti-social phenomena, etc.

The authors believe that the stages of technological progress are associated with the self-organization and with the self-development of technological systems. And all technological revolutions known in history are a change in the stages of self-organization and self-development of technological systems, implementing much more diverse options for the specialization of technological capabilities, their integration and combination. Self-development in this case is associated with a leap due to the emergence of a new technological paradigm, with breakthroughs in theory and in technological solutions of the problems set in the past. But after the stage of self-development the self-organization stage begins associated with the multidimensional organizational use of the capabilities of the new technology until the potential of the existing technological paradigm is completely exhausted (Pilipenko et al., 2021).

Thus, all scientific and technological progress was subordinated to the task of complicating economic systems and increasing GDP as a measure of the socio-economic progress of mankind. At the same time, the exhaustibility of natural resources (natural capital) did not seem to be a significant problem, the solution of which was again associated with the future scientific and technical capabilities of mankind. Thus, humanity has eliminated itself from the need to rationally manage its global portfolio of natural assets, which includes the Earth's biosphere. Estimates show that between 1992 and 2014, produced capital per person doubled, and human capital per person increased by about 13% globally; but the stock of natural capital per person declined by nearly 40% (Dasgupta, 2021). Accumulating produced and human capital at the expense of natural capital is what economic growth and development has come to mean for many people. In other words, while humanity has prospered immensely in recent decades, the ways in which the mankind has achieved such prosperity means that it has come at a devastating cost to Nature. Estimates of our total impact on Nature suggest that we would require 1.6 Earths to maintain the world's current living standards (Dasgupta, 2019; 2021).

And, meanwhile, at the heart of all the processes of system formation in which a person participates, there are universal organizational processes – disintegration (specialization, separation, diversity) and integration (unification, cooperation, inclusion). The infinite generation of elements and mechanisms of their organization in the interweaving of processes of differentiation and integration as mechanisms that spontaneously organize the world, according to Bogdanov (1934) may be associated with the future of humanity itself and the Earth's biosphere, as well as technological progress in the interests of the mankind and of nature. It is a person who associates with technology his goal of maximizing utility both in the economy and in society. However, at the same time, he absolutely ignores the organizational laws of interaction between systems artificially created by man and the Earth's biosphere integrating them into itself.

So the creation of tools and technologies was subordinated to the goal of first facilitating, and increasing the effectiveness of the human labor activity, and then satisfying of people's needs as much as possible. Thus, technological progress has served and serves the desire of people to facilitate their work and make it more productive for the purpose of maximizing human utility (Mokyr, 2005; 2009). At the same time, nature is perceived as a phenomenon given to man for eternal use, regardless of the destructive impact on it by mankind.

Meanwhile, biodiversity allows the nature to be creative, sustainable and adaptable. Just as diversity within a portfolio of financial assets reduces risk and uncertainty, diversity within a portfolio of natural assets – biodiversity – directly and indirectly increases Nature's resilience to shocks, reducing risks to the services on which the mankind relies. As soon as biodiversity decreases, both nature and humanity will suffer. The devastating impacts of COVID-19 and other emerging infectious diseases – of which land-use change and species exploitation are major drivers – could prove to be just the tip of the iceberg if mankind continue on the current path of scientific, technological and socio-economic progress.

So, following the dialectical logic, we can state the fact that the violation of the processes of system formation in the Earth's biosphere and anthropocenosis is associated with the subjective component of this process. In other words, humanity has succeeded in constructing system integrity in the economy, society and technology, having omitted the main point. It lies in the fact that the unlimited success of mankind on this path can only be ensured by implementing the principles of self-organization and self-development of the system and its elements. In other words, it is impossible for mankind to develop human-created systems by decreasing the biosphere of the Earth as their meta-system (Dasgupta, 2004; February 2021). This can be explained in terms of general systems theory (Bertalanffy, 1968; Arnol'd, 1975; 2003.) and based on the patterns of anthropocenosis (Gumilev, 2012a; 2012b; Vernadsky, 1960a; Weber, 2009). From the point of view of systemic categories, a system and an element are a dialectical pair with all the ensuing consequences. They can only develop together by differentiating elements and integrating them as drivers of organizational processes of complicating the system as a whole. Without each other and without observing the laws of their dialectical interaction, they are doomed to complete destruction and death. These logical conclusions are so simple that it seems inexplicable the phenomenon of a well-established practice of man to expand the sphere of his life at the expense of the Earth's biosphere, without the constant restoration and stability of which anthropocenosis is simply impossible. And, meanwhile, this is just the centuries-old result of the triumph of human genius in technology, neglecting a small but fundamentally important truth. It is about a strict requirement to coordinate all stages of scientific and technological progress with an assessment of their contribution to strengthening organizational ties with the Earth's biosphere, the depletion of which will result in the death of the entire planet.

Expansion of Human-Created Systems in the Economy and Society and Homeostasis Destruction of the Earth's Biosphere and Anthropocenosis: Explanation of the Reasons

The authors formulated a paradox associated with scientific and technological progress, which, on the one hand, was an amazing triumph of human genius, and on the other hand, caused monstrous damage to the biosphere, from which mankind emerged many centuries ago. From the point of view of the authors, the reasons for such ignoring anthropocenosis of the dialectics of the interaction of systems artificially created by man in the economy, society and technology, with the Earth's biosphere, should be of an objective nature.

Arguing on the principle from the general to the particular, the authors proceed from the fact that in all the processes of system formation the subjective, object and process components participate, which are subordinate to each other and interdependent. Destruction of homeostasis of the earth's biosphere and anthropocenosis was predetermined by the humans according to the idea of Emile Durkheim (1895) that Homo Sapiens is always and to the same extent also Homo Socius. In this context, the subjective component should be singled out in order to find an explanation for why humanity so recklessly constructed its systemic formations in the economy, society and technology out of dialectical connection with the state of the Earth's biosphere. To solve this puzzle the authors investigated the functions of the subjective component in the processes of system formation. They are based on the dialectics of the phenomena of integration and disintegration, the optimal combination of which at first led to the formation of stable dialectical pairs interacting over the exchange of material objects (Pilipenko et al., 2021). Moreover, the objects of exchange were rather passive in nature, since they changed as needs appeared in them and the desire to produce them, i.e. were due to personalized motivations of participants in market transactions. With the deepening of scientific and technological progress and the introduction of its results into the economic activity of people, the variety of objects of exchange increased, the number of participants in market transactions grew, material objects became more and more intangible (information, intellectual property, etc.). In other words, the object component of system formation processes changed not only quantitatively, but also qualitatively. But it was the subjective driver of the system formation process that remained the main factor of scientific, technological and socio-economic progress of the mankind. However, this does not yet explain the reasons for the destruction of homeostasis of the Earth's biosphere and anthropocenosis.

It all starts with the fact that the economic activity of people gradually began to organize, thanks to the exchange of material objects. It was in this process that the formation of stable system formations took place as a result of the integration of differentiated types of economic activity of people (Pilipenko et al., 2021). As a result of the long evolution of the market economy, economic systems became isolated (atomized) within the national borders of the countries of the world, which, in turn, integrated into themselves economic branches, industrial productions, firms, and etc. Gradually, a person as an employee, consumer, and citizen integrated into the systems artificially created by mankind in the economy and society. Thus, he was more and more separated from nature, less and less felt the need to take into account its requirements in organizing his life.

Such isolation from nature and ignoring the requirements of the dialectical laws of the development of nature intensified as systemic formations in the economy, society and technology acquired the quality of integrity. The point is that as exchange transactions became more complex, not only horizontal market ties of subjects regarding the exchange of material goods and services became stable, but also vertically oriented structural ties mediating the exchange of money, their substitutes, financial instruments, crypto-currencies, and etc. This is how the ability of economic systems to self-movement was formed, i.e., to self-organization in statics and to self-development in dynamics (Figure 2). In other words, human organizations as systemic entities become self-sufficient due to the fact that dialectically related elements are mediated by structural connections that are constantly renewed, stable and unchanged. From a philosophical point of view, the structure of the system embodies "the principle, method, law of the connection of elements within the systemic whole" (Sheptulin, 1975, 1978). In other words, the dialectical laws are implemented through structural connections of exchange participants, regulating their changes. And this happened in economic systems, regardless of nature, because the mechanism of their self-movement was internally embedded into their structural relations.

Further research of systemic integrities in the sphere of human activity led the authors to the conclusion that the change in the structure of the system under the influence of the dialectical laws of its self-movement makes it possible to distinguish two qualitatively different states of it. In the process of self-organization, the system becomes more complex by generating a hierarchy of structural levels that are vertically linked by cause-and-effect relationships (Pilipenko, 2020). While the higher levels in the structural hierarchy solve the problems of lower structural levels of the system, the latter remains stable, although its fragility as the integrity increases. The specificity of the self-organization of the system due to the generation of new levels by the structure is associated with the reproduction of direct and feedback connections by them, i.e., the system becomes more complex due to additional structural levels (Arthur, 2014; 2009; Arthur et al., 1997), repeating and not changing qualitatively. On its stage of self-organization, the system is in statics, changing only organizationally. It means the self-organizing system becomes more complex through the generation of new dialectically interacting structural links (levels). The system in statics becomes more and more complex under the influence of the dialectical law of unity and struggle of opposites operates at the horizontal level of the structure. As for the generation of its new structural levels it occurs under the law of the transition of quantitative changes into qualitative changes (Figure 2, time intervals $t_0 - t_1$ and $t_1 - t_2$).

However, everything has its limits. This is because the structural complication of a static system objectively leads to an increase in its fragility (Taleb, 2012). This is also true for the self-organization of the system in statics. As for its transformation into dynamics, this dialectical leap is realized through the dialectical double negations of horizontal and vertical structural levels of its own static state (Figure 2, time intervals $t_2 - t_3$ and $t_3 - t_4$). With the destruction of the structure, the point of no return is left behind, the direct and reverse structural interdependencies have been destroyed, and the system acquires the qualities of a dynamic one.

The content of such a state of the system is due to the fact that only dialectically complex elements remain from the previous system, which will rebuild structural connections and dialectical interdependencies (Figure 2, time interval $t_2 - t_3$). Actually, this theoretical fragment describes the essence of the future post-coronavirus reality, the specificity of which today is complete uncertainty.

It becomes possible because the system in statics and the system in dynamics represent two forms of its self-movement which are dialectically interconnected: the dynamics of the system cannot be without statics, and statics is intended to form conditions for its dynamic state. Actually, the stages of self-organization and self-development of human-created systems are interconnected in this way. According to G.W.F. Hegel (1892) self-organization and self-development of systems could be treated as characteristics of the objective world. They are inherent in all systemic integrities. With this approach, self-development should be understood as endless changes in the system as a whole, including certain stages (of self-organization) of structural complication of an unchanging system. In other words, it is only about changes in the "left to itself" systemic integrity in the economy, society or technology.

The theoretical structure formed above is complicated due to the fact that it must additionally take into account not only the intersystem connections of the economic and social systems, but also the technological foundation on which they are formed. Figure 2 integrates the technological base when characterizing economic system in statics and in dynamics. If the structural basis of reversible phenomena in technology has not exhausted its potential, then economic systems subordinate to it and cannot count on the implementation of self-development. In fact, it is about the fact that interconnected dialectically systems have not yet exhausted the potential of structural complication in the process of self-organization. Only upon reaching the threshold of complexity by technological systems, all the systemic integrities in the economy and other spheres predetermined by them, can realize their own self-development. In any case the process of self-movement of economic systems in connection with self-organization and self-development should be interpreted as the dialectically interrelated processes, which represent the unity of discontinuous and continuous, relative rest and constant change (Figure 2).

Figure 2. General model of interaction of self-organization and of self-development mechanisms of economic systems and of implementation of dialectical laws
Source: Pilipenko et al., 2021

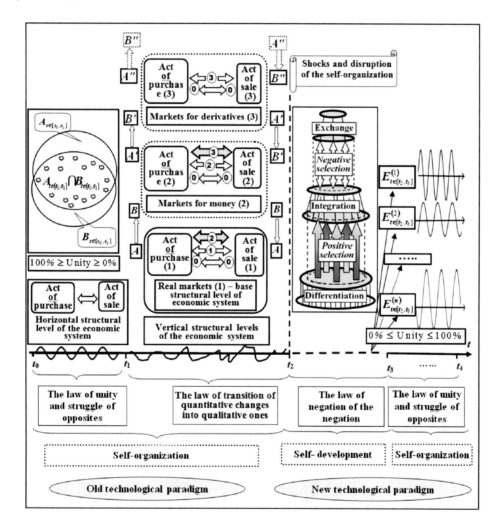

At the same time, it should be emphasized that for each country the limiting states of self-organizing systems will have its own specifics due to the peculiarities of the economy, society and technological basis. Therefore, even if their states coincide in the context of reaching the limits of self-organization, a dialectical leap or discontinuity in the movement of system integrity will have a huge variety of options for different countries and national communities It is possible to concretize these options in modern reality only conditionally, and this largely predetermines the uncertainty of the future post-pandemic reality. In the authors' model (Figure 2), the alternative of self-development of dynamic systems is conditionally reduced to three scenarios – $E^{(1)}, E^{(2)}, ..., E^{(n)}$, in the time intervals $t_2 - t_3$ and $t_3 - t_4$.

The above material gives a theoretical understanding of the process of system formation in the economy and the objective reasons for the separation of systemic integrity, artificially created by man, from the Earth's biosphere. In this process of system formation, people's idea of the objectivity of the disintegration of economic systems from the natural environment was strengthened. Moreover, confidence was formed in the omnipotence of scientific and technological progress, capable of compensating for the exhaustion of natural resources. And the fact that anthropocenosis is dialectically connected by an umbilical cord with the Earth's biosphere, and economic systems must perform certain functions to strengthen this meta-system somehow disappeared from the list of life necessary principles of human existence.

So the authors have discovered objective reasons in this disintegration of the anthropocenosis and the Earth's biosphere. This is due to the fact that a self-organizing economic systems (and before COVID-19, all national economies were static) function by complicating structural links within themselves by multiplying internal structural levels. Moreover, the latter changed in accordance the dialectical laws of unity and struggle of opposites and the transition of quantitative changes into qualitative ones, regardless of connections with the Earth's biosphere. By the way, in a static economy, the mission of a person was reduced to the performance of his professional activity, subject to complete obedience to the requirements of these systemic laws, which made a person more and more passive. This was explained by the fact that the static system was based on the constant repetition of essentially unchanged structural relationships that changed only in connection with the new quality of the objects of exchange. This was typical of the static economy as a whole and was reflected in the perceptions of the political elite, CEOs, and ordinary workers. Such mental constructions also dictated a pattern of thinking and decision-making based on past experience, dictated by constantly repeating, stable feed-forwards and feedbacks and causal dependencies.

And the COVID-19 pandemic and subsequent events are destroying the established assumptions typical of a static economy (Billimoria and Bishop, 2020). And for a person, everything changes according to a shock scenario. If on the established technological base in a static economy a person was needed only as an intermediary in the repetition of structural connections, then the latest technological revolution 4.0 prepared the objective conditions for constructing new structural connections on a new technological foundation, corresponding to a dynamic economy (Figure 2). And so the hour X came, when it became vitally important to rethink the old reality and realize the new quality of systemic formations post the COVID-19 emerging as a result of the law of denial of denial. The authors associated this new quality of dynamic reality with continuity break in its evolution and with its "expansion" due to constructing a new technological base for the economic system, restoring the dialectics of the economy and society, and (Pilipenko et al., 2022a; Pilipenko et al., 2022b) to homeostasis restoration of the biosphere of the Earth and of anthropocenosis. These fundamental transformations demanded the creative role of humans.

Human Beings in the Homeostasis Restoration of the Earth's Biosphere and of Anthropocenosis

The coronavirus pandemic has accelerated the fundamental transformation of anthropocenosis: human-created systemic integrities are objectively ripe for the dialectical leap from statics to dynamics. The new quality of these systemic formations is predetermined by the radically changed structural connections due to (1) constructing a new technological base for the economic system; (2) the need to restore the dialectical interactions of the economy and society, subject to the priority of the social principles of the coexistence; (3) homeostasis restoration of the biosphere of the Earth and of anthropocenosis. The implementation of these cardinal changes in anthropocenosis is possible only if the role of man in the processes of system formation is transformed. It is about the fact that in a static economy its functions were limited by the laws of self-organization (Pilipenko et al., 2021). And in the conditions of a dialectical leap of the human-created economic system from statics to dynamics, and, moreover, of the construction of a dynamic reality after the coronavirus pandemic, the person's functions are changing dramatically. Self-development of systems created by man requires him not to repeat practices already known in the past, but creativity, the ability to form new structural connections that correspond to the new reality. In the new conditions, a person must possess the qualities of a demiurge of the contours of the post-COVID-19 reality both in the economy, society and technology, and of homeostasis of the biosphere of the Earth and of anthropocenosis. The authors associate the potential of such a person with the abilities of an intellectually autonomous individual who can create quite effectively in conditions of complete uncertainty. This implies the ability of talented workers to analyze offline non-trivial problems that have arisen, determine the causes of their occurrence and choose innovative design solutions from a variety of possible alternatives, based on the justifications that have arisen in his head.

However, this is only one side of an intellectually autonomous personality as a complex systemic integrity, which is predetermined by its educational, knowledge component (Pilipenko, 2020). As the modern turbulent situation in the world has shown, this side of a highly intelligent person is undoubtedly important, but the quality of his socialization is no less, and perhaps even more important. Without the dominance of socially necessary norms for the integration of a talented individual into society, the whole integrity can be destroyed. In other words, intellectuals, as the main actors in the self-development of the future dynamic reality, will be able to understand the need for homeostasis restoration of the biosphere of the Earth and of anthropocenosis and restore the destroyed dialectics only in case of their adequate socialization. The point is that the more educated a person is, the more he is predisposed to diversity, however, any creative actions of people are based on their ability, desire and steps towards the consensus, i.e. inclusion in all areas of social activities. Thus, the education and socialization of the individual are the necessary and sufficient conditions for all transformations in the context of imparting dynamic qualities to human-created systems and harmonizing their interaction with nature. It follows that the basis of all modern transformations accelerated by the coronavirus pandemic is the need to rethink the phenomenon of a person (Chardin, 1987), who himself acts as a complex systemic integrity, capable of both self-organization and self-development. Moreover, both in the period of self-organization and in the period of self-development, the formation of a person as an individual with the qualities of intellectual autonomousness is based on two dialectical principles: separation (diversity), which is predetermined by education (in self-organization) and self-education (in conditions of self-development), and socialization (inclusion) – which comes from the society in the process of self-organization and from the individual to the society – in the conditions of self-development. In other words, only adequate socialization will make it possible to form a critical mass of intellectually autonomous individuals who interact and correct socially significant rules of behavior that strengthen the integrity of society.

It is no coincidence that since the mid-2010s, the problems of socialization of individuals (comparable in meaning to the category of "inclusion") have been actively discussed in relation to the macro level (WEF, 2015, 2016, 2017, 2018; Breene, 2016; Pilipenko et al., 2019), and in connection with the formation of company teams (The business case for inclusive growth, 2018; Hunt et al., 2020). However, the identification of the fundamental principles of system formation of structural integrity by humans - diversity and inclusion (integration/cooperation and disintegration/specialization, according to Bogdanov (1934) is only half of the problem. In the economy, these processes lead to system formation through the exchange, and in society diversity and inclusion should be mediated by subjective parameters of ethics in the sense of rules generated by joint human activity, norms that unite society and help to overcome individualism and aggressiveness. In other words, without understanding human psychology, it is impossible to understand the patterns of socialization in relation to modern societies. In this context, the developments of Russian scientists in the field of human psychology turned out to be extremely useful (Slobodchicov et al., 1995; 2000; 2013; Ananiev, 1977; Ushinsky, 2005; Elkonin, 1989; Piaget, 2008; Vygotsky, 1960). It is also about the psychology of human development (Slobodchicov et al., 2000) and psychology of education (Slobodchicov et al., 2013). So, for the purposes of constructing a post-COVID reality with dynamic qualities, education should be aimed at the formation of an intellectually autonomous personality, and socialization under the influence of society should allow students to acquire the ability to reflection. This is what determines the formation of a future talented personality at his initial stage of self-organization. For these purposes, bridging the gap between "what to teach" (on the part of the teacher) and "what to learn" (on the part of the student) is of paramount importance. In this rapprochement, participants of the educational process form individual abilities to overcome their own psychological and cognitive barriers (PCBs) (Pilipenko, 2020; Romer, 1988; 1990; UNESCO, 2016a; 2016b; World Bank, 2018), which will distinguish intellectually autonomous individuals from all other people.

In other words, in the process of self-organization, a person, first, gets an education and acquires the ability to solve any of the most complex problems in the future, if he has learned to overcome his PCBs. And, second, in the same process of self-organization, the student receives socialization skills dictated by society, which allow him to learn how to structure his inner subjectivity. It is about his awareness of the elementary rules of social communication as a product of joint human activity, uniting society, helping to overcome disunity and, moreover, aggressiveness. In this context, the words of N.A. Berdyaev (1952), which he attributed to the state, are very relevant: the socialization of a person will not make him happier it will only not allow him to turn his life and the lives of others into hell.

From the point of view of psychology, a person structures his internal subjectivity as a result of socialization in the process of his self-organization. Having learned to overcome the PCBs in the process of education and having acquired the ability to reflect as the ability to structure one's inner subjectivity, a person makes a dialectical leap "into adulthood" into the stage of his self-development. As an intellectually autonomous personality, he becomes capable of self-reflection, i.e. to understanding the boundaries of one's own subjectivity and to understanding the need for integration (socialization) into society. In the latter case it is about his ability to transcend, i.e. about the desire and ability to expand the boundaries of one's own subjectivity for the purpose of expanding the possibilities for realizing the individual self of a person outside in society. In other words, the socialization of a self-developing personality is the process of objectification of his personified internal subjectivity outside through transcending.

From a philosophical point of view, the crisis of modern education (World Bank, 2018) can be assessed due to its impact on the integrity of society and the misunderstanding of the human phenomenon as a dialectical integrity of unity in diversity. If an individual in education has not acquired the ability to transcend, then his socialization in the structure of society does not strengthen it (at best, does not change). As a result, the society is ready for a quick break of structural ties in extreme conditions, since only a self-developing personality with a high ability to transcend is able to quickly solve the extraordinary problems of strengthening systemic integrity. As a result, the societal crisis caused by the coronavirus pandemic has actually revealed the main cause of all social problems that are associated with a growing gap between a person's ability to self-reflection and to transcend.

Although theoretically it is quite simple to structure the process of self-development of a personality: an intellectually autonomous person is distinguished by the ability to self-education, since this deepens his potential to overcome the PCBs, and the desire for transcending in order to realize all his creative potential. However, both self-education and self-socialization of a person will serve society only if the individual ethical ideas of a person (realization of self-reflection) coincide with the ethical norms of behavior accepted in society (realization of transcending). Thus, ethical standards of behavior come to the fore. Only their convergence in the individual representation of citizens and in the institutionally fixed norms of behavior in society will make it possible to achieve the strengthening of systemic integrity. In addition, the structuring of the post-COVID-19 reality and homeostasis restoration of the biosphere of the Earth and of anthropocenosis completely depend on the understanding by ordinary citizens, and not just the intellectual and political elite, of the importance of inclusion, given the huge diversity of people in the context of ensuring at least a minimum consensus to avoid a general collapse. This applies both to socio-economic problems and to the dialectic of homeostasis of the biosphere of the Earth and of anthropocenosis.

The Global Community and 17 SDGs: Their Subordination and the Sequence of Steps for their Implementation in the Context of Restoring the Homeostasis of the Earth's Biosphere and Anthropocenosis

So, the authors prove that modern and potential future disasters, in fact, are the result of systemic ignorance by humanity of the dialectical laws of self-movement (self-organization and self-development) of human-created systemic integrity in the structure of the Earth's biosphere as a meta-system that integrates them as systemic elements (as part of anthropocenosis). As a result of the gradual rupture of the dialectical interaction of the Earth's biosphere with the anthropocenosis, more and more destructive natural disasters and humanitarian crises are ripening. It is in this context that the phenomenon of a human being becomes paramount (Chardin, 1987) as the first impetus of system formation and the potential beginning of the possible end of anthropocenosis and the Earth's biosphere. From a theoretical point of view, all crises in the economy and society accelerated by the pandemic are forms of the implementation of the broken dialectic of relationships in systemic formations created by mankind in the economy, technology, society and with nature. As a result, the main acting figure of the anthropocenosis - man, in the role of the main driver of the modern technological revolution 4.0, lost his self-sufficiency, and began to be subjected to the destructive effects of the forces of nature. From a theoretical point of view, this is due to the violation by mankind of the dialectic of the interaction between the Earth's biosphere and anthropocenosis. Over a thousand-year history, humanity has managed not only to destroy the integrity of its own artificially created systems in the economy and society, but also to destroy the dialectic of their interactions with

Human Being and the Homeostasis Restoration of the Biosphere of the Earth and of Anthropocenosis

the Earth's biosphere. It all started with the fact that for a long time nature was ignored as an asset, like produced capital (roads, buildings and factories) and human capital (as part of health care, education, socialization). Meanwhile, nature is more than an economic good, since biodiversity allows nature to be creative, sustainable and adaptive. Then diversity in the portfolio of natural assets increases the resilience of the Earth's biosphere to shocks, reduces risks in relation to its ability to recover (Dasgupta, 2004).

Dasgupta (2021) identified the main manifestations of the broken dialectics of the interaction between anthropocenosis and the Earth's biosphere. First, it is about the "inequality of interaction" between human-created systems and nature, which is expressed in the discrepancy between the needs of mankind and the possibilities of nature. Meanwhile, the benefits of nature used to meet human needs are limited by the reserves of natural assets and their ability to recover. In other words, the exhaustion of the Earth's biosphere by mankind must be compensated by his ability to restore the expended natural resources. The lack of an institutional mechanism for the protection of the natural environment leads to the fact that the Earth's biosphere is destroyed by its system-forming elements as part of antopocenosis. Second, the violation of the homeostasis of the Earth's biosphere and antopocenosis is predetermined by the criteria of economic and technological success, regardless of the cost of natural capital and the price of its restoration. Hence the majority of models of economic growth and development in which Nature was considered as a source of a limited supply of resources, the depletion of which must be overcome by technological progress. Meanwhile, an adequate indicator of socio-economic and scientific and technological progress should be the criterion of "comprehensive well-being", assessing the current state of Nature and the well-being of future generations. And, third, based on the theoretical constructions of the authors, the main beneficiary as a result of the restoration of homeostasis of the Earth's biosphere and anthropocenosis is a human being. By the way, he also acts as the main opportunist, violating the dialectic of interaction between the systems artificially created by him and the eco-natural system. Therefore, it is necessary to start with its education and socialization, which would allow to form a hierarchy of social values, which would organically include the principle of "do no harm" to Nature. Only an understanding of the Earth's biosphere as a global public good will allow humans to unite the efforts of the current generations to ensure the endless progress of mankind.

It is no coincidence that since the beginning of the 2000s, research and scientific developments have been intensified within the framework of the UN, which in 2015 led to the definition of 17 sustainable development goals (SDGs) (United Nations Framework Convention on Climate Change, 2021; United Nations, May 2012; Economist Impact, 2022; Uitto and Batra, 2022). Their achievements, according to UN experts, will transform life on the planet in the interests of peace and prosperity. In fact, it was about restoring the broken dialectical interaction between human-created systems in the economy, technology and society and the Earth's biosphere. With all the depth of this idea, structured in the 17 SDGs, it lacks a systemic principle that would make obvious the subordination of steps on the way to their implementation and provide a multiplicative effect.

Following the dialectical logic, the authors conditionally structured all 17 SDGs, based on the priority of solving all problems related to man as a systemic integrity capable of restoring the stability of the economy, society, technological base and the dialectic of interaction between them, as well as reviving harmony in the mutual influence of the anthropocenosis and the Earth's biosphere. Specifically, with a person as a phenomenon in itself, only one SDG4 (qualitative education) is associated, which is important, but absolutely not enough. The lack of adequate socialization among highly educated representatives of the management elite is fraught with irreversible planetary catastrophes, an illustration of which is the massive intensification of the threat of armed clashes of nuclear powers in various parts of the world. As for the restoration of the dialectics of the interaction between the anthropocenosis and the Earth's biosphere, with a high degree of conventionality the following can be distinguished from the 17 SDGs: SDG13 (climate action), SDG14 (life below water), SDG15 (life on land). As a result, out of 17 SDGs, only one goal (SDG4) is associated with the formation of the human of the future, and three goals (SDGs 13-15) relate to the Earth's biosphere. All other 13 SDGs are associated with the dialectics of anthropocenosis, i.e. with restoration of the integrity of economic, social and technological systems and interactions between them:

- the economy can conditionally include such goals as SDG2 (zero hunger), SDG3 (good health and well-being), SDG6 (clean water and sanitation), SDG8 (decent work and economic growth), SDG12 (responsible consumption and production);
- to the society can be attributed the following goals: SDG1 (no poverty), SDG5 (gender equality), SDG0 (reduced inequalities), SDG11 (sustainable cities and communities), SDG16 (peace, justice and strong institutions), SDG17 (partnerships for the goals);
- the technological progress is related to such goals as SDG7 (affordable and clean energy); SDG9 (industry, innovation and infrastructure).

Even with such a more than conditional differentiation of 17 SDGs, there is clearly no hierarchy in terms of significance, no dialectics of their interconnections and no mutual influence.

And, meanwhile, undoubtedly, the most important top-SDG is a human being, or rather, the formation of an intellectually autonomous personality, which must not only be highly educated, but also adequately socialized in the context of its ability to strengthen the integrity of the Earth's biosphere. It is a human being (and only he) who is able at any structural level to find approaches to solving the paramount problems of restoring the dialectics of the interaction between the anthropocenosis and the Earth's biosphere provided that the systemic integrity in the economy, society and technology is strengthened. As noted above, the transformation of a highly educated person into a guarantor of the prosperity of socio-economic systems lies in the coincidence of the hierarchy of social values, ethical norms that should strengthen the societal integrity as well as restore both the anthropocenosis and the biosphere of the Earth.

It turns out that a person's quality education (SDG4 in conjunction with SDGs3 and 6) should be accompanied by the formation of a hierarchy of such social values that would allow a highly educated person to integrate organically into society and direct all their abilities to multiplying the global public good (SDGs1, 2, 8, 12), as well as homeostasis restoration of the biosphere of the Earth and of anthropocenosis (SDGs13-15). Meanwhile, only SDGs 5 and 10 out of 17 SDGs are specifically devoted to ethical standards.

The UN Convention on Biological Diversity (COP15) and the UN Climate Change Conference (COP26) intend to formulate a new, ambitious direction of global actions for the coming decade to restore the homeostasis of the anthropocenosis and the Earth's biosphere, as well as the appropriate institutional structure of the mechanisms necessary to guarantee the fulfillment of the obligations set before the states. However, if their priorities do not include the goal of high-quality education of a person and his adequate socialization, then all their undoubtedly good intentions will pave the way to hell. And another fundamental problem of our time, which must be solved first of all, is the formation of global cooperation, without which the achievement of all the above goals is simply impossible.

RESULTS AND DISCUSSION

The main findings of the study are based on a theoretical construction, according to the logic of which mankind has systematically destroyed the nature and biosphere of the Earth, following a long experience of technological progress and socio-economic prosperity, which makes it possible to abstract from the state of the external environment. However, the purely practical, based on established postulates, the position of mankind in relation to Nature in the status of a free public good, did not suit the authors. They tried to do justice to the human mind and questioned its denial of the Nature, only based on the centuries-old experience of mankind.

Dialectical logic and a systematic approach allowed the authors to restore the theoretical structure of system formation in the economy, subject to the priority importance of the subjective component in this process. It turned out that humanity's non-dialectical denial of Nature has its own objective roots, and they are associated with systemic changes. So, as the structural ties, mediated by man in exchange, strengthened, structural levels arose, which, as they became more complex, endowed economic systems with the quality of integrity and the ability for self-movement. Moreover, this mechanism generated a complication of the systems, regardless of the Nature, since it was embedded into the internal structure of the human-created systemic integrities and ensured its relative isolation from the Earth's biosphere.

However, in this mechanism of self-movement of systemic integrities created by man in the economy and society, two of its modifications are provided. One of them ensures the progress of the system due to its internal organizational links, which optimize the interaction of differentiated (separated) subjects within the framework of integration associations. In this case, the self-movement of the system is realized due to its self-organization – the complication of numerous internal structural connections. It is at this stage that the static economy non-dialectically denies Nature as natural capital, realizing the internal mechanisms of self-movement. As a result, in self-organizing static economic systems, the opportunistic attitude of mankind towards Nature has become stronger.

The second mechanism of self-movement of systems artificially created by man is their self-development. And here the significance of dialectical interaction with Nature increases many times, and not only because of its rapid exhaustion, but in connection with the new quality of a dynamic economy at the stage of its self-development. After the dialectical leap of static economies, accelerated by the pandemic, into an uncertain dynamic reality, they become able to progress only if a number of conditions are met. The authors called this new quality of human systems "expansion", which should expand their dialectical connections by harmonizing the interactions of the economy and society, replacing the old technological base with the latest, generated by technological revolution 4.0, and restoring the homeostasis of the Earth's biosphere and anthropocenosis. Moreover, even the general contours of such a rapprochement have already been outlined by the UN in the structurally presented 17 SDGs.

However, such an "expansion" of economic systems, their acquisition of dynamic qualities with harmonization of interconnections with the Earth's biosphere, rests on a problem that has not been solved, and has become even more complicated due to the transformation of static economies and the acceleration of technological revolution 4.0. And again, the authors are trying to identify the objective reasons for this phenomenon, following the dialectical logic and relying on a systematic approach. It is about a person whose education has been in crisis for several years, according to international think tanks, and whose low level of socialization has more than manifested itself during the pandemic. The authors proved that the main problem here lies in the misunderstanding of the phenomenon of a human being as a complex systemic integrity, self-organization and self-development of which is mediated by education and socialization. At the same time, in primary and secondary education, as well as in the process of socialization at the stage of self-organization of the young generation, it is the society that forms its personality. And self-development of a person occurs as he acquires the qualities of intellectual autonomy and ways of his successful transcending into society. At the same time, all the problems of a dynamic economy based on a new technological base, dialectical interaction with society and the restored homeostasis of the Earth's biosphere and anthropocenosis can be solved if intellectually autonomous individuals with high ethical standards of social integration are formed. Thus, the ethical norms of behavior of intellectuals begin to mediate their self-development, i.e. their desire to ensure technological and socio-economic progress, as well as to restore the homeostatic state of the Earth's biosphere and anthropocenosis.

The authors illustrate this position with a model of restoration of the Earth's biosphere and anthropocenosis by self-sufficient companies with the involvement of a physical and mathematical model of the finite motion of a particle in the central field. At the same time, the authors invest in the physical parameters the generalized economic content. Appealing to the visibility of the developed model representations, it is advisable to form a flat image of the model and operate with polar coordinates r and φ (Figure 3)

Figure 3. Polar coordinates **r** *and* φ
Source: authors' development

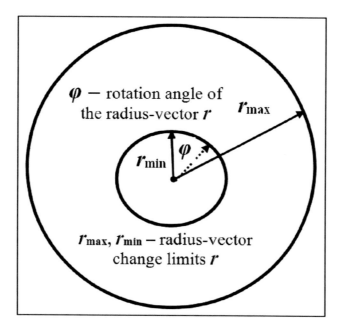

In Figure 3, the inner circle of radius r_{min} reflects the problems of anthropocenosis, the outer circle of the ring of radius r_{max} reflects the problems of the Earth's biosphere. The activities of a self-sufficient company focused on solving these problems reflect the generalized trajectory of the model movement from one problem area to another and back. Obviously, it is about expanding the field of activity of the company.

Since the range of r has two boundaries r_{min} and r_{max}, the motion is finite and the generalized trajectory lies entirely inside the ring bounded by the circles $r = r_{max}$ and $r = r_{min}$. This, however, does not mean that the trajectory is necessarily a closed curve (Figure 4). Omitting mathematical calculations, we note that during the time during which *r* changes from r_{max} to r_{min} and then to r_{max}, i.e. for one loop of the generalized model motion, the radius vector will turn through an angle $\Delta\varphi$, equal to

$$\Delta\varphi = 2\int_{r_{min}}^{r_{max}} \frac{\frac{M}{r^2}dr}{\sqrt{2m(E-U)-\frac{M}{r^2}}}. \tag{1}$$

It passes through the minimum and maximum distance countless times (as, for example, in Figure 3) and fills the entire ring between the two boundary circles in infinite time.

If to count the angle *φ* от направления радиус-вектора, проведенного в точку поворота, then the sections of the generalized trajectory adjacent to this point on both sides will differ only in the sign of *φ* for each identical value; this means that the trajectory is symmetrical about the specified direction.

Figure 4. Finite generalized movement (activity) of a self-sufficient company - fills the entire space between the boundaries of the anthropocenosis (area of minimum size) and the biosphere (area of maximum size)
Source: compiled according to: Landau, L.D. and Lifshitz, (1988). Theoretical Physics. In 10 volumes. V.1. Mechanics. 4th ed. Moscow: Nauka

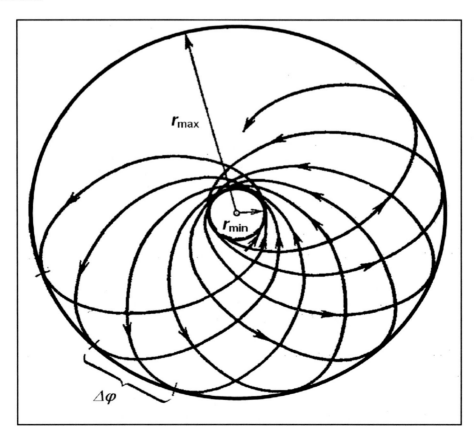

Starting, say, from any of the points $r = r_{max}$, it will be possible to go through a section of the trajectory to a point with $r = r_{min}$, then we will have the same section symmetrically located to the next point with $r = r_{max}$ and so on. Thus, the entire trajectory is obtained by repeating the same sections of the trajectory in the forward and reverse directions. In a sense, it is about the consistent expansion of the field of activity of a self-sufficient company.

These theoretical conclusions, which are largely supported by statistical data and analytical studies of practitioners, reveal the objective reasons for the rupture of the dialectical links between human-created systems and the natural environment. There are many issues on the agenda of the practical realization of the potential of talented individuals for the purposes of forming a dynamic economy and restoring the Earth's biosphere. The authors highlight only one aspect of this problem, which was the first to be solved by self-sufficient companies (Pilipenko, 2022a). First, it is they who, at the micro level of the economic system, are a typical form of organizing the professional activities of gifted individuals. And second, they had to solve problems at their own level not only to create conditions for the unlimited potential of employees with a high intellectual level, but also to correct their socialization, performing the functions of organizational creators of the external environment for the functioning of self-sufficient companies. For these purposes, CEOs of these companies began to actively rebuild their internal organization on the principles of "diversity - equity - inclusion" (Boston Consulting Group, 2021; Hansen, 2022). In other words, the optimization of the forms of organizing the activities of professionals is aimed at adequately transcending intellectually autonomous individuals into the composition of intra-firm teams. And this makes these companies the most important drivers of all system formation processes in the context of the pandemic and post COVID-19. Moreover, these companies introduce breakthrough technologies into their business processes, thereby mediating the formation of a new technological base for the entire economic system. Already today, this is manifested in the growth of corporate ecosystems on the latest technological platforms. As for the problem of the dialectical interaction between the economy and society, as well as the restoration of the ecological environment, the CEOs of these companies are increasingly implementing ESG principles in their long-term development strategies (O'Brien, 2021; Economist Impact, 2022). In other words, self-sufficient companies at the micro level of economic systems mediate the effect of their "expansion", including through homeostasis restoration of the biosphere of the Earth and of anthropocenosis, which is a necessary condition for constructing a dynamic economy.

EMPIRICAL EVIDENCE

The authors used the model of dynamical sets capable of self-affiliation within the framework of the theory of sets with self-affiliation (Mirimanoff, 1917; Chechulin, 2012). The dynamic sets A and B represent the values of the individual and the values (social norms) of society. In this example, the change in the segments of the intersection of dynamic sets illustrates how the structural connections of dialectically related individuals and systemic integrity change. As long as the talented person is satisfied with the "social behavioural norms" in economic and social systems, their interests vary from 100% coincidence to 50% (Figure 5).

Under this condition, the structures of these systems strengthen. In Figure 5 such a state is described by a white arrow with parameters 0. Otherwise, the divergence of interests of individuals and society as well as the integrity of systemic formations become more and more fragile. It is in this context the authors link the system formation in the economy and society with the subjectivization of their structural ties. This means that the value parameters of individuals and their agreement / disagreement with society as a whole play a cardinal role in strengthening the integrity of systems or in their destruction. Moreover dialectical logic forced the authors to consider a person as a polysyllabic phenomenon. It is in this context that the subjective component of the processes of self-movement of systemic integrity in the economy, society and technology performs the function of synchronization of the processes of their self-movement. Actually the same idea was brilliantly formulated by E. Durkheim (1895) in the sense that Homo Sapiens is always and to the same extent also Homo Socius.

The significance of such a coincidence of ethical norms on the part of the individual and society can be illustrated using the category "trust", which describes the result of a successful integration of a person into society. According to experts of the Trust Outlook (2021), especially now, every decision a leader makes will either increase or decrease trust in themselves, in their team, and in their brand. The only way to sustain, or even grow economic system is with trust. They highlight 8 Pillars of Trust which are especially magnified in times of crisis and recovery: clarity; compassion; character; competency; commitment; connection; contribution; and consistency (The Trust Outlook, 2021). 85% of people believe a high trust work environment helps them perform at their best (Trust Outlook 2020TM).

Figure 5. A model representation of the interaction of individuals and society as a systemic integrity, taking into account the segments of the intersection of the individual ethic principles of and social norms that dominate in society
Source: the authors' development

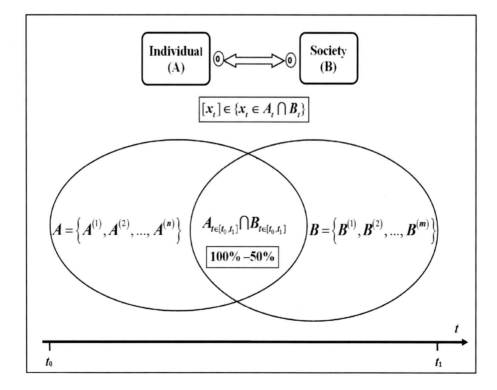

The authors propose analytical models that make it possible to concretize two dialectical components of a human self-organization. First, it is about education, and then the assessment of socialization is given.

Considering the problem of constructing an educational process focused on the formation of an intellectually autonomous personality, the authors proceed from the following provisions (considerations). First, we offer the following, more modern, interpretation of the triad of the methodological system proposed in the 1950th by Pinsky (1978) (Figure6).

Figure 6. Methodical triad of Pinsky (1978)
Source: the authors' development

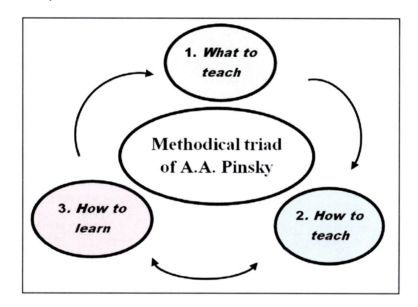

1. "What to teach" should be considered as an object component of the process of self-organization of the intellectual and cognitive sphere of a person-learner;
2. "How to teach" should be considered as a subjective component of the process of self-organization of the intellectual and cognitive consciousness of a person-learner, which is based on reflection when processing information structured by the teacher;
3. "How to learn" should be considered as a process component of the self-organization of the intellectual sphere of a person-learner, based on self-reflection.

In the opinion of the authors, the active role of students in the creation or construction and structuring of knowledge is based on their own cognitive activity through reflection of the methods (samples) of structuring educational and scientific information by the teacher. Through self-reflection of their own mental strategies students form the basis of the content of the element of the triad "how to learn". Moreover, self-reflection should be understood as a formula: "I know not only what I know, but I also know why I know it and how I know it." (How I think about my thinking).

So in the process of self-movement of a student, ultimately, his transition from self-organization to self-development takes place. The main role in this process is played by the timely organization of overcoming numerous PCBs, objectively functioning in the educational consciousness of the student / the learning person and in the professional consciousness of the teacher. Therefore, the next mandatory point is the application of the theory of PCB in teaching (Pilipenko et al., 2019; Pilipenko, 2020). It should be noted that it is the self-reflection formula that opens up the opportunity to independently overcome psychological and cognitive barriers. It is characteristic that PCBs, on the one hand, are the point of discrepancy between the planned and actual level of knowledge assimilation. And on the other hand, their successful overcoming mediates the action of the above mentioned dialectical laws of development.

Figure 7. The spiral of self-organization inherent in the methodological triad of Pinsky, taking into account the functioning of psychological and cognitive barriers in the expanding educational consciousness of the student (view from above)
Source: the authors' development

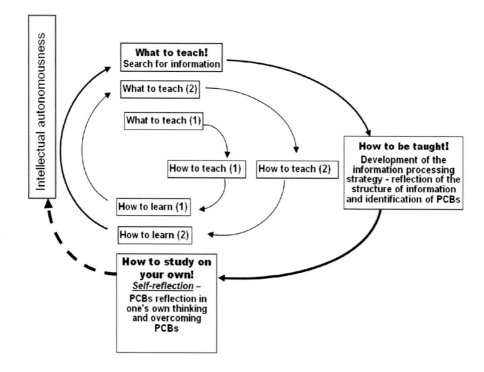

Application of the technology of overcoming the PCBs in the cycles of the triad of Pinsky provides the transition of the learning personality from the cycles of the triad to the spirals of the expanding cognitive consciousness. This model is presented in Fig 7. It contains the principles responsible for the processes of self-organization in the consciousness of the student / the learner.

The logic result of the educational activity described in Figure 7 is the formation of an intellectually autonomous personality, capable of independently overcoming PCBs both in professional activity and in the field of social communication.

Figure 7 corresponds to the temporal interpretation (Figure 8) of the development of the "spiral galaxy" of the reflective consciousness of the student in the direction of the intellectually autonomous personality and of a high degree of social responsibility. Thus, in a model presentation, the authors summarized the main principles of human self-organization and self-development in order to understand the main approaches to solving the fundamental problem of homeostasis restoration of the biosphere of the Earth and of anthropocenosis.

Figure 8. Time scan of the development of the "spiral galaxy" of the reflective consciousness of a student in the direction of the intellectually autonomous personality
Source: the authors' development

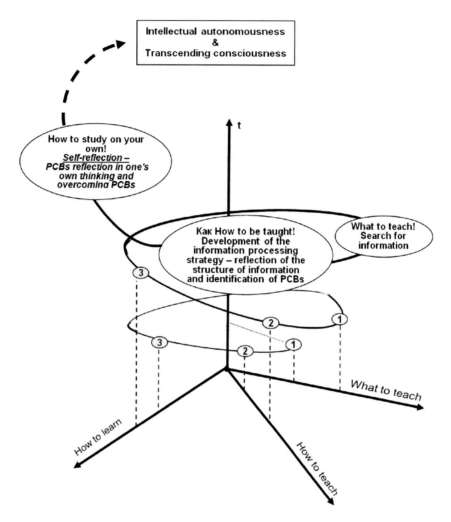

CONCLUSION

The mankind now lives in the anthropocenosis, a new era in which human activity has a dominant impact on planetary processes (Uitto and Batra, 2022). Climate change and other environmental challenges, such as chemical pollution and the mass extinction of species, have become defining challenges of our time. The COVID-19 pandemic that began in 2020 has tragically demonstrated how closely human health and ecosystem health are intertwined. The 2030 Agenda recognizes that sustainable development depends equally on three interlinked pillars: social, economic, and environmental (Uitto and Batra, 2022). All 17 Sustainable Development Goals (SDGs) incorporate each of these dimensions to a varying degree. If one of the dimensions fails, the goal is not achievable. However, it is still necessary to form a hierarchy of interdependence of these goals, which should begin with fundamental changes in a human being – the main beneficiary and the main driver in solving the fundamental problem of homeostasis restoration of the biosphere of the Earth and of anthropocenosis.

The chapter proposes a theoretical approach to solving the problem of the formation of an intellectually autonomous personality, which, in its capacity, is able to freely understand the significance of the goals set for humanity, optimize the order of their implementation and find mechanisms for performing the specific steps to fulfill the definite goal. Only by starting with the preparation of a person capable of self-development and adequately socialized can one save time. Although this statement is still relative, since the education of a person and his socialization takes at least 12 years in the primary and secondary levels of national education systems. And the problems of the adequate functioning of these links today, both in terms of education and in terms of socialization, are only getting worse.

Surprisingly, the problem of adequate socialization of the individual as a separate UN SDG is not even singled out. And, meanwhile, from the point of view of the authors, the phenomenon of a human is due to his multifaceted essence. From the point of view of anthropocenosis, his significance as a thinking being is increasing in modern conditions due to the ultimate stage of violation of the dialectic interaction between the Earth's biosphere and anthropocenosis. In this context, it becomes necessary to understand the significance of a person as a systemic integrity capable of self-organization and self-development. To become self-developing as an intellectually autonomous person capable of solving problems of virtually any difficulty, he must be placed at the forefront and interests of the entire ruling intellectual elite of the planet today otherwise it will be too late. At the stage of self-organization of a person, it is necessary not only to provide him with a qualitative education, but, no less important, with adequate socialization. The socialization of a poorly educated person unambiguously slows down progress, but an asocial highly educated person can destroy not only the anthropocenosis, but the entire biosphere of the Earth. Following Arthur C. Clarke (1954), it is safe to say that the COVID-19 pandemic marked the "Childhood's End" of humanity. Now the main thing is that people realize this truth, build an adequate hierarchy of sustainable development goals and begin to fulfill them with the top one - the creation of a human being as an intellectually autonomous and adequately socialized personality.

REFERENCES

Ananiev, B. G. (1977). *On the problems of modern human science.* Science.

Arnol'd, V. I. (1975). Critical points of smooth functions and their normal forms. *Russian Mathematical Surveys, 30*(5), 1–75. doi:10.1070/RM1975v030n05ABEH001521

Arnol'd, V. I. (2003). *Catastrophe theory.* Springer Science & Business Media.

Arrow, K. J. (2012). *Social choice and individual values* (Vol. 12). Yale University Press.

Arthur, B., Durlauf, S., & Lane, D. (1997). *The Economy as an Evolving Complex System II.* Addison-Wesley.

Arthur, W. B. (2009). *The nature of technology: What it is and how it evolves.* Simon and Schuster.

Arthur, W. B. (2014). *Complexity and the Economy.* Oxford University Press.

Becker, G. S. (1976). *The economic approach to human behavior* (Vol. 803). University of Chicago Press. doi:10.7208/chicago/9780226217062.001.0001

Berdyaev, N. A. (1952). *My Philosophic World-Outlook*. Retrieved from: http://www.berdyaev.com/berdiaev/berd_lib/1952_476.html

Berger, P. L., & Luckmann, T. (1966). *The Social Construction of Reality: a treatise on sociology of knowledge*. Penguin Book.

Bertalanffy, L. V. (1968). *General system theory: Foundations, development, applications*. G. Braziller.

Bhattacharya, A., & Stern, N. (2021). Our Last, Best Chance on. *Finance & Development*.

Billimoria, J., & Bishop, M. (2020). How COVID-19 can be the Great Reset toward global sustainability. In *Sustainable Development Impact Summit*. World Economic Forum.

Bogdanov, A. A. (1934). *Tectology or General Organizational Science*. Conjuncture Institute publishing house.

Boston Consulting Group. (2021). *Use AI to Measure Emissions Exhaustively, Accurately, and Frequently Carbon Measurement*. Survey Report 2021. BCG.

Breene, K. (2016). *Why does inclusive growth matter?* The World Economic Forum. Retrieved from https://www.weforum.org/agenda/2016/01/why-does-inclusive-growth-matter/

Chechulin, V. L. (2012). *Set Theory with Self-affiliation (foundation and some applications)* [Теория множеств с самопринадлежностью (основы и некоторые приложения] (2nd ed.). Perm State University.

Clarke, A. C. (1954). *Childhood's End*. Sidgwick & Jackson Ltd.

Dasgupta, P. (2004). *Human Well-Being and the Natural Environment*. Oxford University Press.

Dasgupta, P. (2019). *Time and the Generations: Population Ethics for a Diminishing Planet*. Columbia University Press. doi:10.7312/dasg16012

Dasgupta, P. (2021a, February). *The Economics of Biodiversity: The Dasgupta Review*. London: HM Treasury. Retrieved from: www.gov.uk/official-documents

Dasgupta, P. (2021b). Economics Nature's Way. Good economics demands that we manage Nature better. *Finance & Development*, 58(3).

de Chardin, P. T. (1987). *Phenomenon of Human Being* [Феномен человека]. Science Publishers.

Doumba, M. (2020). The Great reset must place social justice at its centre. In *Sustainable Development Impact Summit*. World Economic Forum.

Durkheim, E. (1895). *The Rules of Sociological Method*. Sage Publications.

Elkonin, D. B. (1989). *Selected Psychological Works*. Pedagogy Publisher.

Gumilev, L. N. (2012a). *Ethnosphere: History of People and History of Nature*. Eksmo Publishing House.

Gumilev, L. N. (2012b). *Ethnogenesis and the Earth's Biosphere*. Eksmo Publishing House.

Hansen, Moore, & Reiter. (2022). *Decarbonizing the built environment: Takeaways from COP26*. McKinsey & Company.

Hegel, G. W. F. (1892). The science of logics. In *The Logic of Hegel* (2nd ed.). Oxford University Press. http://links.jstor.org/sici?sici=0022-3808(198610)94:5<1002:IRALG>2.0.CO;2-C

Hunt, V., Prince, S., Dixon-Fyle, S., & Dolan, K. (2020). *Diversity wins*. McKinsey.

Impact, E. (2022). The ESG conundrum: how investors and companies can find common purpose in ESG. *Economist Impact*.

Kant, I. (1781). *Critic of Pure Reason* [Critic der Reinen Vernunft]. Verlegts Johann Friedrich Hartknoch.

Koehring, M., & Cornell, P. (2021, December). Delivering on a greener future: What to expect in 2022. *Economist Impact*. Retrieved from https://impact.economist.com/sustainability/net-zero-and-energy/delivering-on-a-greener-future-what-to-expect-in-2022?utm_medium=cpc.adword.pd&utm_source=google&ppc campaignID=17210591673&ppcadID=&utm_campaign=a.22brand_pmax&utm_content=conversion. direct-response.anonymous&gclid=CjwKCAjws--ZBhAXEiwAv-RNL0YqvjUlcowUpazyaL4STGYz7 JHtdVtHZHukHXbYx5g0DJnFqMqHKxoCg_8QAvD_BwE&gclsrc=aw.ds

Lem, S. (1964). *Summa Technologiae*.

Malone, T.W. (2018). *Superminds. The Surprising Power of People and Computers Thinking Together*. Little, Brown Spark.

Marx, K. (1995). *Capital: A New Abridgement*. Oxford University Press.

Mirimanoff, D. (1917). The antinomies of Russel and Burali-Forti and the fundamental problem of set theory [Les antinomies de Russel et de Burali-Forti et le probleme fundamental de la théorie des ensembles]. *L'Enseygnement Mathematiques*, *19*, 37–52.

Mokyr, J. (2005). The intellectual origins of modern economic growth. *Journal of Economic History*, *65*(2), 285 – 351.

Mokyr, J. (2009). *The Enlightened Economy*. Yale University Press.

O'Brien, J., & Stoch, H. (2021). ESG: Getting serious about decarbonisation: From Strategy to Execution. Deloitte Insights.

OECD. (2021). *Government at a Glance 2021*. OECD Publishing. doi:10.1787/1c258f55-

Piaget, J. V. F. (2008). *The Psychology of Intelligence*. Direct-Media.

Pilipenko, A., Pilipenko, O., & Pilipenko, Z. (2019). Education and inclusive development: puzzle of low-learning equilibrium. In Modeling Economic Growth in Contemporary Russia. Bradford, UK: Emerald Publishing Limited. doi:10.1108/978-1-78973-265-820191006

Pilipenko, A. I. (2020). Education and theory of psychological and cognitive barriers: human capital as driver of stable economic growth. In Social, Economic, and Environmental Impacts Between Sustainable Financial Systems and Financial Markets. IGI Global. doi:10.4018/978-1-7998-1033-9.ch011

Pilipenko, A. I., Pilipenko, Z. A., & Pilipenko, O. I. (2022a). Patterns of Self-Sufficient Companies' Network Interactions Reorganization due to COVID-19: Dialectics of Organizational Structures Optimization. In Challenges and Emerging Strategies for Global Networking Post COVID-19. IGI Global.

Pilipenko, A. I., Pilipenko, Z. A., & Pilipenko, O. I. (2022b). Dialectics of Self-Movement of Resilient Companies in the Economy and Society post COVID-19: Patterns of Organizational Transformations of Networking Interactions. In Challenges and Emerging Strategies for Global Networking Post COVID-19. IGI Global.

Pilipenko, O. I., Pilipenko, Z. A., & Pilipenko, A. I. (2021). Theory of Shocks, COVID-19, and Normative Fundamentals for Policy Responses. IGI Global.

Pilipenko, O. I., Pilipenko, Z. A., & Pilipenko, A. I. (2022). COVID-19 Shock and Subsequent Crisis: How It Was. In M. Khosrow-Pour (Ed.), *Research Anthology on Business Continuity and Navigating Times of Crisis* (Vols. 1–4). Information Resources Management Association. doi:10.4018/978-1-6684-4503-7.ch073

Pinsky, A.A. (1978). Methodology as a science. *Soviet Pedagogy, 12*.

Prigogine, I., & Stengers, I. (2018). *Order out of chaos: Man's new dialogue with nature.* Verso Books.

Romer, P.M. (1988). Increasing returns and long-run growth. *The Journal of Political Economy, 94*(5), 1002-1037.

Romer, P. M. (1990). Endogenous technological change. *Journal of Political Economy, 98*(5), S71-S102.

Schwab, K., & Malleret, T. (2020, July 14). *COVID-19's legacy: This is how to get the Great Reset right.* World Economic Forum. Retrieved from https://www.weforum.org/agenda/2020/07/covid19-this-is-how-to-get-the-great-reset-right

Shchedrovitsky, G.P. (1997). *Philosophy. Science. Methodology* [Filosofija, nauka, metodologija]. Shkoly kulturnoi politiki.

Sheptulin, A. P. (1975). Dialectics of the Special and of the General. Krasnoyarsk.

Sheptulin, A. P. (1978). *Marxist-Leninist Philosophy.* Progress Publishers.

Slobodchicov, V. I., & Isaev, E. I. (1995). Human Psychology. Academic Press.

Slobodchicov, V. I., & Isaev, E. I. (2000). Psychology of Human Development. Academic Press.

Slobodchicov, V. I., & Isaev, E. I. (2013). *Psychology of Human Education. Formation of Subjectivity in Educational Processes.* Academic Press.

Sutcliffe, H. (2020). COVID-19: The 4 building blocks of the Great Reset. *Sustainable Development Impact Summit.*

Taleb, N. N. (2012). *Antifragile: Things That Gain from Disorder.* Random House Trade Paperbacks.

The business case for inclusive growth. (2018) *Deloitte Global Inclusive Growth Survey.*

Uitto, J. I., & Batra, G. (2022). Transformational Change for People and the Planet: Evaluating. *Environmental Development.*

UNESCO. (2016a). *Contemporary learning crisis: education' failure in "creating sustainable future for all".* In The Global Education Monitoring Report (2nd ed.). UNESCO Publishing.

UNESCO. (2016b). *Education for People and Planet: Creating Sustainable Futures for All.* Global Education Monitoring Report 2016, Paris: UNESCO Publishing. Retrieved from: https://unesdoc.unesco.org/images/0024/002457/245752e.pdf

United Nations. (2012). *Back to Our Common Future Sustainable Development in the 21st Century (SD21) Project.* UN.

United Nations Framework Convention on Climate Change. (2021). *Nationally determined contributions under the Paris Agreement: Synthesis report by the secretariat.* https://unfccc.int/sites/default/files/resource/cma2021_02_adv_0.pdf

Ushinsky, K. D. (2005). Selected Works (vol. 4). Academic Press.

Vernadsky, V. I. (1960). Biosphere. Selected works. Academic Press.

Vernadsky, V. I. (2018). *Philosophy of Science. Selected Works.* Youwright Publisher.

von Mises, L. H. E. (1998). Human Action: A Treatise on Economics. Yale University Press.

Von Schelling, F. W. J. (1993). *System of transcendental idealism (1800).* University of Virginia Press.

Vygotsky, L. S. (1960). *Development of Higher Mental Functions.* Nauka Publisher.

Weber, M. (2009). *The theory of social and economic organization.* Simon and Schuster.

WEF. (2020). *The Future of Nature and Business.* New Nature Economy Report II. World Economic Forum.

WEF. (2018). *The Inclusive Growth and Development Report.* The World Economic Forum.

World Bank. (2018). *World Development Report 2018: Learning to Realize Education's Promise.* World Bank.

Chapter 11
Return to the Basics:
Vernacular Architecture as a Tool to Address Climate Change

Seyda Emekci
https://orcid.org/0000-0002-5470-6485
Ankara Yildirim Beyazit University, Turkey

ABSTRACT

Buildings use a substantial amount of primary energy, and as a consequence, they are a significant source of greenhouse gas emissions, which contribute to global warming and climate change. A considerable change in building design ideas, methods, technologies, and design and construction systems is necessary in light of the oncoming worldwide energy and environmental concerns. Vernacular architecture is a valuable resource that has the ability to contribute significantly to the definition of sustainable design principles. It is distinguished by a particular style of formal expression that has evolved in reaction to a variety of elements—geographical, climatic, social, and economic—that describe the local location or region in question. The purpose of this chapter is to examine instances of vernacular architecture that have been used to combat climate change and to qualitatively appraise the bioclimatic design solutions that have been applied.

INTRODUCTION

The preservation of the natural environment is essential to the continuous survival of human civilization, which has expanded and grown throughout the course of time. Every region or area has, over the course of time and as a result of the intricate interaction between the natural process of evolution and human adaptation to its surrounding environment, developed particular characteristics that set it apart from others.

The only way for a location to keep its one-of-a-kind identity and characteristics for an extended period of time is for its residents to have an in-depth understanding of the natural systems and a thorough absorption in the time-tested cultural reactions to the environment's assets and liabilities. Both of these factors are essential for long-term sustainability.

DOI: 10.4018/978-1-6684-4829-8.ch011

The preservation of the environment is the most pressing issue confronting humankind today, and one of the most effective strategies is to reduce energy usage. As is commonly known, buildings use around 40% of all energy consumed worldwide and are responsible for 36% of greenhouse gas emissions (EC, 2020). Increasing urbanization, expanding population, and better living circumstances will all lead to the most energy-consuming constructions being located in metropolitan centers (Zhao & Zhang, 2018). With the depletion of energy resources and the danger of climate change, it is imperative to pursue an environmentally friendly and sustainable growth route. Due to the impending global energy and environmental problems, a significant shift in building design ideologies, strategies, innovations, and design and construction systems is required.

Since the beginning of time that has been documented, people in communities and civilizations all across the globe have been putting up structures. These buildings range in terms of size, form, and architectural style. The primary function of buildings is to offer a comfortable and secure living environment that is sheltered from the effects of adverse weather conditions. In this sense, vernacular architecture developed gradually to meet the environmental, social, and sociological characteristics of the society as well as the changing lifestyle over a period of time to accommodate the changing lifestyle (Coch, 1998). The materials and architectural components that have been used have also been tailored to the individual terrain and environment of each location, making them climate-responsive and site-specific. It has a significant capacity to adapt to the environment and provide good thermal comfort since it makes extensive use of natural materials and has bioclimatic qualities. Aside from that, it is a design that is both economically and socially sound, as well as ecologically self-sufficient. These structures achieved a high degree of compatibility with their surroundings and had a minimum environmental effect on the environment. However, modernization, combined with the necessity for rapid and timely shelter supply for an ever-increasing population, has resulted in a flood of innovative building designs, technologies, and materials flooding the market. These are quickly embraced by customers who show a desire for such designs as well as higher expectations for thermal comfort in their houses. From an environmentally benign approach to a succession of quite high-tech and costly replies, the notion of sustainability, which emerged in the 1980s, progressed extremely quickly (Ş Emekci, 2021). The group of new construction specialists often applies new designs without taking the local climatic variables into consideration. As a consequence, traditional structures are being forgotten, and knowledge of their construction processes is fading away. However, it is possible to develop new and more sustainable design strategies for the rapidly expanding building sector by learning from the past. These strategies should take into consideration the local climatic conditions while aiming to reduce the use of energy-intensive and expensive artificial means to provide comfort in the future. Because of these efforts, it is possible that the repercussions of climate change may be reduced. The objective of this chapter is to investigate examples of vernacular architecture that have been utilized as a tool in tackling climate change, as well as to qualitatively assess which bioclimatic design strategies have been used. The chapter also discusses how vernacular architecture can contribute to social sustainability and decrease climate change impacts on society.

VERNACULAR ARCHITECTURE

Since the dawn of recorded history, groups and civilizations all over the world have been constructing buildings. These buildings have varied in size, shape, and design. The development of vernacular architecture was an inherently human reaction to the challenge of fashioning habitations that were appropriate for the surrounding environment and made using materials that were readily available in that environment. Examples of vernacular architecture vary from little caverns that our ancient human ancestors remodeled to make it possible for them to live there to stunning adobe haciendas that were constructed from the mud under our feet. The worldwide supply chain of items that are transported all over the globe is of critical importance to contemporary building practices (Vrijhoef & Koskela, 2000). On the other hand, vernacular traditions were developed as a result of local knowledge of the particular and contextual circumstances that existed close to the place.

Vernacular architecture has its origins in a tradition that makes use of locally sourced and already existent materials because of restricted access to resources. Additionally, vernacular architecture promotes environmentally responsible behaviors and ideas (Hantash, 2017).

Vernacular architecture integrated environmentally friendly techniques while also integrating the region's distinctive cultural values. These practices are now incorporated into conceptual ideas for buildings in the modern day. These principles are reflected in the methods of building that are developed in response to the local environment, way of life, traditions, and cultural norms. The procedures were simple, straightforward, and not harmful to the natural environment in any way. The peculiarities of vernacular architecture play a significant part in the process of modifying the climate inside structures (Hantash, 2017).

The term "vernacular architecture" refers to a simple type of structure that is unique to a particular time and place. When it comes to the construction of structures, it often does not include the supervision of a qualified architect and makes use of the resources and expertise that are available locally (Glassie, 2000). It might be challenging to specify particular parameters for all types of vernacular architecture. A great deal is determined by the resources and traditions of each region. Traditional cultural construction practices that have been handed down over a number of generations are included in the realm of vernacular architecture. Although the forms and methods change throughout time, vernacular architecture is consistently simple, economical, and environmentally friendly (Coch, 1998).

How Vernacular Architecture Helps to Cope with Climate Change

The most pressing challenge in today's society is how to cut energy usage while yet protecting the environment. According to a recent study, roughly 40% of the world's energy is used in buildings (Topfer, 2009). The majority of the world's energy-consuming structures will be found in developing countries due to rising populations, increased urbanization, and rising living standards. Demands for a sustainable development path based on renewable energy sources and energy efficiency are growing because of energy depletion and climate change (Motealleh et al., 2018).

Sustainable buildings are quickly becoming the norm in the construction industry as a new kind of structure that is not only environmentally conscious but also aesthetically remarkable. The long-term viability of these kinds of structures makes a contribution to the long-term viability of society as a whole, cultural traditions, and, most crucially, the economy. In addition, the relevant technical developments in the construction sector are anticipated to be impacted as a result of the provision of adequate solutions to achieve sustainability measures.

In the past, the built environment was designed to be more closely aligned with the unique characteristics of each individual terrain. Construction approaches were created to interact with their environment, focusing on a variety of locally sourced materials wherever possible (Salman, 2018). In spite of this, the spread of globalization has resulted in a gradual uniformity of architectural styles. Identical skyscrapers, airports, shopping malls, and petrol stations became contemporary city symbols thanks to orthogonal design and concrete construction decades later. It's becoming more necessary for architecture to reclaim its local vernacular in light of climate change's diverse impacts. Cities are under danger from a variety of natural disasters, including more frequent and more powerful rains, droughts, floods, and hurricanes.

Most people associate vernacular architecture with buildings like English thatched cottages and clapboarded New England saltboxes, African mud huts, or Brazilian favela tin and concrete-block ziggurats. However, it has been defined as items from the rural past as well as items from other lands that are related with the identity of the people who created and lived, or lived, in the structures in which they were found (Brown & Maudlin, 2012). Since ancient times, inhabitants of every location in the world have developed bioclimatic approaches and strategies for reducing the consequences of bad weather conditions caused by the environment's climate. The types of clothes worn, the diurnal and nocturnal labor patterns, the construction of structures, and the activities carried out inside them all serve as direct indicators of the solutions found in each location(Al-Hinai et al., 1993). In architecture, the origins of bioclimatic design can be dated back to the design concepts that are used in the majority of vernacular or traditional structures across the globe. Throughout history, vernacular architecture has changed and developed in response to the environmental, cultural, technical, and historical context of the place on which it was erected. As a result, vernacular architecture is considered to be well-suited to the local climate and natural environment, and hence to express a complete knowledge (Nguyen et al., 2019). The increased strain imposed on by current global environmental challenges has caused architectural designers to embrace regionalism and the knowledge of traditional buildings, believing that these structures are energy efficient and sustainable.

Based on an analysis of the effect of climate and the optimization of building environmental performance, vernacular architecture is the foundation of its practice. In other words,architects are attempting to reduce our environmental footprint and resource consumption by working with the external climate. Because of this, vernacular architecture can significantly reduce building energy usage without sacrificing current living standards (Singh et al., 2010). A building's occupants' well-being is directly impacted by their ability to adapt to external climatic conditions like wind and sunlight.mAs a result, vernacular architecture and sustainability have a common goal of reducing energy use while also improving the quality of life for building occupants.

Vernacular Architecture: Design Strategies to Achieve Sustainability

Vernacular architecture were created as a response to the needs of societies prior to the industrial era. They were also developed as a result of the insurmountable constraints imposed by the region and climate, as well as as a result of a unique interaction between the human mind and the knowledge gained from observing natural phenomena (Engin et al., 2007). The term "vernacular architecture" refers to a style of local or regional construction that makes use of the conventional building materials and resources that are indigenous to the region in which the structure is situated. As a consequence of this, the building in question has a tight connection to the environment in which it was built, is aware of the distinct geographical characteristics and cultural elements of its surrounds, and is profoundly impacted by these factors. Because of this, they are distinctive to many locations around the globe and have even developed as a method of reiterating one's identity. Vernacular structures, whether as a single structure or as part of a larger community, are the greatest illustrations of the harmony that exists between human behavior, architecture, and the natural environment. It has unwritten knowledge on how to maximize the energy efficiency of buildings at a cheap cost by using locally sourced materials in their construction (Zhai & Previtali, 2010). Furthermore, it is consistent with the values, economy, way of life, and cultures that were originally there (Nabakov, 1999).

Settlement Patterns

The design of a building settlement has an impact on how the prevalent climate will be adjusted to generate mesoclimates in the area around the structure. For example, the direction of streets has an effect on the passage of wind through a city. Furthermore, the width and direction of roadways have an impact on the amount of solar radiation that is received by people and surfaces in the surrounding landscape. Thus, a growing body of research indicates that there is a substantial link between a sustainable city and notions like as density or compaction, which are important in ensuring climatic comfort while also conserving energy (Golkar, 2000). In a hard and cold environment, the settlement patterns are quite compact. Alleyways are formed between the buildings, shielding the residents of a city from the cold weather and snowstorms (Bodach et al., 2014). However, in hot environment, because of the sun's angle, the passageways were constructed in the style of vernacular architecture. Therefore, people who were exposed to direct sunlight did not experience any discomfort on the local streets to the greatest extent feasible. In other words, vernacular architects took into consideration the low width to height ratio and designed paths that provided excellent thermal comfort for people to walk on (Mazraeh & Pazhouhanfar, 2018).

Vernacular Human settlements are a reflection of the people' desire and ability to engage in commerce and communal living, as well as their desire to maintain the circumstances of social cohesiveness in order to coexist as peacefully as possible in spite of opposing interests. This is more than the straightforward design of the structures that make up the towns; rather, these settlements are a reflection of the aspirations and capabilities of the people who live there.

Building Form and Orientation

The orientation of the structure should be chosen by taking into account the topography aspects of the location, as well as bioclimatic factors such as the calculated sun and wind direction. The building should be well-protected from passive solar energy in the summer and well-gained in the winter. While doing so, the building's transparent (window, winter garden, etc.) and opaque (facade, roof, etc.) surfaces may be utilised. Instead of focusing just on heat gain or heat prevention in building orientation, lighting of building areas and offering aesthetic comfort should also be considered. The activities in the building (work, entertainment, meetings, etc.) should be used as a guideline.(Şeyda Emekci, 2021). Therefore in vernacular architecture strategies, Solar radiation for passive heating is an useful design method in cold temperate climates, particularly during the winter months. In cold environment, the building has a more compact floor layout, and it has been found that the southern direction is the most appropriate orientation. Even in hot and humid climates, right orientation of the structure may be beneficial in shifting air flow for natural ventilation of the surrounding area. Because the objective of creating such buildings was to collect the least amount of solar energy possible throughout the warmer seasons of the year, the north direction was determined to be the most acceptable orientation for the structure in the hot and humid climate (Mazraeh & Pazhouhanfar, 2018). The lengthier facade is often oriented north–south, which lowers the amount of time the building is exposed to direct sunlight.

The majority of vernacular buildings are one-room constructions. Since just a single fire pit was utilized to provide heat in cold regions, it was appropriate to designate a single, big, central chamber as the preferred location (Zhai & Previtali, 2010). Multi-room examples are also available. The building of towering vaulted ceilings to allow for air stratification in warmer regions, when fire was not necessary to heat the home, is made easier by the use of single rooms. Besides that, the primary activities in a housing are resting or sleeping activities, which are divided into two groups of rooms, such as a parent's bedroom and a child's bedroom. It is placed according to the sun. Supporting activities in the home are for gathering and are specified by two groups of rooms, such as the family room and the stairwell room, respectively. Family room, which serves as a social space, is positioned in the center of the building as an axis point. Though it is placed in the rear of the building, the service room is designed to be accessible from any direction due to the sun (Fadli & Thamrin, 2017).

Furthermore, floor plans with rectangular forms were somewhat more common than floor plans with round shapes in the samples of single family houses that were examined, with aspect ratios ranging between 1 and about 1.5. It should be emphasized, however, that many vernacular customs are adapted to big extended families or many families living in close proximity to one another. This is most likely due to the fact that fire was the major source of heat in northern latitudes, and it was more effective with square floor designs(Zhai & Previtali, 2010).

Semi-Open Spaces in Buildings

Semi-open areas are key architectural and typological aspects of vernacular architecture because they allow for natural ventilation. They has been designed to be inviting and comfortable in response to the moderate climatic conditions and to the introverted way of life. As circulation spaces and transitional zones between inside and outdoors, they also accommodate domestic and agricultural tasks as well as serving as social meeting places for the community. Additionally, they function as microclimate regulators, giving protection from the sun and rain and therefore moderating the external environmental conditions. As a vast number of studies have discovered, semi-open areas are primarily found on the south side of buildings, where they provide shade during the summer cooling season. At the same time, the suitable width of their openings allows for ideal direct solar gains throughout the heating season (Philokyprou et al., 2015). Studies exploring the thermal efficiency of semi-open spaces have examined the relationship between typological elements of these spaces, such as their architectural arrangement, the amount of exposed regions and the degree of exposure, and their direction (Chun et al., 2004; Thravalou et al., 2015). Material choices, in addition to orientation and architectural plan, have a significant influence on the environmental performance of semi-open areas (Potvin, 1997). More precisely, the use of locally sourced materials with high thermal mass is associated with the environmental significance of these places. When it comes to both cooling and heating, the thermal inertia of the building materials is an excellent passive technique to use throughout both periods(Cardinale et al., 2013).

Open Spaces

The courtyard building form is an excellent architectural choice for hot climates. This is the primary reason why it is considered to be one of the most important aspects of vernacular design in the rural areas of hot and arid regions. They are often located in the geographic center of the area, and they are either totally exposed to the unobstructed view of the sky or, in certain instances, slightly covered by overhangs (Verma et al., 2022).

In contrast to the majority of contemporary structures, which are designed without paying sufficient attention to their potential effects on the natural environment, sustainable design is an approach to architecture that has been implemented to improve both the quality of the natural environment and the quality of the indoor environment of buildings. This is accomplished by minimizing the adverse effects that buildings and their natural surroundings have on one another (Memarian & Brown, 2003).

Traditional courtyard buildings in the hot environment are one of the most effective examples of climate-responsive architecture. These buildings were constructed with meticulous consideration paid to the climatic needs as well as the socio-cultural settings in which they were built. Furthermore, the central courtyard itself is one of the most useful components. It is possible to think of it as a room in the middle of the house; it does not have a roof, but it typically has a paved section, a pool, many trees and flowers to create an independent microclimate; and it is one of the most cost-efficient elements (Soflaei et al., 2017).

A section of the outdoors that is constructed in the midst of the home and is encircled by the rooms of the house. In addition to offering users the benefits of privacy and safety, it also brings natural light and fresh air into the spaces that surround it. Also, the courtyard in the middle of the house enables the houses in a traditional urban neighborhood to attach to each other, representing the strong social solidarity amongst the members of the society, granting them security and safety, and most importantly, providing them with protection from the weather conditions represented in the harsh sun and dusty wind (Hantash, 2017).

Courtyards are beneficial to the climate because they provide shade, ventilation, and most significantly, air movement. This air movement creates a draft of cold air falling descending into the courtyard at night, and a draft of hot air moving upwards during the daytime. One can argue that it will have a favorable influence on the temperature within the building if the orientation of the courtyard is done correctly, and if the openings of the rooms that surround the courtyard are done properly as well (Dunham, 1961). In addition, the inclusion of arcades inside the courtyard will make the facades of the building get a greater amount of shade, which, once again, will help to reduce the impact of heating on the walls.

The following describes the operation of a central courtyard as a passive cooling strategy: the air inside the courtyard gradually grows warmer as the day goes toward evening. The courtyard is able to maintain its cool temperature because of the laminar layers of air that collect cool air, which then flow into the rooms that surround the courtyard. The temperature in the courtyard gradually rises in the morning, which enables the courtyard to keep its cool temperature until solar radiation is cast directly onto it. During the day, a warm breeze blows over the home, but it does not enter the courtyard and instead only generates eddies inside it, unless baffles are built to divert airflow (Sahebzadeh et al., 2017).

Construction Techniques

There have been reports on a number of vernacular construction approaches that have been used to contemporary sustainable buildings for passive designs, as well as to traditional structures. Several vernacular building approaches have been adopted in the construction of structures in hot desert settings where cooling and daylighting are desired. Roof ponds, domes, air vents, and cooling towers, among other things, are among the ways that have been discovered. The selection of construction methods and materials for such structures is largely dictated by the intended advantages, as well as the availability of construction supplies and skilled labor in the local area, among other factors (Alrashed et al., 2017). The literature also has extensive coverage of traditional methods such as adobe and rammed earth building, as well as the wattle and daub method of construction (Niroumand et al., 2013).

Building Materials

Building materials have also had a significant impact on vernacular building traditions. Ecological knowledge is most prominently shown in traditional folk housing construction via the effective utilization of local resources. Because of a lack of funds and cumbersome transportation, residential structures in old communities can only be created using materials found in the surrounding area. These materials most of the time have a lower thermal conductivity coefficient, which allows for significant energy savings while also improving the comfort of the interior environment (Wang & Liu, 2013).

The use of alternative materials and methods, such as those found in vernacular architecture, can help to minimize the total embodied energy of a structure (Fernandes et al., 2015). In order to obtain optimal interior thermal comfort, researchers have hypothesized that vernacular architecture should be used since it uses less energy and has low environmental effect. This type of architecture is believed to be very useful in the development of contemporary sustainable architectural design (Chiou & Elizalde, 2019). The use of vernacular construction materials and methods has been successfully implemented in a number of sustainable building projects across the globe (Hamard et al., 2020; Hashemi, 2018).

In accordance with the ASHRAE Handbook of Fundamentals and a few research publications, the thermal characteristics of vernacular materials were determined. Stone (slate), tile, thatch, bark, skin, felt, wood, and grass were some of the roofing materials used in traditional buildings. Thatch has exceptional insulating properties (Zhai & Previtali, 2010).

CONSTRUCTING SOCIAL SUSTAINABILITY

According to Morin, it is important to analyze what kind of structure housing ought to take if it is to recover the dimensions of both tangible and immaterial values, as well as if it is to make a contribution to the production of stronger social cohesiveness in the environment in which we live (Morin, 2006). Because of the legacy of social and cultural importance left by vernacular structures, the lessons that we've learned from them bring us right back at the center of this dilemma. Is there not room in today's more global society, which has a propensity to trivialize culture, for the recreation of social relationships and the transformation of our social and cultural distinctions into a richness of importance that may be reintegrated into the production of our habitats? (Guillaud, 2014).

Environmental crises, globalization, contacts and conflicts between cultures as well as rapid technological advancements, are some of the other major issues that have had a significant influence on how the world is viewed, structured, and lives in throughout the twenty-first century. Each of these problems has major social and cultural repercussions, and they are all linked in some manner to vernacular traditions.

Vernacular architecture incorporates both tangible and intangible qualities that are a testament to mankind's power of adaptability in its living surroundings, as well as mankind's profound respect for nature, regardless of the particular aspects of its environment that are being taken into consideration. Nature and culture, as well as the identity of the local community, the capacity to create the finest living circumstances possible, as well as knowledge and expertise, are inextricably intertwined.

Construction of vernacular architecture is a testament to the skills and expertise of the artisans or nameless builders that created the structure. In the landscape, these 'traces' of vernacular building cultures can be seen in the visual aspect of materials such as earth, stone, wood, plants, farmhouses and outbuildings (such as granaries and hay barns), roofs, building and decoration details, relationships between buildings and their surroundings, as well as relationships between buildings and their surroundings (paths, ponds, streams) (Dipasquale et al., 2014). All of these characteristics illustrate human's ability to adapt to a new environment, to suit his or her own requirements, and to address the social and cultural identity of a given region or country.

The word "vernacular architecture" has evolved in the past two decades or so as a phrase that has gained currency not just within the field of cultural heritage studies, but also within the discourse on sustainability and sustainable development. Non-arrogant, linked, serene, minimalist building that is in organic relationship with its site and topographies is a common description of this kind of architecture (Olukoya & Atanda, 2020). Despite its humble origins, vernacular architecture has a startling level of inventiveness that reflects a high degree of communal intelligence as well as a process of building experimentation that progressively transformed into experience. It contains both intangible and physical information, and it is a testament to the human being's remarkable ability to adapt to a variety of environmental restrictions and possibilities. Taking this into consideration, it is argued that vernacular architecture demonstrates cultural sustainability lessons through its approach, which demonstrates the capacity to transform locally available materials (such as earth, stone, plants, and wood) into construction elements and buildings (such as stables, houses, mosques, and churches) that are organic with their surroundings and address the social and cultural identity of the specific locality (Dipasquale et al., 2014). In this context, vernacular architecture serves as a breeding ground for regional identities, traditional skill and technique, human creativity, and collective memory, among other things.

LESSONS LEARNED FROM VERNACULAR ARCHITECTURE

The human civilization has developed and advanced over the course of time, and it is imperative that the natural environment be protected in order to ensure that it will continue to exist in perpetuity. There is a uniqueness to each region or location that has emerged through time and via a complex interaction of natural and human development and adaptation to the surrounding environment. Different logical answers for ensuring human comfort conditions are represented by vernacular architecture in conjunction with distinct weather conditions. In this study, the findings revealed that traditional vernacular architecture was a climate-responsive architecture, which is a key aspect in achieving long-term sustainability. Residents of this region's vernacular architecture employed several climatically adapted techniques to keep their structures protected from the severe outside environment. They designed their buildings to allow for the least amount of sunlight penetration and the most efficient utilization of wind currents possible, using a variety of techniques.

There are many important lessons to be learned from vernacular architecture, including a holistic approach to climate and environment that takes into consideration both the overall urban form and the particular building form. Because of the compact design of structures, cities and towns were able to be equally small, while still giving shade and shelter from adverse weather to pedestrians in the surrounding area.

Another lesson is regarding building level. For example, the primary tactics for coping with its subtropical environment in winter include solar passive heating, low thermal mass, decreasing direct solar gains by building orientation and shading, enhancing airflow, and rain protection, among other things. In hot and humid weather, the use of light (low thermal mass) and air-permeable materials for the building envelope, together with the proper location of openings, increases the amount of natural ventilation that must be provided to maintain comfort. The dwellings are angled to the north and south, as recommended, in order to prevent direct solar gains from the facades of the buildings. All of the housings in the neighborhood that were inspected had enough shade, semi-open exteriors that were suitable for all types of activities, and a high roof overhang that protected them from severe rains. The majority of roofs are built of thatch, which is a lightweight and highly insulating material. Passive solar heating is the only bioclimatic design technique that has been established that cannot be found in the native architecture of the subtropical climate. The use of such strategies results in a significant reduction in energy consumption. Consequently, reducing greenhouse gas emissions from energy is a viable option.

Vernacular architecture reveals a startling level of originality, which reflects both a high degree of communal intelligence and a process of building experimentation that progressively evolved into experience. Both of these factors contribute to the creation of vernacular architecture. This social and cultural history demonstrates a high degree of inventiveness in terms of adjusting to available resources and making effective use of those resources. It may be implemented in a variety of ways that are practical, inventive, aesthetically pleasing (for example, colored coatings, frescos, plantings), and artistic. These creative expressions ought to serve as a source of motivation for the day that will come.

The social and cultural aspects of vernacular architecture are also expressed in a building language that conveys the immaterial values of individuals who constructed and lived in the space. This is because vernacular architecture is produced by people who share those values. The many expressions of sacredness (religious or agnostic, myths and legends), expressions of symbolism and identity connected with the building systems, and apotropaic protective measures are all examples of this collective memory's expression of place attachment. These ideals may also be expressed via communal rituals or in private settings that are enjoyable to occupy as a way of life.

Vernacular Human settlements reflect the inhabitants' desire and capacity to trade and live together, as well as to preserve the conditions of social cohesion in order to live as peacefully as possible in spite of competing interests. This is more than the simple architecture of the buildings that make up the settlements; instead, these settlements reflect the inhabitants' desire and capacity.

CONCLUSION

In comparison to modern structures, researchers and industry professionals have started making suggestions about, and engaging in debates about, the vernacular architecture's inherent capacity for sustainability. In many people's minds, modern structures are synonymous with a wide variety of traits that are harmful to the environment. Some of these features include pollution, unmanageable energy consumption, inefficient use of resources, and carbon footprints. Vernacular architecture has been under scrutiny and discussion in recent years as a possible solution to the ever-increasing number of environmental problems. Researchers and professionals in the field have started to propose and debate the topic.

Human civilization has developed throughout the course of history and has been able to maintain its continuity via integration with the environment. Humanity's continued existence is reliant on the protection of nature. As a result of the multifaceted interaction of evolution and human adaptability to the surrounding environment, over the course of time, every location and region developed distinctive characteristics that set it apart from other locations; these distinctive characteristics are the "essence" of "identity."

The construction industry has a substantial influence on the environment and natural resource availability. Building design philosophies, strategies, technologies, and construction processes must undergo a significant transformation in order to meet the world's increasing energy and environmental concerns.

The huge range of climate, topography, and culture found around the globe has a significant impact on vernacular architecture. It has unwritten knowledge on how to maximize the energy efficiency of buildings at a low cost by using locally sourced materials in their construction. A specific place's vernacular buildings have developed through time to meet the difficulties posed by the environment, construction materials, and cultural expectations of the period.

The concept of sustainability has often been seen as an essential component in the formation of both tangible and intangible cultural resources. The concepts of sustainability and the maintenance of cultural identity are complimentary to one another. Vernacular architecture is a type of traditional architecture that has developed over time by making use of the resources and technologies available in the surrounding natural and cultural environment. These elements of sustainable design are an integral part of vernacular architecture. Vernacular architecture fosters the best possible relationships between people and the locations in which they live.

To promote sustainable and climate-specific building design, vernacular architecture, which have been developed with the help of local builders who are knowledgeable about their location on the planet, can help contemporary architects learn from their mistakes and adapt to their surroundings.

REFERENCES

Al-Hinai, H., Batty, W. J., & Probert, S. D. (1993). Vernacular architecture of Oman: Features that enhance thermal comfort achieved within buildings. *Applied Energy, 44*(3), 233–258. doi:10.1016/0306-2619(93)90019-L

Alrashed, F., Asif, M., & Burek, S. (2017). The Role of Vernacular Construction Techniques and Materials for Developing Zero-Energy Homes in Various Desert Climates. *Buildings, 7*(1), 17. Advance online publication. doi:10.3390/buildings7010017

Bodach, S., Lang, W., & Hamhaber, J. (2014). Climate responsive building design strategies of vernacular architecture in Nepal. *Energy and Building, 81*, 227–242. doi:10.1016/j.enbuild.2014.06.022

Brown, R., & Maudlin, D. (2012). Concepts of vernacular architecture. In The SAGE Handbook of Architectural Theory (pp. 340–368). doi:10.4135/9781446201756.n21

Cardinale, N., Rospi, G., & Stefanizzi, P. (2013). Energy and microclimatic performance of Mediterranean vernacular buildings: The Sassi district of Matera and the Trulli district of Alberobello. *Building and Environment, 59*, 590–598. doi:10.1016/j.buildenv.2012.10.006

Chiou, Y.-S., & Elizalde, J. S. (2019). Thermal Performances of Three Old Houses: A Comparative Study of Heterogeneous Vernacular Traditions in Taiwan. *Sustainability*, *11*(19), 5538. Advance online publication. doi:10.3390u11195538

Chun, C., Kwok, A., & Tamura, A. (2004). Thermal comfort in transitional spaces—basic concepts: Literature review and trial measurement. *Building and Environment*, *39*(10), 1187–1192. doi:10.1016/j.buildenv.2004.02.003

Coch, H. (1998). Chapter 4—Bioclimatism in vernacular architecture. *Renewable & Sustainable Energy Reviews*, *2*(1), 67–87. doi:10.1016/S1364-0321(98)00012-4

Dipasquale, L., Correia, M., Mecca, S., Achenza, M., Guillaud, H., Mileto, C., ... Zanini, L. (2014). *VerSus: Heritage for Tomorrow Vernacular Knowledge for Sustainable Architecture*.

Dunham, D. (1961). The Courtyard House As A Temperature Regulator. *Ekistics*, *11*(64), 181–186.

EC. (2020). *In focus: Energy efficiency in buildings*. Retrieved January 18, 2022, from https://ec.europa.eu/info/news/focus-energy-efficiency-buildings-2020-feb-17_en

Emekci, Ş. (2021). Sustainability in Architecture: Low-tech or High-tech? In Proceedings Article (pp. 107–111). Alanya Hamdullah Emin Paşa University. doi:10.38027/ICCAUA2021216n14

Emekci, Ş. (2021). Determination of Sustainable Architectural Design Criteria in Protected Areas: Focus group method. *Journal of Design+Theory*, *17*(33), 229–242. doi:10.14744/tasarimkuram.2020.82687

Engin, N., Vural, N., Vural, S., & Sumerkan, M. R. (2007). Climatic effect in the formation of vernacular houses in the Eastern Black Sea region. *Building and Environment*, *42*(2), 960–969. doi:10.1016/j.buildenv.2005.10.037

Fadli, C., & Thamrin, H. (2017). Room Arrangement of Vernacular Dwelling in Mandailing, Indonesia. *IACSIT International Journal of Engineering and Technology*, *9*(4), 3427–3434. doi:10.21817/ijet/2017/v9i4/170904185

Fernandes, J., Mateus, R., Bragança, L., & Correia da Silva, J. J. (2015). Portuguese vernacular architecture: The contribution of vernacular materials and design approaches for sustainable construction. *Architectural Science Review*, *58*(4), 324–336. doi:10.1080/00038628.2014.974019

Glassie, H. (2000). *Vernacular architecture* (Vol. 2). Indiana University Press.

Golkar, K. (2000). Sustainable urban designing in borders of deserts. *J. Fine Arts*, *3*, 43–52.

Guillaud, H. (2014). Socio-cultural sustainability in vernacular architecture. In *Versus: heritage for tomorrow* (pp. 48–55). Firenze University Press. Retrieved from https://hal.archives-ouvertes.fr/hal-01159772

Hamard, E., Cammas, C., Lemercier, B., Cazacliu, B., & Morel, J.-C. (2020). Micromorphological description of vernacular cob process and comparison with rammed earth. *Frontiers of Architectural Research*, *9*(1), 203–215. doi:10.1016/j.foar.2019.06.007

Hantash, T. F. A. (2017). *Building A Zero Energy House For UAE : Traditional Architecture Revisited*. Presented at the *5th International Conference on Zero Energy Mass Customised Housing*.

Hashemi, F. (2018). *Adapting vernacular strategies for the design of an energy efficient residential building in a hot and arid climate*. City of Yazd.

Mazraeh, H. M., & Pazhouhanfar, M. (2018). Effects of vernacular architecture structure on urban sustainability case study: Qeshm Island, Iran. *Frontiers of Architectural Research*, 7(1), 11–24. doi:10.1016/j.foar.2017.06.006

Memarian, G., & Brown, F. E. (2003). Climate, Culture, And Religion: Aspects Of The Traditional Courtyard House In Iran. *Journal of Architectural and Planning Research*, 20(3), 181–198.

Morin, E. (2006). Organization and Complexity. *Annals of the New York Academy of Sciences*, 879(1), 115–121. doi:10.1111/j.1749-6632.1999.tb10410.x

Motealleh, P., Zolfaghari, M., & Parsaee, M. (2018). Investigating climate responsive solutions in vernacular architecture of Bushehr city. *HBRC Journal*, 14(2), 215–223. doi:10.1016/j.hbrcj.2016.08.001

Nabakov, P. (1999). Encyclopedia of Vernacular Architecture of the World. Traditional Dwellings and Settlements Review, 10(2), 69–75.

Nguyen, A. T., Truong, N. S. H., Rockwood, D., & Tran Le, A. D. (2019). Studies on sustainable features of vernacular architecture in different regions across the world: A comprehensive synthesis and evaluation. *Frontiers of Architectural Research*, 8(4), 535–548. doi:10.1016/j.foar.2019.07.006

Niroumand, H., Zain, M. F. M., & Jamil, M. (2013). A guideline for assessing of critical parameters on Earth architecture and Earth buildings as a sustainable architecture in various countries. *Renewable & Sustainable Energy Reviews*, 28, 130–165. doi:10.1016/j.rser.2013.07.020

Olukoya, O. A. P., & Atanda, J. O. (2020). Assessing the Social Sustainability Indicators in Vernacular Architecture—Application of a Green Building Assessment Approach. *Environments*, 7(9), 67. Advance online publication. doi:10.3390/environments7090067

Philokyprou, M., Michael, A., Savvides, A., & Malaktou, E. (2015). Evaluation of bioclimatic design features of vernacular architecture in Cyprus. *Case studies from rural settlements in different climatic regions*.

Potvin, A. (1997). The arcade environment. *Architectural Research Quarterly*, 2(4), 64–79. doi:10.1017/S1359135500001603

Sahebzadeh, S., Heidari, A., Kamelnia, H., & Baghbani, A.-N. (2017). Sustainability Features of Iran's Vernacular Architecture: A Comparative Study between the Architecture of Hot–Arid and Hot–Arid–Windy Regions. *Sustainability*, 9(5), 749. doi:10.3390u9050749

Salman, M. (2018). Sustainability and vernacular architecture: Rethinking what identity is. In *Urban and Architectural Heritage Conservation within Sustainability*. IntechOpen.

Singh, M. K., Mahapatra, S., & Atreya, S. K. (2010). Thermal performance study and evaluation of comfort temperatures in vernacular buildings of North-East India. *1st International Symposium on Sustainable Healthy Buildings*, 45(2), 320–329. 10.1016/j.buildenv.2009.06.009

Soflaei, F., Shokouhian, M., & Soflaei, A. (2017). Traditional courtyard houses as a model for sustainable design: A case study on BWhs mesoclimate of Iran. *Frontiers of Architectural Research*, *6*(3), 329–345. doi:10.1016/j.foar.2017.04.004

Thravalou, S., Philokyprou, M., Michael, A., & Savvides, A. (2015). *The role of semi-open spaces as thermal environment modifiers in vernacular rural architecture of Cyprus*. Biocultural.

Topfer, K. (2009). *Energy efficiency in buildings: transforming the market*. WBCSD.

Verma, T., Kamal, M. A., & Brar, T. (2022). An Appraisal of Vernacular Architecture of Bikaner. *Climatic Responsiveness and Thermal Comfort of Havelis*, *9*, 41–60.

Vrijhoef, R., & Koskela, L. (2000). The four roles of supply chain management in construction. *European Journal of Purchasing & Supply Management*, *6*(3), 169–178. doi:10.1016/S0969-7012(00)00013-7

Wang, T., & Liu, Y. P. (2013). Ecological Design Strategies of Vernacular Architecture in Shanxi, China. *Applied Mechanics and Materials*, *275–277*, 2773–2776. . doi:10.4028/www.scientific.net/AMM.275-277.2773

Zhai, Z., & Previtali, J. M. (2010). Ancient vernacular architecture: Characteristics categorization and energy performance evaluation. *Energy and Building*, *42*(3), 357–365. doi:10.1016/j.enbuild.2009.10.002

Zhao, P., & Zhang, M. (2018). The impact of urbanisation on energy consumption: A 30-year review in China. *Urban Climate*, *24*, 940–953. doi:10.1016/j.uclim.2017.11.005

Chapter 12
Turkey's Financial Alignment With the European Green Deal

Meryem Filiz Baştürk
Uludag University, Turkey

ABSTRACT

Signatory countries of the Paris Climate Change Agreement have committed to decreasing carbon emissions and reducing global warming by at least two degrees Celcius for fighting against climate change. Moreover, the European Union has declared to European Green Deal, and it has been taken one step further from the Paris Climate Change Agreement. European Green Deal aims to transform Europe into the first carbon-neutral continent in the world in 2050. EU member countries have prioritized achieving the reduced emission level and then reaching carbon-neutral societies in the next phase. European Green Deal has not only been related to EU member countries but also has affected Turkey as a prominent commercial partner and candidate member country. Ministry of Trade – Republic of Turkiye has released the Green Deal Action Plan to comply with the changes that will emerge with European Green Deal and related transformation policies. This study has been aimed to evaluate the third section of the Green Deal Action plan, which involves green finance.

INTRODUCTION

Global climate change, the damage caused by solid-fossil wastes to the environment, has increased the environmental awareness of economies. Reflections of the general concerns about environmental issues lead to global level initiatives such as the Paris Climate Change Agreement (COP21), which was signed in December 2015 with the participation of 195 countries. Following the COP21 statement, signatory countries confirmed reducing carbon emissions and targeted falling the global temperature by 2 degrees (REN21, 2016, pp. 110-111). Turkey ratified the Paris Climate Change Agreement in October 2021.

The European Union has taken the Paris Climate Change Agreement further with the European Green Deal. The European Green Deal was announced by the EU on 11 December 2019. According to this agreement, the European continent will be the first carbon-neutral continent by 2050. The European Union aims to implement a new growth strategy to reduce global climate change with the European

DOI: 10.4018/978-1-6684-4829-8.ch012

Green Deal (European Commission, 2019). According to this new strategy, reducing foreign dependency and ensuring energy supply security becomes a priority by increasing renewable energy investments, which stand out with their environmental friendliness, in all member countries (Cengiz & Kutlu, 2021).

To adapt to the changes that will arise with the implementation of the European Green Deal, the Green Deal Action Plan was published on 16 July 2021 by the Ministry of Trade - Republic of Turkiye. In this study, Green Finance, the third main title of the action plan consisting of nine main titles, is evaluated. The increase in green investments depends on the distribution and accessibility of financial resources. This makes financial compliance a strategic issue. Today, the share of investments made in green is increasing; at the same time, investors' interest grew, and financial instruments diversified. Stabilizing practices become vital, as some involve some form of financial innovation. It also requires regulation in many areas, from calculating returns to the parties' sharing the risks. These conditions make financing strategies vital for the development of green investments. In this direction, Turkey's financial compliance with the European Green Deal is discussed in detail in this study.

This study will follow those sections; firstly, will be defined the concept of green finance. Then, the second section will have been tried to summarise previous studies as a literature review. The third section will have been focused on the economic rationale behind green finance. And in the last section will have been tried to evaluate the green finance roadmap for Turkey.

DEFINITION OF GREEN FINANCE

Green finance has not been clearly defined. Although climate finance, green finance, and sustainable finance have been used interchangeably mainly, there have been scope differences between the concepts. The concept of climate finance has only been related to the effects of climate change. Due to the narrow meaning of climate finance, green finance has been dealt with in a broader scope. This concept has been related to prevalent environmental issues. In more broadly terms, green finance is a generic term for financial instruments, methods, and policies associated with the developing projects, policies, and technologies for sustainable development targets. On the other hand, sustainable finance includes environmental, social, economic, and governance factors (Berrou, Dessertine & Migliorelli, 2019, p. 13; UNEP, 2016, p. 10).

There have been at least three fundamental reasons for the lack of definitional clarities of green finance. Firstly, identifying the green sector and activities has been taken some difficulties. Although defining renewable energy or energy productivity investment has not been taken as a clarifying difficulty; clarifying green finance has not been an easy task. Some approaches have included electric and hybrid cars in the definition of green finance. But some others have embraced a more rigid framework and have been interested in only renewable and sustainable instruments. Consequently, this kind of definitional complication has been visibly in the clarifies made by diverse international institutions. Secondly, defining specific criteria for green finance instruments has been so difficult. And thirdly, the definitional ambiguity of green finance has undermined the trust and hindered the development of this emerging market (Berrou, Ciampioli & Marini, 2019, pp. 32 – 33).

LITERATURE REVIEW

There have not been rich empirical studies in the green finance literature yet. The limited number of research has dealt with more about some developments in recent years. The ambiguous nature of the green finance definition and the finite data explain why limited research has been done. Nevertheless, the spread of green finance instruments in financial markets and the popularity of topics related to green finance will trigger the research soon. This section of the study will explain the results of some research in the green finance literature.

One of the limited studies on green finance has been released by Khan et al. (2021), and they have been focused on researching the effects of green finance on the ecological footprint in the 26 Asian countries from 2011-to 2019 periods. After the analysis, they concluded that green finance had reduced ecological footprints in Asian region countries.

Another study by Rasoulinezhad & Taghizadeh-Hesary (2022) has dealt with complex relationship variables between carbon emissions, green finance, green energy index, and energy efficiency in the ten most supportive countries of green finance in the 2002-2018 period. The conclusion of this study indicated that green bonds have contributed to green investments, reducing carbon emissions levels.

Also, the research of Zheng et al. (2021) has been examined bankers' attitudes in private commercial banks towards green finance and green finance instruments in the 2014 – 2019 period in Bangladesh. Their analysis results have illustrated that private commercial banks have played a more active role in achieving sustainable development goals via green finance than the other banks and non-bank financial institutions in Bangladesh.

Additionally, Lee & Lee (2022) have been evaluated the relationship between green finance and green factor productivity in 30 provinces in China in the term of 2016-2018. The results of their analysis have shown that the development of green finance has increased green factor productivity.

Wang, Cai & Elahi (2021) has been indicated that green finance plays a reductive role in carbon emission levels in a study of 125 cities in China in the 2015 – 2017 period. Also, their findings have shown that green credits are more effective tools than green venture capital as a green finance investment. For authors, this finding may be related to the fact that green credits have been used before green venture capital, and green credits have been the lion's share in financial investment resources in China. In addition, high-level financial costs have encouraged the firms to use green credits. In conclusion, their study has been stated that green credits are more effective financial instruments in reducing carbon emissions due to scale and costs advantages.

Some studies have been interested in the role of green finance in environmental quality. Zhou & Zhang (2020) have been highlighted the high-level correlation between developments in green finance and economic development. Furthermore, green finance has positively affected the environmental quality and reduced or eliminated environmental degradation.

Likewise, the study by Li, Fan & Zhao (2022) has handled the role of green finance in the transition to a low carbon economy in 30 provinces in China. Their analysis has been based on data from the 2001-and to 2019 periods. Results of the research have confirmed the positive role of green finance in the transition to the low carbon economy.

THE ECONOMİC RATIONALE FOR GREEN INVESTMENT

Achieving sustainable economic growth depended on increasing green investment, founded on environmentally–friendly characteristics. The green investment is based on a broad scope. This concept has included renewable energy, energy efficiency, pollution prevention and control, environmentally sustainable management of land use, conservation of biodiversity, clean transport, sustainable water and wastewater management, climate change adaptation, eco-efficient and adapted products, and green buildings (CMBT, 2021, p. 6). Although the scope of green projects and investments is wide, renewable energy investments constitute the lion's share among them.

There are strong economic motives behind the increasing importance of green investment in contemporary economies. The primary of that is environmental damage related to solid fossil fuels. Many economies have recently increased green investment (especially renewable energy investment) to reduce solid fuel use. Numerous countries have declared transition plans to a green economy. For example, European Union declared The European Green Deal, which has chosen a primary target to make Europe the first carbon-neutral continent in 2050 in world history. Also, Turkey has targeted reducing to zero-emission level by 2053. The second motivation for increasing green investment in the contemporary world could be related to reaching an equilibrium point between rising energy needs and reducing environmental damages. The third motivation is related to energy security concepts. Domestic conventional energy resources are insufficient for many economies. Accordingly, for filling the demand-supply gap, economies tend to energy import. Cyclical fluctuations in energy markets damage sustainability and create security issues related to accessibility. In most contemporary economies, investing in green energy sources has been understood as escaping energy security risks. The last is about concrete links between green investment as a newly developing sector and creating employment in the labor market. Growing the green investment sector could be created employment expansion opportunities in the predominantly youth population societies especially (Kaygusuz, 2007, pp. 79-80).

Although green investment includes a broad window of opportunity for many economies, has been faced with some financial and non-financial barriers. These kinds of barriers bear mutual impacts, which have both related lenders and borrowers. For an analytical insight, barriers to green investment could divide into three different categories. The first one could relate to the obstacles which impact on lender's side. As seen in the table.1 could be possible to define five different sub-categories about the lenders' sides related barriers. These named that, financial modeling challenges, lack of collateral, high up-front investment, longer paybacks and availability of investment-ready projects. On the other side, barriers related to borrowers could be split up into three different sub-categories: behavior patterns, providers' weakness, and limited access to capital. In addition, barriers that impact both sides – lenders' and borrowers'- could be defined as lack of awareness and/or lack of technical capacity, adverse regulatory environment, legal framework issues, and limited experience in the financial sector (Aldana, Bray-Cartillier, & Shuford, 2014, p. 7).

The most decisive barriers as seen in the table.1 are lack of awareness and/or lack of technical capacity. Because of that, a lack of technical infrastructure will cause a choice problem for lenders, and they will have some trouble selecting suitable projects. Similarly, project developers who search for lenders will have a matching problem with available lenders. Choice and matching problems between lenders and borrowers are especially prevalent in emerging markets (IRENA, 2016, p. 31). The policies that target the elimination of deficiencies in technical infrastructure contribute to the removal of other barriers to green investment. Elimination of deficiencies in technical infrastructure will create a facilitating effect on

Table 1. Barriers to green investment

Affects the lender	Affects the borrower
Financial modeling challenges	Behavior patterns
Lack of collateral	Technology providers weaknesses
High up-front investments	Limited Access to capital
Longer paybacks	
Availability of investment ready projects	
Lack of awareness and/or lack of technical capacity	
Adverse regulatory environment	
Legal framework issues	
Limited experience in the financial sector	

Sources: Aldana, Bray – Cartillier, & Shuford, 2014; IRENA, 2016:27).

developing legal frameworks. Furtherly, the financial sector will be more experienced in green investment operations, and following then, the entry cost will decrease compared to the starting level. Some issues related to the long-term refunds and limited accessibility to start-up capital needs will have dissolved.

The public sector should be leading in eliminating deficiencies in technical infrastructure and encouraging private sector funds for green investment. The high entry costs in the green investment justify the public intervention via support policies at the start-up level (Mazzucato, 2021, p. 165). Here, elaborating on the main motives behind the public and private sector investment is crucial because both sides have totally different purposes. The primary motive behind the green investment is the high marginal costs of capital and technologies for the public sector. High marginal costs discourage private-sector lenders, and the public sector has tried to compensate for the cost loadings. Also, creating investors friendly climate in the green investment by the public sector is vital for drawing the interest of private funds. In summary, the public sector encourages lenders to invest in renewable energy projects which have borne some difficulties and immaturities compared to private sector funds (IRENA & CPI, 2004, pp. 40; 45).

Characteristics burdens of green investment have created a need for "patient capital". Patient capital could be in different forms. Typical forms of patient capital are feed-in tariffs, and development banks play a crucial role there (Mazzucato, 2021, pp. 188 – 189). Many countries have adopted feed-in tariffs as a prevalent form of patient capital for increasing renewable energy investments. A policy instrument in Turkey called YEKDEM is a typical example of feed-in tariffs applications. After the matureness of market conditions replacing a kind of support policies mechanisms such as "green tariffs ("yeşil tarife"- YETA)" (TSKB, 2021, p. 35). Development banks play a pivotal role in green investment, especially renewable energy investments. National development banks also have a considerable share in financing renewable energy projects in developing countries. In addition, multinational development banks have collaborated with some national development banks to finance renewable energy investments (Brown et al., 2012: 18). For example, 83% of public investment had been funded by national, multilateral, and bilateral development finance institutions in 2013-18 (IRENA & CPI, 2020, p. 45).

INCENTIVE MECHANISM IN TURKEY

Turkey ratified the Paris Climate Agreement in October 2021. Also, Turkey has been stated by 2030 on Intended Nationally Determined Contribution (INDC), greenhouse gas emissions will be reduced by 21% from the business as usual level and targeted zero-emission in 2053 (MTF, 2021). According to Turkey's National Renewable Energy Action Plan (TREAP), Turkey has been targeted that supply at least 30% of total electric energy demand and 10% of transportation sector needs with renewable energy sources until 2023.

Renewable energy investments play a vital role in reaching TREAP targets. Although Turkey's renewable energy investments have been grown in the last decade and are one of the countries that upgrading investment rates, renewable energy usage volume is not at the targeted level. Substantially, this situation is related to the high start-up initial cost. In the investment theory, high start-up initial costs have been supposed to reduce the number of projects, and some incentive mechanisms may tolerate investment risks. Turkey has created some investment mechanisms related to renewable energy projects following that policy idea. Two different prominent incentive mechanisms could be mentioned. The first called YEKDEM (Yenilenebilir Enerji Kaynaklarını Destekleme Mekanizması – Renewable Energy Support Mechanism) and other is YEKA (Yenilebilir Enerji Kaynak Alanı – Renewable Energy Resource Areas). The first one, YEKDEM, was conducted by the Use of Renewable Energy Resources for the Purpose of Electricity Generation Act no.5346 which regulated price, volume, duration, and payments incentives for renewable energy start-ups. YEKDEM also included that fixed price purchased guarantee for start-ups for a ten-year duration. After the duration period was closed in YEKDEM in June 2021, a new implementation was released on 29th January-2022, which is some differences from first-period mechanisms, such as fixed price purchased guarantee currency basis replaced U.S. dollar to Turkish Lira (SHURA, 2021, p. 20).

Also, the YEKA model has been aimed at raising renewable energy share in total electric energy production and, again, raising the percentage of the domestic output in renewable energy technologies (MENR, 2022). In addition, the model has implemented a market-based tendering process that supports solar and wind capacity setups (SHURA, 2020b, pp. 30-31).

Turkey has implemented some mechanisms which may be an example of "patient capital". TURSEFF, financial investment support has been developed by EBRD (European Bank for Reconstruction and Development) has envoy some characteristics of patient-capital. This program aims to create investment channels for start-ups in renewable energy production and support the energy productivity applications in SMEs. Investments have been provided for users with national commercial banks, which include, Akbank, Garanti Bankası, Denizbank, İş Bankası, Vakıf Bank ve Yapı Kredi Bankası according to the program. TURSEFF was started with Phase I in 2010; then Phase II was conducted in 2013, and finally, Phase III began in 2017. Starting at the Phase III process, some leasing institutions have provided investments for users, such as AkLease, Garanti Leasing, and QNB Finance Leasing (SHURA, 2020a, p. 42). TURSEFF has supported the 2277 projects and provided funds for investment of 726 million euros since the beginning of 2010, the last 11 years. Renewable energy installations support in the program's scope has been close to 625 MW (TurSEFF, 2022).

These incentive mechanisms have contributed to increasing green investment in general and renewable energy investments specifically. Especially, these support mechanisms accelerated long-term investment projects and provided predictability for investors who were anxious about the unique risks of renewable energy production. The incentive mechanisms in Turkey could be assessed as a facilitator factor

Table 2. Development of renewable energy resources in installed energy power (MW)

Resources	2013	2014	2015	2016	2017	2018	2019	2020	2021/9
Hydroelectric	22.289	23.643	25.868	26.682	27.273	28.291	28.503	30.984	31.447
Wind power	2.760	3.630	4.498	5.751	6.516	7.005	7.591	8.832	10.168
Solar power	0	40	310	833	3.421	5.063	5.995	6.667	7.534
Biomass	224	288	345	467	575	739	1.163	1.485	1.782
Geothermal	311	405	624	821	1.064	1.283	1.515	1.613	1.650
Total Renewable Energy	**25.583**	**28.006**	**31.645**	**34.554**	**38.849**	**42.381**	**44.768**	**49.581**	**52.581**

Source: (TSKB, 2021, p. 31).

on investment requests of projects in the scope of renewability. Effects of the facilitating characteristics of these mechanisms have been traceable in the share of renewable energy the installed energy power. Table.2 indicated that the share of renewable energy in the installed energy power had increased since 2013. The most considerable increase has been in the share of hydroelectric power, wind power, and solar power in the installed energy power, respectively. Although the share of wind power and solar power has significantly increased, hydroelectric power resources are still the most significant share of the installed energy power in Turkey. The program of YEKDEM and the tender process of YEKA have played a decisive role in the decreasing initial costs of solar and wind energy projects, and these developments have produced a growing share of renewable energy resources in the installed energy power (TSKB, 2021, pp. 31; 51).

ROADMAP OF GREEN FINANCE IN TURKEY

Green finance has been a third main part of the Green Deal Action Plan published by the Ministry of Trade – Republic of Turkiye. This acting plan has involved increasing green finance resources at the national level and determining a roadmap for benefiting from the growing volume of worldwide green investment flows. This section has been focused on determining the implementation of the roadmap and has emphasized applications in Green Deal Action Plan.

Sustainable Finance Framework

The sustainable Finance Framework of Turkey has been released in November-2021 by the Ministry of Treasure and Finance – Republic of Turkiye. This framework has been aimed at implementing bond issue operations such as green bonds, sustainable bonds, Sukuk, credit operations, and other debt instruments in international markets by the Ministry of Treasure and Finance. The sustainable finance framework has involved four main elements. The first element has been related to the use of proceeds; the second one has been comprised of the evaluation criteria of projects and the selection process. The third of them has been focused on the management of proceeds, and the last one has been relevant to the reporting. Ministry of Treasure and Finance has also been committed that each bonds issue operation will be compatible with these basic principles (MTF, 2021, p. 12).

Table 3. Renewable energy financing current situation and forecasts

	2002 - 2018	2019 - 2030
Credits	**98**	**70**
International Finance Institutions	50	35
National Commerical Banks	45	30
Leasing Institutions	3	5
Alternative Borrowing Instruments	***2***	***30***
Risk / Venture Capital	2	10
Bonds	0	10
Crowd funding	0	5
Others	0	5

Source: SHURA, 2019, p. 97.

Sustainability Strategy in the Turkish Banking Sector

The banking sector's share of financial transactions has been significant in Turkey. Naturally, actors in the banking sector have been playing prominent roles in green investment. This decisive role of the sector may be seen easily in table.3 which contains splitting two-different periods of finance for renewable energy investments. Renewable energy investments could be seen as indicators of green investments, and these figures may be given hints at the size of investments in Turkey.

According to the table.3 saw that 98% of finance of renewable energy investments had been via credits, and assets had provided only 2% of them with alternative debt instruments. International finance institutions and national commercial bank credits have been predominant in renewable energy investment in this period (SHURA, 2019, p. 97). SHURA (2019) estimated the transformation of borrowing channels for renewable energy investment in Turkey in 2019-2030 projections. According to the projections, the alternative borrowing instruments' share will rise to 30%, and the percentage of credits in renewable energy investments will decline to the 70% level. Alternative debt instruments, except for credits, will be more functional investment tools and will be detecting more diversified debts instrument in the renewable energy projects in Turkey in the near future.

In addition, BDDK (2021) has prepared a "Sustainable Banking Strategy Document", which had been committed to Green Deal Acting Plan. This document has involved some policy proposals for developing sustainable finance and has offered solutions to current and possible problems for the finance process for investment based on the reality of the predominant role of banks in green finance in Turkey.

The Turkish banking sector has encountered some structural and institutional problems in providing funds for green investments. Some structural problems in the banking sector have been related to the macroeconomic uncertainties in the Turkish economy, low saving rates, and the short-term funding structure of bank operations. Also, the Turkish banking sector should tackle some issues linked to weak institutional frameworks. One of the leading weaknesses has been the absence of green categorization. Green categorization could aid banks in defining categories of green investments and could be given comparing tools for alternative green borrowing instruments. Banks have been deprived of coherent policy development opportunities without these kinds of categorization tools (BDDK, 2021, p. 11). Structural problems of banks in green finance operations have been needing long-term policy tools and development frameworks. Paris Climate Change Agreement and the European Green Deal that takes this agreement one step further could be seen as an opportunity to solve institutional deficits. For example, according to the European Green Deal, the EU has been committed to transforming The European Investment Bank into The European Climate Bank. The bank will be devoted half of its actions to environmentally friendly projects (European Commission, 2020, p. 9). Although green investment banks are not a worldwide phenomenon, it is so possible said that the number of green banks would be increasing due to the incremental development of green investments in the near future. The establishment of A Green Bank of England, Green Finance Organization in Japan, and Clean Energy Finance Cooperation in Australia may be evaluated as the indicators of these institutional trends (Taranto & Dinçel, 2019, p. 41). Like these international trends, it may be considered to establish a green bank that will only give funds to green investment projects in Turkey.

Green Bonds and Green Lease Certificates Guideline

Green bonds have offered one of the best suitable alternative tools for sustainable development to governments. Green bonds were first issued by the EIB (European Investment Bank) in 2007 (Griffith-Jones et al., 2012, pp. 28 – 29). Initially, green bonds were issued only by international development banks; then, some national-based public and private institutions followed these trends and issued their own green bonds. The terms "green" in the definition of "green bonds" have indicated a commitment that bonds have only used on funding for the environmentally friendly investment (IRENA, 2020, pp. 6-8).

Green bonds have getting played more role in green investments. The increasing prevalence of green bonds has caused some problems in definitions. Then, the need for specific standards has been debated among financial actors. Two different international standards have stood out in bonds markets. The first is named "green bond principles", and the latter is called "climate bonds standard". Both of these standards are based on voluntarily implemented (IRENA, 2020, p. 6). Except for these foremost standards, the EU Commission has announced the implementation of a "European Green Bond Standard" in the European Green Deal Investment Plan. The green bond standards of EU Commissions have been again voluntarily based implemented, involving four elements. The first one is that funds have been compatible with EU taxonomy principles. The second is related to the transparency principle, and parties should be sharing the information about how bonds incomes are used. The third one might be called the "audit principle", which involves that green bonds have to audit by an external auditor. And the last one offered an external audit architecture and indicated that European Securities Markets Authority (ESMA) should control external auditors (European Commiesion, 2021). European green standards have not only comprised the EU member countries but are also suitable for non-EU bond issuers. The standards will have been prevented the problem of greenwashing, which vital issue for green investment.

The first green bonds were issued by TSKB in 2016 for Turkey. After the first, TSKB continued the bond issue, and other banks have followed. Table.4 indicates some details of the green bond issue in Turkey. As seen in the table.4, green bonds are first issued by development and investment banks, and then commercial banks are involved in the process. Involving the banks in the green bonds issued has been expanded with the participation of the real sector, and interest in the process has been increased. Arçelik has been the first real sector actor who issued green bonds. The most significant bond issue was made in July 2017 by Aydem Energy. Although the green bond has been an emerging market, it is a notable increase in recent years. The increasing interest has been predicted in both banking sector bond issues and real sector bond issues in the coming years.

Table 4. Green and sustainable bonds issued in Turkey

Date of issue	Issuing institutions	Amount	Expiry	Type	Currency
18.05.2016	TSKB	300	5	Green/ environmental bond	dolar
28.03.2017	TSKB	300	10	Green/ environmental bond	dolar
30.06.2017	TSKB	150	5	Green/ environmental bond	TL
2017	Garanti Bankası	150	5	Green bond	dolar
21.08.2019	TSKB	50	10	Green/ environmental bond	dolar
20.12.2019	TSKB	50	5	Green/ environmental bond	dolar
2019	Garanti Bankası	50	5	Green bond	dolar
21.01.2020	TSKB	50	5	Green/ environmental bond	dolar
22.01.2020	Yapı Kredi	50	5	Green/ environmental bond	dolar
05.08.2020	TSKB	50	5	Green/ environmental bond	dolar
08.12.2020	TSKB	750	5	Sustainable bond	dolar
14.01.2021	TSKB	350	5	Sustainable bond	dolar
28.01.2021	TSKB	600	5	Sustainable bond	dolar
28.05.2021	Arçelik	350	5	Green bond	euro
27.07.2021	Aydem Enerji	750	5,5	Green bond	dolar

Source: TSKB, 2021: 53; Garanti Bankası, 2022; Yapı-Kredi Bankası, 2022; Arçelik, 2022; Aydem Enerji, 2022.

The growth and development of green bond markets in Turkey as a developing country will have been provided support for transparency. Furthermore, increasing transparency will have attracted foreign investors in green bonds, and it will have been easier to find investors for environmentally friendly projects (CBI, 2017, p. 19). SPK (Sermaye Piyasası Kurulu – Capital Markets Board of Turkey) has been released guidance about green debt instruments and green lease certificates in November-2021 in line with these goals. This guidance has been involved some definitions of green debt tools and frameworks related to the green lease certificate mechanisms. The guidance has been based on green bond principles by International Capital Market Association (ICMA) and has been comprised of four main elements. The first one is "the use of proceeds obtained from the issue and corresponded funds from green debt tools should be invested in green projects. The following is the project evaluation and selection process, and the third is the management of proceeds obtained from the issue. The last one is about reporting.

Sukuk has been an environmentally friendly financial tool in Islamic finance. It may call as "green Sukuk" which was firstly issued by Tadau energy company in Malaysia in July 2017 (CBI, 2007, p. 19). Green Sukuk has been one of the widespread financial instruments in Islamic countries. The total amount of green Sukuk issue is close to 4.3 billion dollars in the UAE, Saudi Arabia, Indonesia, and Malaysia in 2019 (CBI, 2020b). The first green Sukuk was issued an amount of 50 million dollars by TSKB on behalf of Zorlu energy company in Turkey (TSKB, 2021, p. 54).

CONCLUSION

Economies have been aimed at transforming into a low-carbon economy to reduce climate change in the 21st century. EU has been leading role in transforming into a low-carbon economy, and the union has declared a target of reaching the first carbon-neutral continent in the world in the year 2050 in the European Green Deal. European Green Deal has not been related only to member countries, and also close commercial partner economies have been affected by the new environmental deal. Turkey, a prominent commercial partner for European economies and a nominated country of the EU, will have been one of the most affected countries by the Green Deal process.

Transform to a green economy has been a concrete phenomenon with the Paris Climate Change Agreement, and the European Green Deal takes this agreement one step further. Yet, the financing problem of the low-carbon green economy is still being Achilles tendon in current economic circumstances. Conventional finance tools have been very insufficient for investing low-carbon and environmentally friendly economy. The effectiveness of conventional financial tools and harmonization of them with green economy needs has been raised the importance of green finance. In addition, new and innovative instruments such as green bonds or green Sukuk will have been widespread in financial markets. Growing green investments are primarily dependent on deepening the green finance markets. The lack of financial resources has been hindered green investment projects. Adverse effects of lack of financial resources have been primarily felting at the start-up level of green investments.

Ministry of Trade – Republic of Turkiye has declared Green Deal Action Plan and emphasized green finance in the third main section. The plan has been involved some steps for building secure green finance architecture. Some of the policy steps are operational now, and other parts of the measures have been planned and will be operating in the near future. It will be possible to reach a regulatory framework for the Turkish economy. Some other stakeholders have been declared some additional strategic papers for addition to the Green Deal Action Plan such as the Ministry of Treasure and Finance – Republic of Turkiye has announced that Sustainable Finance Framework in November 2021; BDDK has implemented the "Sustainable Banking Strategy Documents at again November 2021; SPK (Sermaye Piyasası Kurulu – Capital Markets Board of Turkey) has been released Green Debt Instruments and Green Lease Certificate. Different stakeholders have been tried to fulfill their responsibilities for aiming targets related to the Green Deal Action Plan. These applications in the context of green investment will have been diversified to green finance instruments and have been facilitated to reach sufficient financial resources. Furtherly deepening green finance tools will have accelerated the harmonization of the Turkish economy to the European Green Deal.

REFERENCES

About TURSEFF (2022). Retrieved from https://www.turseff.org/sayfa/facility

Aldana, M., Braly-Cartillier, I., & Shuford, L. (2014). *Guarantees for Green Markets Potential and Challenges. Inter-American Development Bank*. IDB.

BDDK. (2021). *2022-2025 Sustainable Banking Strategic Plan* [2022-2025 Sürdürülebilir Bankacılık Stratejik Planı]. Retrieved January 15, 2022, from https://www.bddk.org.tr/KurumHakkinda/EkGetir/5?ekId=36

Berrou, R., Ciampoli, N., & Marini, V. (2019). Defining green finance: Existing standards and main challenges. In *The Rise of Green Finance in Europe* (pp. 31–51). Palgrave Macmillan. doi:10.1007/978-3-030-22510-0_2

Berrou, R., Dessertine, P., & Migliorelli, M. (2019). An overview of green finance. *The rise of green finance in Europe*, 3-29.

Cengiz, Ç., & Kutlu, E. (2021). The budget of the European Union and green energy: Perceptions, actions and challenges. *International Journal of Social Inquiry*, *14*(1), 171–197. doi:10.37093/ijsi.950563

CMBT. (2021). *CMB Green Debt Instrument and Green Lease Certificate Guide* [SPK Yeşil Borçlanma Aracı ve Yeşil Kira Sertifikası Rehberi]. Retrieved January 10, 2022, from https://www.spk.gov.tr/Sayfa/Dosya/1350

European Commission. (2019). *The European Green Deal.* Communication from the Commission COM(2019) 640 Final, 11.12.2019. Retrieved June 25, 2020, from https://eur-lex.europa.eu/legal-content/EN/TXT/?qid=1596443911913&uri=CELEX:52019DC0640#document2

European Commission (2020). *Sustainable Europe Investment Plan, European Green Deal Investment Plan.* Retrieved December 10, 2021, from file:///C:/Users/User/Downloads/Commission_Communication_on_the_European_Green_Deal_Investment_Plan_EN.pdf.pdf.

European Commission. (2021). *Commission puts forward new strategy to make the EU's financial system more sustainable and proposes new European Green Bond Standard.* Retrieved December 10, 2021, from https://ec.europa.eu/commission/presscorner/detail/en/ip_21_3405

Griffith-Jones, S., Ocampo, J. A., & Spratt, S. (2012). *Financing renewable energy in developing countries: Mechanisms and responsibilities.* European Report on Development. Retrieved March 5, 2018, from https://policydialogue.org/files/publications/Financing_Renewable_Energy_in_Developing_Countries.pdf

IRENA. (2016). *Unlocking renewable energy investment: The role of risk mitigation and structured finance.* IRENA. Retrieved January 25, 2018, from https://www.res4med.org/wp-content/uploads/2017/11/IRENA_Risk_Mitigation_and_Structured_Finance_2016.pdf

IRENA. (2020). *Renewable energy finance: Green Bonds.* International Renewable Energy Agency. Retrieved March 3, 2021, from https://www.irena.org/publications/2020/Jan/RE-finance-Green-bonds

IRENA & CPI. (2018). *Global Landscape of Renewable Energy Finance, 2018.* International Renewable Energy Agency. Retrieved March 5, 2021, from https://www.irena.org/publications/2018/jan/global-landscape-of-renewable-energy-finance

IRENA & CPI. (2020). *Global Landscape of Renewable Energy Finance, 2020.* International Renewable Energy Agency. Retrieved March 5, 2021, from https://www.irena.org/publications/2020/Nov/Global-Landscape-of-Renewable-Energy-Finance-2020

Kaygusuz, K. (2007). Energy for sustainable development: Key issues and challenges. *Energy Sources. Part B, Economics, Planning, and Policy*, *2*(1), 73–83.

Khan, M. A., Riaz, H., Ahmed, M., & Saeed, A. (2021). Does green finance really deliver what is expected? An empirical perspective. *Borsa İstanbul Review*. doi:10.1016/j.bir.2021.07.006

Lee, C. C., & Lee, C. C. (2022). How does green finance affect green total factor productivity? Evidence from China. *Energy Economics*, *107*, 1–15.

Li, W., Fan, J., & Zhao, J. (2022). Has green finance facilitated China's low-carbon economic transition? *Environmental Science and Pollution Research*. doi:10.1007/s11356-022-19891-8

Mazzucato, M. (2021). *Debunking the Myth of Entrepreneurial State Public Sector-Private Sector Antagonism* [Girişimci Devlet Kamu Sektörü-Özel Sektör Karşıtlığı Masalının Çürütülmesi]. Koç Üniversitesi Yayınları.

MENR. (2022). *YEKA Modeli*. Retrieved January 11, 2022, from https://enerji.gov.tr/eigm-yenilenebilir-enerji-uretim-faaliyetleri-yeka-modeli

MT. (2021). *Yeşil Mutabakat Eylem Planı*. Retrieved October 1, 2021, from https://ticaret.gov.tr/data/60f1200013b876eb28421b23/MUTABAKAT%20YE%C5%9E%C4%B0L.pdf

MTF. (2021). *Republic of Turkey Sustainable Finance Framework*. Retrieved January 11, 2022, from https://ms.hmb.gov.tr/uploads/2021/11/Republic-of-Turkey-Sustainable-Finance-Framework.pdf

Rasoulinezhad, E., & Taghizadeh-Hesary. (2022). Role of green finance in improving energy efficiency and renewable energy development. *Energy Efficiency*, *15*(14), 1–12.

REN-21. (2016). *Renewables – Global Status Report: 2016*. Retrieved June 4, 2017, from http://www.ren21.net/wp-content/uploads/2016/05/GSR_2016_Full_Report_lowres.pdf

SHURA Enerji Dönüşümü Merkezi. (2019). *Türkiye'de Enerji Dönüşümünün Finansmanı*. Retrieved July 25, 2020, from https://www.shura.org.tr/wp-content/uploads/2019/10/Turkiyede_Enerji_Donusumunun_Finansmani.pdf

SHURA Enerji Dönüşümü Merkezi. (2020a). *Enerji verimliliği çözümü: Finansman mekanizmaları*. Retrieved June 20, 2021, from https://www.shura.org.tr/wp-content/uploads/2020/10/SHURA_FinansmanMekanizmalari.pdf

SHURA Enerji Dönüşümü Merkezi. (2020b). *Türkiye enerji dönüşümünü hızlandırmak için 2020 yılı sonrası düzenleyici politika mekanizması seçenekleri: Şebeke ölçeğinde ve dağıtık güneş ve rüzgar enerjisi kapasite kurulumları*. Retrieved January 10, 2022, from https://www.shura.org.tr/wp-content/uploads/2021/01/2020_yili_sonrasi_duzenleyici_politika.pdf?_ga=2.202753873.1740824367.1645079503-919185812.1644914363

SHURA Enerji Dönüşümü Merkezi. (2021). *Türkiye'de Yenilenebilir Enerji Tedariki ve Belgelenmesi*. Retrieved January 10, 2022, from https://shura.org.tr/turkiyede-yenilenebilir-enerji-tedariki-ve-belgelenmesi/

T.C. EPDK. (2005). *Law on the Use of Renewable Energy Sources for Electrical Energy Production* [Yenilenebilir Enerji Kaynaklarının Elektrik Enerjisi Üretimi Amaçlı Kullanımına İlişkin Kanun]. Kanun No: 5346. Retrieved June 20, 2020, from https://www.epdk.gov.tr/Detay/Icerik/3-0-0-122/yenilenebilir-enerji-kaynaklari-destekleme-mekanizmasi-yekdem

T.C. SBB. (2019). *Eleventh Development Plan (2019-2023)* [On Birinci Kalkınma Planı (2019-2023)]. Retrieved June 20, 2020, from https://www.sbb.gov.tr/wp-content/uploads/2021/12/On_Birinci_Kalkinma_Plani-2019-2023.pdf

TSKB. (2020). *Energy Outlook 2020* [Enerji Görünümü 2020]. Retrieved July 10, 2021, from https://www.tskb.com.tr/i/assets/document/pdf/enerji-sektor-gorunumu-2020.pdf

TSKB. (2021). *Energy Outlook 2021* [Enerji Görünümü 2021]. Retrieved January 20, 2022, from https://www.tskb.com.tr/i/assets/document/pdf/enerji-sektor-gorunumu-2021.pdf

UNEP. (2016). *Definitions and concepts.* Background note. Inquiry working paper 16/13. https://www.unep.org/resources/report/definitions-and-concepts-background-note-inquiry-working-paper-1613

Wang, F., Cai, W., & Elahi, E. (2021). Do green finance and environmental regulation play a crucial role in the reduction of Emissions? An empirical analysis of 126 Chinese Cities. *Sustainability*, *13*(23), 1–20.

Zheng, G. W., Siddik, A. B., Masukujjaman, M., Fatema, N., & Alam, S. S. (2021). Green Finance Development in Bangladesh: The role of private commerical banks (PCBs). *Sustainability*, *13*(2), 1–17.

Zhou, X., Tang, X., & Zhang, R. (2020). Impact of green finance on economic development and environmental quality: A study based on provincial panel data from China. *Environmental Science and Pollution Research International*, *27*, 19915–19932.

Chapter 13
Urban Transition via Municipal Transformation

Konstantinos Asikis
Urban Planning Laboratory, Archtecture School, National Technical University of Athens, Greece

Alexia Spyridonidou
European Public Law Organisation, Greece

Sofia Aivalioti
Bax & Company, Spain

ABSTRACT

Municipalities have to play a substantial role for urban resilience and sustainable development not only as local administrators, but also as strategists for integrated territorial planning, enablers by generating the urban activity and growth, servants by providing public quality-of-life amenities, and investors (public investments, infrastructures, and soft projects). Cities need further specialization to deal with complexities of the modern challenges, including pandemics, climate change, resources depletion, and socio-economic issues. They need to become more responsive and effective to emergencies as well as the long-term sustainability. They have to keep up with social and technological innovation and transit to a new model of operation. In order to achieve these, municipalities have to be organized like the contemporary companies; plan intelligent policies regarding resources, social capital, and economic growth, with respect to SDGs; apply smart and healthy city projects; involve cities to the international programmes, initiatives, and networks; and be familiar with the new era funding tools.

BACKGROUND

There is no doubt that the human settlements (cities, towns, villages) and their administrators are in the threshold of some significant and fast changes regarding their status and sustainable development.
    These changes are:

- Challenges, like the climate change, the urban shrinkage, the rural depopulation, the housing, the social inclusion, etc.
- Opportunities, like the Smart Cities applications, the Energy Transition, the participatory governance, the greening, etc.

The sooner the municipal structures be aware of this critical threshold and adapt properly their priorities and function, the more effective policies are going to plan and implement, for the cities and towns transition towards to sustainability.

METHODOLOGY

This chapter aims to:

- Underline the critical position of the cities and towns regarding the strategic planning and the implementation of the European policies.
- Explore the average duties and competencies of the local administrative structures, both as authorities and development engines.
- Expose the current significant and complicated challenges, and the need for the cities to be ready not only to face them, but also to grab the development opportunities.
- Encourage the administrative authorities to transform themselves, in order to be able to lead their cities to a thriving future that includes all the citizens.
- Refers the contemporary Agenda and the Tools that are available to support the efforts for the creation of some more sustainable urban models
- Suggest some useful Guidelines as a Roadmap overview, in order to set some steps which are necessary in this municipal transformation and urban transition procedure

INTRODUCTION

With Europe being one of the most urbanised continents in the world, the development of cities will determine the future economic, social and territorial development of the European Union (European Union. European Comission. Directorate-General for Regional Policy, 2011). The green and digital transition has to be driven by the municipalities – especially the urban ones.

The same stands for combating climate change, building communities resilient to crises, such as Covid-19 pandemic. Municipalities have to play a substantial role regarding urban resilience - sustainable development.

This central role is recognized across a number of EU and global initiatives and policy documents, i.e., the EU Green Deal ("A european green deal", n.d.) and the concept of local Green Deals (European Commission, Directorate-General for Internal Market, Industry, Entrepreneurship and SMEs, Executive Agency for Small and Medium-sized Enterprises, Durieux, Hidson, 2021), Fit for 55 (Plan, 2011) the Recovery and Resilience Facility (RRF) ("Recovery and resilience facility", 2022), the EU Urban Agenda ("The urban agenda for the EU, n.d.), the EU plans of circular economy, sustainable urban mobility, the UN-backed campaign "Cities Race to Zero" (C40) ("Race to Zero Campaing", n.d.), etc.

Moreover, urban municipalities are in the center of "100 Climate Neutral and Smart Cities Mission" (European Commission, Directorate-General for Research and Innovation, Gronkiewicz-Waltz, Larsson, Boni, 2020), one of the five ambitious EC Missions.

Why is it so complex to create realistic transition pathways? Municipalities have to be not only the local "administrators", but also the:

1. "strategists" for integrated territorial planning,
2. "enablers", by generating the urban-human-economic-cultural-etc. activity and growth,
3. "servants" by providing public services and quality-of-life features and amenities,
4. "investors", considering the public investments, infrastructures and soft projects.

They need further specialization to deal with complexities of the modern challenges, including pandemics, climate change, resources depletion and socio-economic issues. They also need to become more responsive and effective to emergencies as well as the long-term sustainability. They have to keep up with social and technological innovation and transit to a new model of operation. The evolving and increasing challenges as well as their multifaceted nature, calls for a better understanding of the different roles they play in the twin transition ecosystems. The stakeholders concerned must understand the different 'hats' of the cities, while the municipalities need to be internally prepared to serve all these roles, adapt, evolve and be flexible.

In order to achieve these, Municipalities have to:

1. be organized like the contemporary companies,
2. plan intelligent policies regarding resources, social capital and economic growth, with respect to SDGs,
3. apply smart and healthy city projects
4. be involved and active to international programmes, initiatives and networks,
5. be familiar with the new era - innovative funding tools.

THE FRAMEWORK

Municipalities in Europe face tremendous socio-economic and environmental challenges and they are called to take action to some of the most tough problems of humanity, namely, the climate crisis, overpopulation, pandemics, inequality and resource depletion among others. They are the implementers of projects and activities that can contribute to a more sustainable future. However do not always take this role and miss out on the opportunity to create better environments for people and the planet.

Traditionally, municipal governments function in a political, financial and administrative domain and act in a relative autonomy in accordance with the country's laws and constitutions. They can establish their own local policies, manage their own budget with received income from municipal taxation or other funding sources, be responsible for the territorial development and provide public services. However, these functions can remain limited due to lack of internal capacity, funding, human resources and entrepreneurship. Political changes do not permit in various cases for long term implementation and continuity of envisioned projects and programmes and frequent changes in the structure and priorities of municipalities hampers future development and resilience.

Figure 1.
Source: European Union. European Commission. Directorate-General for Regional Policy (2011).

Figure 2.
Source: European Union. European Commission. Directorate-General for Regional Policy (2011).

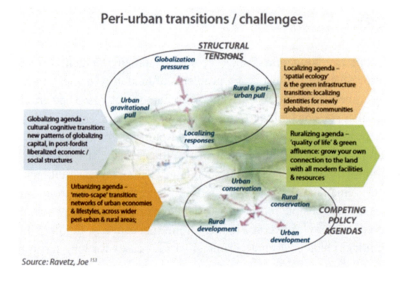

Figure 3.
Source: *Recovery and Resilience Facility (2022).*

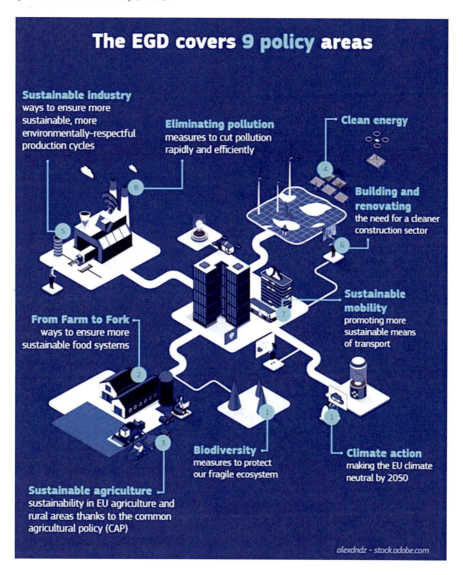

Oftentimes, municipal governments have to deal with unpredictable situations such as the case of the recent pandemic or disasters due to the climate change. In addition, lots of municipal infrastructure is getting old, obliging municipalities to channel significant funds and resources towards their replacement or repair.

According to the EISMEA, Duriex, and Hidson (2021):

Ultimately, all of the policy areas of the EGD are implemented in cities" [...] "They implement 70% of EU legislation, handle one-third of public spending, and manage two-thirds of public investment. In addition, they are often responsible for the direct provision of many services for their citizens".

Cities are places of economic activity, knowledge generation, community engagement, and cradles of innovation, and socio-economic transformation. They bring together local businesses, entrepreneurs, social enterprises, educational establishments, and citizens to find solutions to help tackle the biggest challenges of the twenty first century, whether it is delivering on climate ambitions, or building sustainable recovery and resilience post-COVID.

According to the European Commission, Gronkiewicz-Waltz, Larsson, and Boni (2020):

Cities play a pivotal role in achieving climate neutrality. They take up only 4% of the EU land area, but are home to 75% of EU citizens. This number is expected to rise to 85% by 2050. Worldwide, cities account for more than 65% of energy consumption and for more than 70% of CO2 emissions. The targets of the European Green Deal – reducing emissions by 55% by 2030 and becoming the first climate neutral continent by 2050 – will be impossible to achieve without cities in the vanguard of concerted efforts.

Therefore, we understand well that municipalities can have a pivotal role in reversing many of today's challenges. Their power over local developments can provide substantial opportunities regarding urban resilience and sustainable development. Their connection with citizens can support the co-creation of better neighborhoods and resilient communities.

THE URBAN TRANSITION

The Cities of the new era have to be Green, Smart and Healthy, in order to be resilient and follow a sustainable trajectory. What this means or includes, is given by several contemporary definitions. To select three of the most popular and characteristic:

1. On the IEEE web page ("Smart Cities Community", n.d.) As world urbanization continues to grow with the total population living in cities forecast to increase by 75% by 2050, there is an increased demand for intelligent, sustainable environments that offer citizens a high quality of life. This is typically characterized as the evolution to Smart Cities. We believe a Smart City brings together technology, government and society and includes but is not limited to the following elements:
   - A smart economy
   - Smart energy
   - Smart mobility
   - A smart environment
   - Smart living
   - Smart governance
2. A smart sustainable city is an innovative city that uses information and communication technologies (ICTs) and other means to improve quality of life, efficiency of urban operation and services, and competitiveness, while ensuring that it meets the needs of present and future generations with respect to economic, social and environmental aspects ("Smart sustainable cities: An analysis of definitions", 2014).
3. A WHO healthy city commits to leading with the social, physical, cultural environments aligned to create a place that is actively inclusive and non-exploitative and facilitates the pursuit of health and

Figure 4.
Source: European Innovation Council and SMEs Esecutive Agency (EISMEA), Duriex, and Hidson (2021)

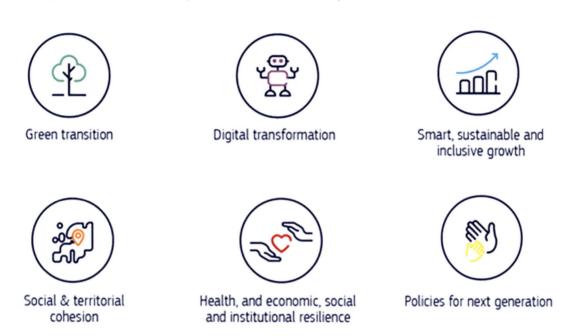

well-being for all. In relation to the place theme, through the adoption of the Consensus, members of the WHO Network commit to ("Compendium of tools, resources and networks, Support package for implementation", 2020).
- Ensure human-centred urban development and planning
- Make the healthiest choice the cheapest, easiest, and most accessible choice
- Create places that support health and well-being throughout the life-course
- Ensure community and participatory governance of places
- Enhance community resilience.

From the above given definitions, we can easily extract some key-words that generally (but not strictly) compose the urban issues "complex":

human – living – society – environment – governance – economy – resilience

Which in urban transition process outlines the focus of the policies aimed the SDGs (United Nations, n.d.a). The sooner its adoption by the municipalities, the better their adaptation to the SDGs Agenda2030 (United Nations, n.d.b) challenges.

Figure 5.
Source: European Commission, Directorate-General for Research and Innovation, Gronkiewicz-Waltz, Larsson, Boni (2020).

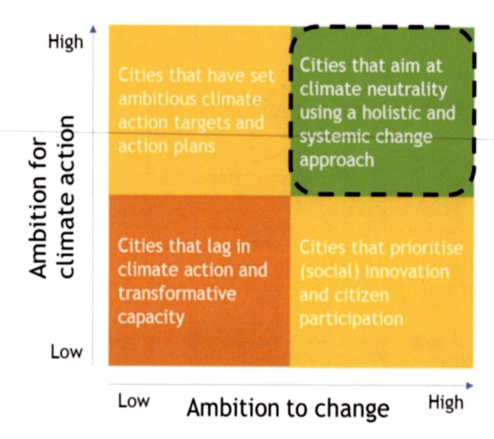

RACE OR CONCURRENCE?

During last decade, a newly-emerged, non-typical "race" is occurring:

Urban Transition needs **VERSUS** Municipal Transformation **ability**.

The first part of this relation became mandatory, due to the recent severe challenges regarding resilience that cities face, the main of which are the combination of the climate, energy and health crises. But not surprisingly, the main issue is the second part: the municipal capacity building is necessary in order for cities to acquire a critical tool:

Urban Transition **VIA** Municipal Transformation **capacity.**

The above mentioned relations create different level of evolution rate among cities, which is able determine their future. The European Commission, recognizes that good governance requires good administrative capacities. Thus, administrations need to be "*efficient and effective in implementing the policies and/or tasks entrusted*" to them. For this purpose, the European Commission has published a Practical Toolkit: Roadmaps for Administrative Capacity Building" (European Commission, n.d.) putting the skills in the core of any required transition. Moreover, as mentioned in the same document

…in the draft 2021-2027 regulations proposed by the Commission, include new provisions that place more emphasis on capacity building. Administrative capacity is also identified as key to effective imple-

Urban Transition via Municipal Transformation

Figure 6.
Source: European Commission, Directorate-General for Research and Innovation, Gronkiewicz-Waltz, Larsson, Boni (2020).

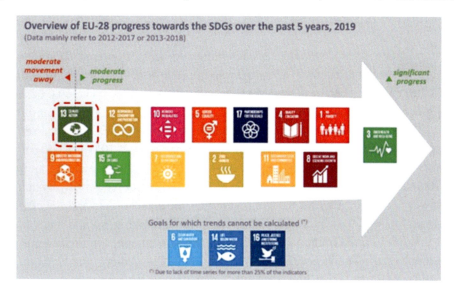

mentation of the funds, especially in the investment guidance of the 2019 European Semester country reports (Annex D). (European Commission, n.d.)

MUNICIPALITIES IN TRANSFORMATION

After the above mentioned, it is obvious that the municipalities have to be transformed from the traditional and strict role of the "administrators" as local authorities are, to the contemporary and wide role of:

Figure 7.
Source: European Commission, Directorate-General for Research and Innovation, Gronkiewicz-Waltz, Larsson, Boni (2020).

"Strategists" for Integrated Territorial Planning

Go Beyond Fragmented Planning

The human, natural and economic resources are restricted, while the challenges that local communities face became more intense. This fact don't allow the waste of time or resources, as well as the lack of focus to some specific targets which can be achieved via an integrated planning. The plan has to be produced by bottom-up procedures, feasible, financial balanced, with time milestones and a measurable impact. A representative example is the "100 climate neutral and smart cities by 2030" that requires the successful mobilization of investments and funds of many million euros (approx. 10.000 EUR per citizen), making use of different financing and funding instruments and tools and implementing approximately 100 projects by 2030. It is more than clear that such and ambitious endeavor requires well planned, integrated processes and a very good capacity for working with multiple stakeholders and the citizens ("Mission climate-neutral and Smart Cities: Info kit for cities now available. Research and innovation," 2021).

The impact that has to be examined is not only the "traditional" one, environmental and financial. New needs have emerged, like the health impact assessment. The society of a "healthy city" is much more rigorous and subsequently capable to achieve an integrated and long-term growth.

Prioritise and Create a Common Vision that Reflects the Local Needs, along with the National and EU Values and Directions

A necessary substance for a successful plan is the specific description of a vision. This gives to the stakeholders the sense of a common goal that has to be achieved. The vision and the main pillars of the local policies have to be adapted with the contemporary needs, the national planning and the EU values.

The above mentioned linkages have the potential to make the implementation of the planning more feasible and effective, more aligned with the general planning and much more eligible for the funding tools.

Coordinate Departments and Stakeholders, Integrate with all Levels of Government, Work in Collaboration and Networking with Other Organizations

The internal and external coordination and synchronization are critical factors that can determine the effectiveness of a territorial planning applying.

A long-term aimed, well-structured, balanced and relatively stable, sources-time feasible, inputs-outputs measurable and community-embraced plan is not only a condition for the emerging of the local dynamics and the eligibility for public funding, but also very attractive for private investments.

Additionally, in a world of much information, the networking can offer not only the opportunities for collaborations, but also the peer-learning via successful paradigms, as well as the proper dissemination channels for ideas and products.

"Enablers" by Generating the Urban-Human-Economic-Cultural-etc. Activity and Growth

Creation of more Social Cohesion, Equal Opportunities

Recent crises, pandemic and climate crisis, made a sometimes sub-estimated factor to be emerged: no one society is isolated, no one society can to achieve its resilience and sustainability without a focus to the social cohesion and equity.

Opportunities for Personal and Community Growth

Personal and community status have interactive bonds that have the potential to boost or prevent the local trajectory. Additionally, in a globalized economy, European countries have not the quantity, but mainly the quality factors as their strong assets.

These include Human Skills, Social Values, Culture of Innovation, Cultural Heritage. Contrary to the massive production and consumption with the disruptive consequences to the global human and environmental health, European municipalities have to prioritize local values, anchored with the territory, as the main pillars of their policies.

Affordable Housing, Transportation, Energy Independence, Culture, etc.

The New European Bauhaus (Von der Leyen, n.d.) initiative calls people to imagine and build together a sustainable and inclusive future that is beautiful for our eyes, minds, and souls. Beautiful are the places, practices, and experiences that are:

- Enriching, inspired by art and culture, responding to needs beyond functionality.
- Sustainable, in harmony with nature, the environment, and our planet.
- Inclusive, encouraging a dialogue across cultures, disciplines, genders and ages.

It is about improving our daily lives, focusing on better living together in more beautiful, sustainable and inclusive places. It is about bridging global challenges with local solutions to achieve our climate targets and support a broader transformation on the ground. A triangle of three inseparable values guides it ("Welcome to the new European Bauhaus prizes 2022", 2022).

- sustainability, from climate goals, to circularity, zero pollution, and biodiversity,
- aesthetics, quality of experience and style, beyond functionality,
- inclusion, valorizing diversity, equality for all, accessibility and affordability.

Support Local Businesses and Entrepreneurship: Provide Services/Coaching/Incentives

The global competition and unstable conditions, like the energy crises due to the wars, make more necessary than ever the most entrepreneurship-friendly environment that a local authority could provide and ensure. This includes as a basis a fair tax-system, but mainly a complex of strong incentives for research, entrepreneurship and innovation.

"Servants" by Providing Public Services and Quality-of-life Features and Amenities

Harness Digitalization, Smart Cities, Healthy Cities

Smart cities policies, followed the integrated approach as it mentioned in the 3nd sub-chapter, have the potential to boost local communities to their sustainable growth. But there is a point that it must be noted: no one growth would be sustainable by omitting the other mentioned above factors, as the social cohesion.

As mentioned also on the 3nd sub-chapter, Healthy Cities is a perquisite for rigorous communities, able to produce and interactive effectively. Subsequently, the quality-of-life features of the urban environment as well as the provided amenities, are not some additional assets, but the main indexes of the local prosperity.

The elaboration of the smart and healthy cities policies, taking on consideration that they have their basic values in common, has the potential to offer the most achievable attractive and creative environment for permanent residence and professional activity.

Local Democracy, Inclusive Governance

Local democracy seems to be a simple notion but is not. Due to the quantity of the stakeholders and their different features, local democracy has to include a simultaneously broad and specific frame of local action. Last years these efforts are easier due to the digital tools expansion.

Further the "traditional" bottom-up procedures, like the co-creative plans and public consultation regarding measures, new, more direct forms of local democracy have to start be emerged, by using the new tools, like the participatory budgeting or investing.

Crowd funding for instance (with other words Citizens Financing), is a relatively new financial tool that could be very powerful, aiming the local growth.

"Investors" Considering the Public Investments, Infrastructures and Soft Projects

The last reference to the prior paragraph, emerges one of the most important municipal features that indicates a successful municipality.

Combine Investments, Inclusion of Nature-based Solutions, be Familiar with the New Era - Innovative Funding Tools

As referred above, financial sources are restricted. On the other side, the need for public investments is more necessary than ever, in order to create the proper conditions for a competitive territory. The "cure" for this fact could be the applying of some new financial instruments, like the "Green Bonds", Public-private sector partnerships, etc. Once more, this aspect is pivotal in the "100 climate neutral and smart cities" mission, that clearly refers to "combination of multiple funding sources" in order to achieve efficiency gains by exploiting synergies and the mobilization of a wider range of actors and resources. Some examples of funding sources integration are the ESI (European Structural and Investment) funds, with domestic funds, the involvement of private investors and the third sector ("Mission climate-neutral and Smart Cities: Info kit for cities now available. Research and innovation", 2021).

Access to European Funds

Beyond the national funding programmes, there is a variety of European ones. They follow the European declared policies and they can offer some excellent financing opportunities, regarding e.g. networking, culture, social values, entrepreneurship, climate action and innovation.

Horizon2020, Erasmus+, CERV, LIFE, ("Find calls for proposals and tenders", n.d.) URBACT III (About us, n.d.)., IURC (About IURC, n.d.), are only some of them. Unfortunately, only a minority of the municipalities can manage to be an active part of these. The problem is more intense regarding small municipalities, especially those in isolated, rural areas.

Create a Productive Internal Environment

The municipal politicians, executives and employees have to adopt a completely different approach regarding their role or job respectively. A just internal regulation, a well-structured framework, a collaborative work-flow procedure, a high-skilled personnel and a collective policy-applying could make a municipality as effective as a contemporary company is.

Unique Ability to Develop Economies of Scale

ADDITIONAL TIPS FOR MUNICIPALITIES

Deal with future uncertainties / Future proofing municipalities

- Plan for more extreme climate change scenarios, prepare the population, the businesses adapt, the city infrastructure

    Keep up with innovation and be adaptable/flexible

- Provide training for new required skills: accessing communities (especially marginalised), deliver co-creation, define new ways of governance, access/develop new business models
- Human-centred innovation
- Invest in digitalisation for better and more efficient services
- Adaptability to "market": translating in adaptability to crisis/health/environment

Figure 8.
Source: European Commission, Roadmaps for Administrative Capacity Building [15]

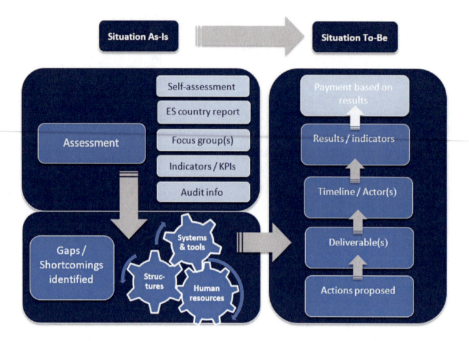

CLOSING REMARKS

Fostering innovation, creating more agile and smarter municipalities will stimulate local economies, increase employment opportunities and support future sustainability.

Municipal governments could rely less on the upper tier governments and show leadership to deal with their local realities and create better places to live, work and create social and economic growth.

REFERENCES

AboutIURC. (n.d.). Retrieved from https://www.iurc.eu/about

About us. (n.d.). Urbact. Retrieved from https://urbact.eu/who-we-are

Compendium of tools, resources and networks, Support package for implementation. (2020). WHO Europe.

Durieux, E., & Hidson, M. (2021). *Local green deals: a blueprint for action: The European Commission's 100 intelligent cities challenge*. Publications Office. https://data.europa.eu/doi/10.2826/94389

Duriex, E., & Hidson, M. (2021). *Local Green Deals: A Blueprint for Action*. Retrieved from https://www.intelligentcitieschallenge.eu/sites/default/files/2021-06/Local%20Green%20Deals-8.pdf

European Commission. (n.d.). *Roadmaps for Administrative Capacity Building*. Retrieved from https://ec.europa.eu/regional_policy/sources/policy/how/improving-investment/roadmap_toolkit.pdf

European green deal. (n.d.). Retrieved from https://ec.europa.eu/info/strategy/priorities-2019-2024/european-green-deal_en

European Union. European Commission. Directorate-General for Regional Policy. (2011). *Cities of tomorrow: Challenges, visions, ways forward.* Publications Office of the European Union.

Find calls for proposals and tenders. (n.d.). European Commission. Retrieved from https://ec.europa.eu/info/funding-tenders/opportunities/portal/screen/home

Gronkiewicz-Waltz, H., Larsson, A., & Boni, A. (2020). *100 climate-neutral cities by 2030 - by and for the citizens: report of the mission board for climate-neutral and smart cities.* Publications Office. https://data.europa.eu/doi/10.2777/46063

Mission climate-neutral and Smart Cities: Info kit for cities now available. Research and innovation. (2021, October 29). Retrieved from https://research-and-innovation.ec.europa.eu/news/all-research-and-innovation-news/mission-climate-neutral-and-smart-cities-info-kit-cities-now-available-2021-10-29_en

Plan, A. (2011). *Communication from the Commission to the European Parliament, the Council, the European Economic and Social Committee and the Committee of the Regions.* European Commission.

Race to Zero Campaign. (n.d.). Retrieved from https://unfccc.int/climate-action/race-to-zero-campaign

Recovery and resilience facility. (2022, July 29). Retrieved from https://ec.europa.eu/info/business-economy-euro/recovery-coronavirus/recovery-and-resilience-facility_en

Smart Cities Community. (n.d.). Institute of Electrical and Electronics Engineers (IEEE). Retrieved from https://www.ieee.org/membership-catalog/productdetail/showProductDetailPage.html?product=CMYSC764

Smart sustainable cities: An analysis of definitions. (2014). ITU-T Focus Group on Smart Sustainable Cities.

The urban agenda for the EU. (n.d.). Retrieved from https://ec.europa.eu/regional_policy/en/policy/themes/urban-development/agenda/#:~:text=The%20Urban%20Agenda%20for%20the%20EU%20is%20an%20integrated%20and,of%20life%20in%20urban%20areas

United Nations. (n.d.a). *The 17 goals | sustainable development.* United Nations. Retrieved from https://sdgs.un.org/goals

United Nations. (n.d.b). *Transforming our world: the 2030 Agenda for Sustainable Development.* Retrieved from https://sdgs.un.org/2030agenda

Von der Leyen, U. (n.d.). *Beautiful, sustainable, together.* New European Bauhaus. Retrieved from https://new-european-bauhaus.europa.eu/index_en

Welcome to the new European Bauhaus prizes 2022. (2022). Retrieved from http://prizes.new-european-bauhaus.eu/home

Compilation of References

Abbaszadeh, M. A., Ghourichaei, M. J., & Mohammadkhani, F. (2020). Thermo-economic feasibility of a hybrid wind turbine/PV/gas generator energy system for application in a residential complex in Tehran, Iran. *Environmental Progress & Sustainable Energy*, *39*(4), 1–12. doi:10.1002/ep.13396

About Citizen Cyberlab . (n.d.). Retrieved February 8, 2022, from https://www.citizencyberlab.org/about

About Journey North . (n.d.). Retrieved February 8, 2022, from https://journeynorth.org/about-journey-north

About Mada Tech . (n.d.). Retrieved February 8, 2022, from https://www.madatech.org.il/ en/about

About TURSEFF (2022). Retrieved from https://www.turseff.org/sayfa/facility

About us. (n.d.). Urbact. Retrieved from https://urbact.eu/who-we-are

AboutIURC. (n.d.). Retrieved from https://www.iurc.eu/about

Adetayo, A. J., & Williams-Ilemobola, O. (2021). Librarians' generation and social media adoption in selected academic libraries in Southwestern, Nigeria. *Library Philosophy and Practice (e-Journal), 4984*. https://digitalcommons.unl.edu/libphilprac/4984

Adetayo, A. J. (2021a). Fake News and Social Media Censorship. In R. J. Blankenship (Ed.), *Deep Fakes, Fake News, and Misinformation in Online Teaching and Learning Technologies*. IGI Global. doi:10.4018/978-1-7998-6474-5.ch004

Adetayo, A. J. (2021b). Leveraging Bring Your Own Device for Mobility of Library Reference Services: The Nigerian Perspective. *The Reference Librarian*, *62*(2), 106–125. doi:10.1080/02763877.2021.1936342

Adger, W. N., Huq, S., Brown, K., Declan, C., & Mike, H. (2016). Adaptation to climate change in the developing world. *Progress in Development Studies*, *3*(3), 179–195. doi:10.1191/1464993403ps060oa

Adimassu, Z., & Kessler, A. (2016). Factors affecting farmers' coping and adaptation strategies to perceived trends of declining rainfall and crop productivity in the central Rift valley of Ethiopia. *Environmental Systems Research*, *5*(1), 1–16. doi:10.118640068-016-0065-2

Aghaei, J., Nezhad, A. E., Rabiee, A., & Rahimi, E. (2016). Contribution of plug-in hybrid electric vehicles in power system uncertainty management. *Renewable & Sustainable Energy Reviews*, *59*, 450–458. doi:10.1016/j.rser.2015.12.207

Ahmad, N. N. N., & Hossain, D. M. (2015). Climate Change and Global Warming Discourses and Disclosures in the Corporate Annual Reports: A Study on the Malaysian Companies. *Procedia: Social and Behavioral Sciences*, *172*, 246–253. doi:10.1016/j.sbspro.2015.01.361

Aien, M., Hajebrahimi, A., & Fotuhi-Firuzabad, M. (2016). A comprehensive review on uncertainty modeling techniques in power system studies. *Renewable & Sustainable Energy Reviews*, *57*, 1077–1089. doi:10.1016/j.rser.2015.12.070

Compilation of References

Aigrain, P., Brugière, F., Duchêne, E., de Cortazar-Atauri, I. G., Gautier, J., Giraud-Héraud, E., & Touzard, J. M. (2016). Work of prospective on the adaptation of the viticulture to climate change: Which series of events could support various adaptation strategies? [Travaux de prospective sur l'adaptation de la viticulture au changement climatique: quelles séries d'événements pourraient favoriser différentes stratégies d'adaptation?]. In *BIO Web of Conferences* (Vol. 7). EDP Sciences.

Aldana, M., Braly-Cartillier, I., & Shuford, L. (2014). *Guarantees for Green Markets Potential and Challenges. Inter-American Development Bank*. IDB.

Aledo-Ruiz, M. D., Ortega-Gutiérrez, J., Martinez-Caro, E., & Cegarra-Navaroo, J. G. (2017). Linking an unlearning context with firm performance through human capital. *European Research on Management and Business Economics*, *23*(1), 16–22. doi:10.1016/j.iedeen.2016.07.001

Algeria's INDC-UNFCCC. (2015). *Algeria's Intended Nationally Determined Contribution (INDC) to Achieve the Objectives of the United Nations Framework Convention on Climate Change (UNFCCC)*. Available online: https://www4.unfccc.int/sites/submissions/indc/Submission%20Pages/submissions.aspx

Algerian Ministry of Energy. (2019). *New Energies, Renewables and Energy Management* [Energies Nouvelles, Renouvelables et Maitrise de l'Energie.]. Algerian Ministry of Energy.

Al-Hinai, H., Batty, W. J., & Probert, S. D. (1993). Vernacular architecture of Oman: Features that enhance thermal comfort achieved within buildings. *Applied Energy*, *44*(3), 233–258. doi:10.1016/0306-2619(93)90019-L

Almadhor, A. (2019, January). Intelligent control mechanism in smart micro grid with mesh networks and virtual power plant model. In 2019 16th IEEE Annual Consumer Communications & Networking Conference (CCNC) (pp. 1-6). IEEE. doi:10.1109/CCNC.2019.8651822

Alrashed, F., Asif, M., & Burek, S. (2017). The Role of Vernacular Construction Techniques and Materials for Developing Zero-Energy Homes in Various Desert Climates. *Buildings*, *7*(1), 17. Advance online publication. doi:10.3390/buildings7010017

Amerasinghe, N., Thwaites, J., Larsen, G., & Ballesteros, A. (2017). *The Future of the Funds: Exploring the Architecture of Multilateral Climate Finance*. World Resources Institute.

Ananiev, B. G. (1977). *On the problems of modern human science*. Science.

Angiz, L. M. Z., Emrouznejad, A., & Mustafa, A. (2010). Fuzzy assessment of performance of a decision making-units using DEA: Non-Radial approach. *Expert Systems with Applications*, *37*(7), 5153–5157. doi:10.1016/j.eswa.2009.12.078

Annunziata, E., Pucci, T., Frey, M., & Zanni, L. (2018). The role of organizational capabilities in attaining corporate sustainability practices and economic performance: Evidence from Italian wine industry. *Journal of Cleaner Production*, *171*, 1300–1311. doi:10.1016/j.jclepro.2017.10.035

Arlt, D., Hoppe, I., & Wolling, J. (2011). Climate change and media usage: Effects on problem awareness and behavioural intentions. *The International Communication Gazette*, *73*(1), 45–63. doi:10.1177/1748048510386741

Arnell, N. W. (2004). Climate change and global water resources: SRES emissions and socio-economic scenarios. *Global Environmental Change*, *14*(1), 31–52. doi:10.1016/j.gloenvcha.2003.10.006

Arnell, N. W., Lowe, J. A., Brown, S., Gosling, S. N., Gottschalk, P., Hinkel, J., Lioyd-Hughes, B., Nicholls, R. J., Osborn, T. J., Osborne, T. M., Rose, G. A., Smith, P., & Warren, R. F. (2013). A global assessment of the effects of climate policy on the impacts of climate change. *Nature Climate Change*, *3*(5), 512–519. doi:10.1038/nclimate1793

Arnol'd, V. I. (1975). Critical points of smooth functions and their normal forms. *Russian Mathematical Surveys*, *30*(5), 1–75. doi:10.1070/RM1975v030n05ABEH001521

Arnol'd, V. I. (2003). *Catastrophe theory*. Springer Science & Business Media.

Arrow, K. J. (2012). *Social choice and individual values* (Vol. 12). Yale University Press.

Arslan, O., & Karasan, O. E. (2013). Cost and emission impacts of virtual power plant formation in plug-in hybrid electric vehicle penetrated networks. *Energy*, *60*, 116–124. doi:10.1016/j.energy.2013.08.039

Arthur, B., Durlauf, S., & Lane, D. (1997). *The Economy as an Evolving Complex System II*. Addison-Wesley.

Arthur, W. B. (2009). *The nature of technology: What it is and how it evolves*. Simon and Schuster.

Arthur, W. B. (2014). *Complexity and the Economy*. Oxford University Press.

Asadu, A. N., Ozioko, R. I., & Dimelu, M. U. (2018). Climate Change Information Source and Indigenous Adaptation Strategies of Cucumber Farmers in Enugu State, Nigeria. *Journal of Agricultural Extension*, *22*(2), 136–146. doi:10.4314/jae.v22i2.12

Asamblea General de las Naciones Unidas. (2015). Transforming our world: the 2030 Agenda for sustainable development. Resolution 15–16301, United Nations.

Asiaei, K., Bontis, N., Alizadeh, R., & Yaghoubi, M. (2021). Green intellectual capital and environmental management accounting: Natural resource orchestration in favor of environmental performance. *Business Strategy and the Environment*, *31*(1), 76–93. doi:10.1002/bse.2875

Asmus, P. (2010). Microgrids, virtual power plants and our distributed energy future. *The Electricity Journal*, *23*(10), 72–82. doi:10.1016/j.tej.2010.11.001

Bahramara, S., Moghaddam, M. P., & Haghifam, M. R. (2016). Optimal planning of hybrid renewable energy systems using HOMER: A review. *Renewable & Sustainable Energy Reviews*, *62*, 609–620. doi:10.1016/j.rser.2016.05.039

Ballerini, L., & Bergh, S. I. (2021). Using citizen science data to monitor the Sustainable Development Goals: A bottom-up analysis. *Sustainability Science*, *16*(6), 1945–1962. doi:10.100711625-021-01001-1 PMID:34316319

Balta Portoles, J., & Dragicevic Sesic, M. (2017). Cultural rights and their contribution to sustainable development: Implications for cultural policy. *International Journal of Cultural Policy*, *23*(2), 159–173. doi:10.1080/10286632.2017.1280787

Bandura, A. (1977). Self-efficacy: Toward a unifying theory of behavioral change. *Psychological Review*, *84*(2), 191–215. doi:10.1037/0033-295X.84.2.191 PMID:847061

Bardon, A. (2020, September 9). *Faith and politics mix to drive evangelical Christians' climate change denial*. https://theconversation.com/faith-and-politics-mix-to-drive-evangelical-christians-climate-change-denial-143145

Barnett, J., & Webber, M. (2010). *Accommodating migration to promote adaptation to climate change* (Policy research working paper No. 5270). The World Bank.

Basak, P., Chowdhury, S., Halder nee Dey, S., & Chowdhury, S. P. (2012). A literature review on integration of distributed energy resources in the perspective of control, protection and stability of microgrid. *Renewable & Sustainable Energy Reviews*, *16*(8), 5545–5556. doi:10.1016/j.rser.2012.05.043

Bayraktaroglu, A., Calisir, F., & Baskak, M. (2019). Intellectual capital and firm performance: An extended VAIC model. *Journal of Intellectual Capital*, *20*(3), 406–425. doi:10.1108/JIC-12-2017-0184

BDDK. (2021). *2022-2025 Sustainable Banking Strategic Plan* [2022-2025 Sürdürülebilir Bankacılık Stratejik Planı]. Retrieved January 15, 2022, from https://www.bddk.org.tr/KurumHakkinda/EkGetir/5?ekId=36

Compilation of References

Becker, G. S. (1976). *The economic approach to human behavior* (Vol. 803). University of Chicago Press. doi:10.7208/chicago/9780226217062.001.0001

Behera, S., Sahoo, S., & Pati, B. B. (2015). A review on optimization algorithms and application to wind energy integration to grid. *Renewable & Sustainable Energy Reviews*, *48*, 214–227. doi:10.1016/j.rser.2015.03.066

Ben Ncir, C.-E., Hamza, A., & Bouaguel, W. (2021). Parallel and scalable Dunn index for the validation of big data clusters. *Parallel Computing*, *102*, 1–10. doi:10.1016/j.parco.2021.102751

Bennett, W. L., & Iyengar, S. (2008). A New Era of Minimal Effects? the Changing Foundations of Political Communication. *Journal of Communication*, *58*(4), 707–731. doi:10.1111/j.1460-2466.2008.00410.x

Berdyaev, N. A. (1952). *My Philosophic World-Outlook*. Retrieved from: http://www.berdyaev.com/berdiaev/berd_lib/1952_476.html

Berger, P. L., & Luckmann, T. (1966). *The Social Construction of Reality: a treatise on sociology of knowledge*. Penguin Book.

Berinsky, A. J. (2017). Rumors and Health Care Reform: Experiments in Political Misinformation. *British Journal of Political Science*, *47*(2), 241–262. doi:10.1017/S0007123415000186

Berrou, R., Dessertine, P., & Migliorelli, M. (2019). An overview of green finance. *The rise of green finance in Europe*, 3-29.

Berrou, R., Ciampoli, N., & Marini, V. (2019). Defining green finance: Existing standards and main challenges. In *The Rise of Green Finance in Europe* (pp. 31–51). Palgrave Macmillan. doi:10.1007/978-3-030-22510-0_2

Bertalanffy, L. V. (1968). *General system theory: Foundations, development, applications*. G. Braziller.

Bhattacharya, A., & Stern, N. (2021). Our Last, Best Chance on. *Finance & Development*.

Billimoria, J., & Bishop, M. (2020). How COVID-19 can be the Great Reset toward global sustainability. In *Sustainable Development Impact Summit*. World Economic Forum.

Bini, B. S., & Mathew, T. (2016). Clustering and regression techniques for stock prediction. *Procedia Technology*, *24*, 1248–1255. doi:10.1016/j.protcy.2016.05.104

Blewitt, J. (Ed.). (2015). Understanding sustainable development (3rd ed.). Routledge (Taylor & Francis Group).

Boafo, Y. A., Saito, O., Jasaw, G. S., Otsuki, K., & Takeuchi, K. (2016). Provisioning ecosystem services-sharing as a coping and adaptation strategy among rural communities in Ghana's semi-arid ecosystem. *Ecosystem Services*, *19*, 92–102. doi:10.1016/j.ecoser.2016.05.002

Bodach, S., Lang, W., & Hamhaber, J. (2014). Climate responsive building design strategies of vernacular architecture in Nepal. *Energy and Building*, *81*, 227–242. doi:10.1016/j.enbuild.2014.06.022

Bogdanov, A. A. (1934). *Tectology or General Organizational Science*. Conjuncture Institute publishing house.

Boiral, O. (2006). Global Warming: Should Companies Adopt a Proactive Strategy? *Long Rabge Planning. International Journal of Strategic Management*, *39*(2), 315–330.

Boiral, O., Heras-Saizarbitoria, I., & Brotherton, M. (2019). Corporate sustainability and indigenous community engagement in the extractive industry. *Journal of Cleaner Production*, *235*, 701–711. doi:10.1016/j.jclepro.2019.06.311

Bonsu, N. O. (2020). Towards a circular and low-carbon economy: Insights from the transitioning to electric vehicles and net zero economy. *Journal of Cleaner Production*, *256*, 120659. Advance online publication. doi:10.1016/j.jclepro.2020.120659

Boons, F., Montalvo, C., Quist, J., & Wagner, M. (2013). Sustainable innovation, business models and economic performance: An overview. *Journal of Cleaner Production*, *45*, 1–8. doi:10.1016/j.jclepro.2012.08.013

Boston Consulting Group. (2021). *Use AI to Measure Emissions Exhaustively, Accurately, and Frequently Carbon Measurement*. Survey Report 2021. BCG.

Bouzid, A. M., Guerrero, J. M., Cheriti, A., Bouhamida, M., Sicard, P., & Benghanem, M. (2015). A survey on control of electric power distributed generation systems for microgrid applications. *Renewable & Sustainable Energy Reviews*, *44*, 751–766. doi:10.1016/j.rser.2015.01.016

Bouznit, M., Pablo-Romero, M. D. P., & Sánchez-Braza, A. (2020). Measures to promote renewable energy for electricity generation in Algeria. *Sustainability*, *12*(4), 1468. doi:10.3390u12041468

Breene, K. (2016). *Why does inclusive growth matter?* The World Economic Forum. Retrieved from https://www.weforum.org/agenda/2016/01/why-does-inclusive-growth-matter/

Broto, V. C., & Bulkeley, H. (2013). A survey of urban climate change experiments in 100 cities. *Global Environmental Change*, *23*(1), 92–102. doi:10.1016/j.gloenvcha.2012.07.005 PMID:23805029

Brown, R., & Maudlin, D. (2012). Concepts of vernacular architecture. In The SAGE Handbook of Architectural Theory (pp. 340–368). doi:10.4135/9781446201756.n21

Brubaker, M., Berner, J., & Tcheripanoff, M. (2013). LEO, the Local Environmental Observer Network: A community-based system for surveillance of climate, environment, and health events. *International Journal of Circumpolar Health*, *2013*(72), 22447. doi:10.3402/ich.v72i0.22447

Brundtland Commission. (1987). *WCED (Report of the world commission on environment and development: our common future)*. Oxford University Press.

Buchholz, B. M., Brunner, C., Naumann, A., & Styczynski, A. (2012, July). *Applying IEC standards for communication and data management as the backbone of smart distribution*. In 2012 IEEE Power and Energy Society General Meeting. IEEE.

Bulkely, H., & Newell, P. (2015). *Governing Climate Change*. Routledge. doi:10.4324/9781315758237

Burton, I. (2011). Adaptation to climate change: Context, status, and prospects. In J. D. Ford & L. Berrang-Ford (Eds.), *Climate Change Adaptation in Developed Nations: From theory to practice*. Springer. doi:10.1007/978-94-007-0567-8_35

Caldon, R., Patria, A. R., & Turri, R. (2004, September). Optimisation algorithm for a virtual power plant operation. In *39th International Universities Power Engineering Conference, 2004. UPEC 2004* (Vol. 3, pp. 1058-1062). IEEE.

Campion, M. A., Mumford, T. V., Morgeson, F. K., & Nahrgang, J. D. (2005). Work redesign: Eight obstacles and opportunities. *Human Resource Management*, *44*(4), 367–390. doi:10.1002/hrm.20080

Candelon, B., Carare, A., & Miao, K. (2016). Revisiting the new normal hypothesis. *Journal of International Money and Finance*, *66*(C), 5–31. doi:10.1016/j.jimonfin.2015.12.005

Cantrell, K. B., Ducey, T., Ro, S. K., & Hunt, P. G. (2008). Livestock Waste-to-Bioenergy Generation Opportunities. *Bioresource Technology*, *99*(17), 7941–7953. doi:10.1016/j.biortech.2008.02.061 PMID:18485701

Compilation of References

Cardinale, N., Rospi, G., & Stefanizzi, P. (2013). Energy and microclimatic performance of Mediterranean vernacular buildings: The Sassi district of Matera and the Trulli district of Alberobello. *Building and Environment*, *59*, 590–598. doi:10.1016/j.buildenv.2012.10.006

Carvalho, A. (2010). Media(ted)discourses and climate change: A focus on political subjectivity and (dis)engagement. *Wiley Interdisciplinary Reviews: Climate Change*, *1*(2), 172–179. doi:10.1002/wcc.13

Castells-Quintana, D., Lopez-Uribe, M. P., & McDermott, T. K. J. (2018). Adaptation to climate change: A review through a development economics lens. *World Development*, *104*, 183–196. doi:10.1016/j.worlddev.2017.11.016

Castillo-Gimenez, J., Montanes, A., & Picazo-Tadeo, A. J. (2019). Performance in the treatment of municipal waste: Are European Union Member States so different? *The Science of the Total Environment*, *687*, 1305–1314. doi:10.1016/j.scitotenv.2019.06.016 PMID:31412464

CCA (Caribbean Conservation Association). (1991). *Dominica Country Environmental Profile*. http://www.irf.org/wp-content/uploads/2015/10/DominicaEnvironmentalProfile.pdf

CDA (Claude Davis and Associates). (2010). *Greenhouse Gas Mitigation Assessment for Dominica, Final Report*. https://unfccc.int/sites/default/files/dominica_mitigation_assessment_final_report%5B1%5D.pdf

CE (Country Economy). (2021). *Dominica National Debt*. https://countryeconomy.com/national-debt/dominica

Cegarra-Navarro, J. G., Wensley, A. K. P., & Sánchez Polo, M. T. (2014). A conceptual framework for unlearning in a homecare setting. *Knowledge Management Research and Practice*, *12*(4), 375–386. doi:10.1057/kmrp.2013.6

CEN/CENELEC/ETSI Joint Working Group. (2011). *Standards for Smart Grids, Final Report*. Author.

Cengiz, Ç., & Kutlu, E. (2021). The budget of the European Union and green energy: Perceptions, actions and challenges. *International Journal of Social Inquiry*, *14*(1), 171–197. doi:10.37093/ijsi.950563

CEREFE (2020). *Energy Transition in Algeria: Lessons, State of Play and Prospects for accelerated development of Renewable Energies, (2020 Edition)* [Transition Energétique en Algérie: Leçons, Etat des Lieux et Perspectives pour un Développement Accéléré des Energies Renouvelables, (Edition 2020)]. Commissariat aux Energies Renouvelables et à l'Efficacité Energétique, Premier Ministre, Alger.

CFR (Council on Foreign Relations). (2013). *The Global Climate Change Regime*. https://www.cfr.org/report/global-climate-change-regime

Chang, C. H., & Chen, Y. (2012). The determinants of green intellectual capital. *Management Decision*, *50*(1), 74–94. doi:10.1108/00251741211194886

Chang, W. Y. (2014). A literature review of wind forecasting methods. *Journal of Power and Energy Engineering*, *2*(04), 161–168. doi:10.4236/jpee.2014.24023

Charles, D. (2019). Untapping the Potential of the Green Climate Fund in Transforming the Caribbean Community Member States' Renewable Energy Nationally Determined Contributions into Action. In *Towards Climate Action in the Caribbean Community*. Cambridge Scholars Publishing.

Charnes, A., Cooper, W. W., Lewin, A. Y., & Seiford, L. M. (Eds.). (1994). *Data envelopment analysis: theory, methodology and application*. Kluwer Academic Publisher. doi:10.1007/978-94-011-0637-5

Charnes, A., Cooper, W. W., & Rhodes, E. (1978). Measuring the efficiency of decision making units. *European Journal of Operational Research*, *2*(6), 429–444. doi:10.1016/0377-2217(78)90138-8

Chechulin, V. L. (2012). *Set Theory with Self-affiliation (foundation and some applications)* [Теория множеств с самопринадлежностью (основы и некоторые приложения] (2nd ed.). Perm State University.

Chen, D., Yin, L., Wang, H., & He, P. (2014). Pyrolysis Technologies for Municipal Solid Waste: A Review. *Waste Management (New York, N.Y.)*, *34*(12), 2466–2486. doi:10.1016/j.wasman.2014.08.004 PMID:25256662

Cheng, L., Zhou, X., Yun, Q., Tian, L., Wang, X., & Liu, Z. (2019, November). A review on virtual power plants interactive resource characteristics and scheduling optimization. In *2019 IEEE 3rd Conference on Energy Internet and Energy System Integration (EI2)* (pp. 514-519). IEEE. 10.1109/EI247390.2019.9061780

Chen, X., Huang, L., Liu, J., Song, D., & Yang, S. (2022). Peak shaving benefit assessment considering the joint operation of nuclear and battery energy storage power stations: Hainan case study. *Energy*, *239*, 121897. Advance online publication. doi:10.1016/j.energy.2021.121897

Chen, Y. (2008). The positive effect of green intellectual capital on competitive advantages of firms. *Journal of Business Ethics*, *77*(3), 271–286. doi:10.100710551-006-9349-1

Chen, Y. (2008a). The driver of green innovation and green image–green core competence. *Journal of Business Ethics*, *81*(3), 531–543. doi:10.100710551-007-9522-1

Chevalier, J. M., & Buckles, D. J. (2009). *SAS2 Guide to Collaborative Research and Social Mobilization* [SAS2 Guía para la Investigación Colaborativa y la Movilización Social]. Plaza y Valdés.

Chiou, Y.-S., & Elizalde, J. S. (2019). Thermal Performances of Three Old Houses: A Comparative Study of Heterogeneous Vernacular Traditions in Taiwan. *Sustainability*, *11*(19), 5538. Advance online publication. doi:10.3390u11195538

Chorafas, D. N. (2011). *Sovereign debt crisis: The new normal and the newly poor*. Palgrave Macmillan. doi:10.1057/9780230307124

Chun, C., Kwok, A., & Tamura, A. (2004). Thermal comfort in transitional spaces—basic concepts: Literature review and trial measurement. *Building and Environment*, *39*(10), 1187–1192. doi:10.1016/j.buildenv.2004.02.003

Chung, H. M., & Gray, P. (1999). Special section: Data mining. *Journal of Management Information Systems*, *16*(1), 11–16. doi:10.1080/07421222.1999.11518231

Clarke, A. C. (1954). *Childhood's End*. Sidgwick & Jackson Ltd.

Clémençon, R. (2016). The Two Sides of the Paris Climate Agreement: Dismal Failure or Historic Breakthrough? *Journal of Environment & Development*, *25*(1), 3–24. doi:10.1177/1070496516631362

CMBT. (2021). *CMB Green Debt Instrument and Green Lease Certificate Guide* [SPK Yeşil Borçlanma Aracı ve Yeşil Kira Sertifikası Rehberi]. Retrieved January 10, 2022, from https://www.spk.gov.tr/Sayfa/Dosya/1350

Cobb, M. D., Nyhan, B., & Reifler, J. (2013). Beliefs Don't Always Persevere: How Political Figures Are Punished When Positive Information about Them Is Discredited. *Political Psychology*, *34*(3), 307–326. doi:10.1111/j.1467-9221.2012.00935.x

Coch, H. (1998). Chapter 4—Bioclimatism in vernacular architecture. *Renewable & Sustainable Energy Reviews*, *2*(1), 67–87. doi:10.1016/S1364-0321(98)00012-4

Collins, G. (2017). *Iran's looming water bankruptcy*. Center for Energy Studies, Rice University's Baker Institute for Public Policy.

Compendium of tools, resources and networks, Support package for implementation. (2020). WHO Europe.

Coombs, W. T. (2015). *Ongoing Crisis Communication* (4th ed.). Sage.

Coombs, W. T., & Laufer, D. (2017). Global crisis management-current research and future directions. *Journal of International Management*, *24*(3), 199–203. doi:10.1016/j.intman.2017.12.003

CPP (Check Petrol Price). (2021). *Welcome to CheckPetrolPrice.com*. http://www.checkpetrolprice.com/global-gasbuddy/Diesel-price-in-Dominica.php

Crompton, J. L., Lee, S., & Shuster, T. J. (2001). A Guide for Undertaking Economic Impact Studies: The Springfest Example. *Journal of Travel Research*, *40*(1), 79–87. doi:10.1177/004728750104000110

CSIRO & Bureau of Meteorology. (2015). *Climate Change in Australia: Information for Australia's Natural Resource Management Regions*. Technical Report, CSIRO and Bureau of Meteorology.

Dai, A. (2013). Increasing drought under global warming in observations and models. *Nature Climate Change*, *3*(1), 52–58. Advance online publication. doi:10.1038/nclimate1633

Dal Mas, F., & Paoloni, P. (2019). A relational capital perspective on social sustainability; the case of female entrepreneurship in Italy. *Measuring Business Excellence*, *24*(1), 114–130. doi:10.1108/MBE-08-2019-0086

Dasgupta, P. (2021a, February). *The Economics of Biodiversity: The Dasgupta Review*. London: HM Treasury. Retrieved from: www.gov.uk/official-documents

Dasgupta, P. (2021b). Economics Nature's Way. Good economics demands that we manage Nature better. *Finance & Development*, *58*(3).

Dasgupta, P. (2004). *Human Well-Being and the Natural Environment*. Oxford University Press.

Dasgupta, P. (2019). *Time and the Generations: Population Ethics for a Diminishing Planet*. Columbia University Press. doi:10.7312/dasg16012

Davenport, T., & Prusak, L. (1998). *Working knowledge: How organizations manage what they know*. Harvard Business Press.

Davidson, D. (2016). Gaps in agricultural climate adaptation research. *Nature Climate Change*, *6*(5), 433–435. doi:10.1038/nclimate3007

Davis, A., & Howard, E. (2005). Spring recolonization rate of monarch butterflies in eastern North America, new estimates from citizen-science data. *Journal of the Lepidopterists Society*, *59*(1), 1–5.

de Chardin, P. T. (1987). *Phenomenon of Human Being* [Феномен человека]. Science Publishers.

De Cuba, K., Burgos, F., Contreras-Lisperguer, R., & Penny, R. (2008). *Limits and Potential of Waste-To-Energy Systems in the Caribbean*. Organization of American States.

De la Vega-Salazar, M. Y. (2011). Possible risk factors for Mexico's epicontinental fish in the face of global warming [Posibles factores de riesgo de los peces epicontinentales de México ante el calentamiento global]. *Avances en Investigación Agropecuaria, México*, *15*(2), 65–78.

Delgado-Verde, M., Amores-Salvadó, J., Martín-de Castro, G., & Navas-López, J. (2014). Green intellectual capital and environmental product innovation: The mediating role of green social capital. *Knowledge Management Research and Practice*, *12*(3), 261–275. doi:10.1057/kmrp.2014.1

Delli Paoli, A., & Addeo, F. (2019). Assessing SDGs: A methodology to measure sustainability. *Athens Journal of Social Sciences*, *6*(3), 229–250. doi:10.30958/ajss.6-3-4

Dempster, A. P., Laird, N. M., & Rubin, D. B. (1977). Maximum likelihood from incomplete data via the EM algorithm. *Journal of the Royal Statistical Society. Series B. Methodological*, *39*(1), 1–38. doi:10.1111/j.2517-6161.1977.tb01600.x

Deng, Y., Mu, Y., Dong, X., Wang, X., Zhang, T., & Jia, H. (2022). Coordinated operational planning for electric vehicles considering battery swapping and real road networks in logistics delivery service. *Energy Reports*, *8*, 1019–1027. doi:10.1016/j.egyr.2021.11.185

DGDC (Dominica Geothermal Development Company Ltd.). (2018). *Geothermal project to get funding from World Bank*. https://www.geodominica.dm/news/geothermal-project-to-get-funding-from-world-bank/

Dhillon, J., Kumar, A., & Singal, S. K. (2014). Optimization methods applied for Wind–PSP operation and scheduling under deregulated market: A review. *Renewable & Sustainable Energy Reviews*, *30*, 682–700. doi:10.1016/j.rser.2013.11.009

Dickinson, T., & Burton, I. (2011). Adaptation to climate change in Canada: A multi-level mosaic. In J. D. Ford & L. Berrang-Ford (Eds.), *Climate Change Adaptation in Developed Nations: From theory to practice*. Springer. doi:10.1007/978-94-007-0567-8_7

Diffenbaugh, N. S., Singh, D., Mankin, J. S., Horton, D. E., Swain, D. L., Touma, D., Charland, A., Liu, Y., Haugen, M., Tsiang, M., & Rajaratnam, B. (2017). Quantifying the influence of global warming on unprecedented extreme climate events. *Proceedings of the National Academy of Sciences of the United States of America*, *114*(19), 4881–4886. doi:10.1073/pnas.1618082114 PMID:28439005

Dimeas, A. L., & Hatziargyriou, N. D. (2007, November). Agent based control of virtual power plants. In *2007 International Conference on Intelligent Systems Applications to Power Systems* (pp. 1-6). IEEE. 10.1109/ISAP.2007.4441671

Dinc, M., & Haynes, K. E. (1999). Sources of regional inefficiency an integrated shift-share data envelopment analysis and input-output approach. *The Annals of Regional Science*, *33*(4), 469–489. doi:10.1007001680050116

Ding, R., Liu, Z., Li, X., Hou, Y., Sun, W., Zhai, H., & Wei, X. (2022). Joint charging scheduling of electric vehicles with battery to grid technology in battery swapping station. *Energy Reports*, *8*, 872–882. doi:10.1016/j.egyr.2022.02.029

Dinh, D.-T., & Huynh, V.-N. (2020). k -PbC: An Improved Cluster Center Initialization for Categorical Data Clustering. *Applied Intelligence*, *50*(8), 2610–2632. doi:10.100710489-020-01677-5

Dipasquale, L., Correia, M., Mecca, S., Achenza, M., Guillaud, H., Mileto, C., ... Zanini, L. (2014). *VerSus: Heritage for Tomorrow Vernacular Knowledge for Sustainable Architecture*.

DNO. (2011). *Government Has No Control Over Gas Prices – PM*. https://dominicanewsonline.com/news/homepage/news/business/government-has-no-control-over-gas-prices-pm/

DNO. (2021). *Dominica has potential to generate over 120 megawatts of electricity – Vince Henderson*. https://dominicanewsonline.com/news/homepage/news/dominica-has-potential-to-generate-over-120-megawatts-of-electricity-vince-henderson/

Döll, P., & Romero-Lankao, P. (2017). How to embrace uncertainty in participatory climate change risk management- A roadmap. *Earth's Future*, *5*(1), 18–36. doi:10.1002/2016EF000411

Dominica, C. S. O. (Central Statistical Office of Dominica). (2021). *Motor Vehicles Licensed During the Year 2002 to 2020*. https://stats.gov.dm/subjects/transport/motor-vehicles-licensed-during-the-year-2002-to-2017/

Dong, H., Li, P., Feng, Z., Yang, Y., You, Z., & Li, Q. (2019). Natural capital utilization on an international tourism island based on a three-dimensional ecological footprint model: A case study of Hainan Province, China. *Ecological Indicators*, *104*, 479–488. doi:10.1016/j.ecolind.2019.04.031

Dorler, D., Fritz, S., Voigt-Heucke, S., & Heigl, F. (2021). Citizen science and the role in Sustainable Development. *Sustainability*, *13*(10), 5676. doi:10.3390u13105676

Doumba, M. (2020). The Great reset must place social justice at its centre. In *Sustainable Development Impact Summit*. World Economic Forum.

Drennan, L., McConnell, A., & Stark, A. (2015). Risk and crisis management in the public sector. Routledge printers, Abingdon.

Druckman, J. N., & Lupia, A. (2016). Preference Change in Competitive Political Environments. *Annual Review of Political Science*, *19*(1), 13–31. doi:10.1146/annurev-polisci-020614-095051

Duda, R. O., Hart, P. E., & Stork, D. G. (Eds.). (2000). *Pattern classification* (2nd ed.). Wiley-Interscience.

Dunham, D. (1961). The Courtyard House As A Temperature Regulator. *Ekistics*, *11*(64), 181–186.

Dunlap, R. E. (2013). Climate Change Skepticism and Denial: An Introduction. *The American Behavioral Scientist*, *57*(6), 691–698. doi:10.1177/0002764213477097 PMID:24098056

Dunn, J. C. (1974). Well-separated clusters and optimal fuzzy partitions. *Journal of Cybernetics*, *4*(1), 95–104. doi:10.1080/01969727408546059

Durieux, E., & Hidson, M. (2021). *Local green deals: a blueprint for action: The European Commission's 100 intelligent cities challenge*. Publications Office. https://data.europa.eu/doi/10.2826/94389

Duriex, E., & Hidson, M. (2021). *Local Green Deals: A Blueprint for Action*. Retrieved from https://www.intelligentcitieschallenge.eu/sites/default/files/2021-06/Local%20Green%20Deals-8.pdf

Durkheim, E. (1895). *The Rules of Sociological Method*. Sage Publications.

EC. (2020). *In focus: Energy efficiency in buildings*. Retrieved January 18, 2022, from https://ec.europa.eu/info/news/focus-energy-efficiency-buildings-2020-feb-17_en

Edvinsson, L., & Malone, M. (1997). *Intellectual Capital: Realizing your company's true value by finding its hidden brainpower*. Oxford University Press, New York.

Eisenhardt, K., & Graebner, M. (2007). Theory building from cases: Opportunities and challenges. *Academy of Management Journal*, *50*(1), 25–32. doi:10.5465/amj.2007.24160888

El Bakari, K., & Kling, W. L. (2012, May). Fitting distributed generation in future power markets through virtual power plants. In *2012 9th International Conference on the European Energy Market* (pp. 1-7). IEEE. 10.1109/EEM.2012.6254692

Elkonin, D. B. (1989). *Selected Psychological Works*. Pedagogy Publisher.

Elliott, J. A. (Ed.). (2013). An introduction to sustainable development (4th ed.). Routledge (Taylor & Francis Group).

EMDAT. (2021). *The International Disaster Database*. Center for Research on the Epidemiology of Disasters-Cred. https://public.emdat.be/data

EM-DAT. (2021). *The International Disaster Database*. https://www.emdat.be/

Emekci, Ş. (2021). Determination of Sustainable Architectural Design Criteria in Protected Areas: Focus group method. *Journal of Design+Theory*, *17*(33), 229–242. doi:10.14744/tasarimkuram.2020.82687

Emekci, Ş. (2021). Sustainability in Architecture: Low-tech or High-tech? In Proceedings Article (pp. 107–111). Alanya Hamdullah Emin Paşa University. doi:10.38027/ICCAUA2021216n14

Enete, I. C. (2014). Impacts of climate change on agricultural production in Enugu state, Nigeria. *Journal of Earth Science & Climatic Change, 5*(9). https://www.cabdirect.org/cabdirect/abstract/20153228679

Engin, N., Vural, N., Vural, S., & Sumerkan, M. R. (2007). Climatic effect in the formation of vernacular houses in the Eastern Black Sea region. *Building and Environment, 42*(2), 960–969. doi:10.1016/j.buildenv.2005.10.037

Ernst & Young. (2010). *The response of companies to climate change* [La respuesta de las empresas al cambio climático. Elegir el camino correcto]. Recovered from: https://www.ey.com/MX/es/Services/Specialty-Services/Climate-Change-and-Sustainability-Services

ETI (Energy Transition Initiative). (2015). *Energy Snapshot – Dominica*. https://www.nrel.gov/docs/fy15osti/62704.pdf

ETI (Energy Transition Initiative). (2020). *Dominica: Energy Snapshot*. Accessed March 22, 2022. https://www.energy.gov/sites/default/files/2020/09/f79/ETI-Energy-Snapshot-Dominica_FY20.pdf

European Commission (2020). *Sustainable Europe Investment Plan, European Green Deal Investment Plan*. Retrieved December 10, 2021, from file:///C:/Users/User/Downloads/Commission_Communication_on_the_European_Green_Deal_Investment_Plan_EN.pdf.pdf.

European Commission. (2016). *Guidance on municipal waste data collection*. https://ec.europa.eu/eurostat/documents/342366/351758/Guidance+on+municipal+waste/3106067c-6ad6-4208-bbed-49c08f7c47f2

European Commission. (2019). *The European Green Deal*. Communication from the Commission COM(2019) 640 Final, 11.12.2019. Retrieved June 25, 2020, from https://eur-lex.europa.eu/legal-content/EN/TXT/?qid=1596443911913&uri=CELEX:52019DC0640#document2

European Commission. (2021). *Commission puts forward new strategy to make the EU's financial system more sustainable and proposes new European Green Bond Standard*. Retrieved December 10, 2021, from https://ec.europa.eu/commission/presscorner/detail/en/ip_21_3405

European Commission. (n.d.). https://ec.europa.eu/environment/topics/waste-and-recycling_en

European Commission. (n.d.). *Roadmaps for Administrative Capacity Building*. Retrieved from https://ec.europa.eu/regional_policy/sources/policy/how/improving-investment/roadmap_toolkit.pdf

European Environment Agency. (2021). *Municipal waste management across European Countries*. https://www.eea.europa.eu/publications/municipal-waste-management-across-european-countries

European green deal. (n.d.). Retrieved from https://ec.europa.eu/info/strategy/priorities-2019-2024/european-green-deal_en

European Union. European Commission. Directorate-General for Regional Policy. (2011). *Cities of tomorrow: Challenges, visions, ways forward*. Publications Office of the European Union.

Eurostat. (n.d.). https://ec.europa.eu/eurostat/cache/metadata/en/cei_wm030_esmsip2.htm

Evans, C., Abrams, E., Reitsma, R., Roux, K., Salmonsen, L., & Marra, P. P. (2005). The neighborhood nestwatch program: Participant outcomes of a citizen-science ecological research project. *Conservation Biology, 19*(3), 589–594. doi:10.1111/j.1523-1739.2005.00s01.x

Everitt, B. (1980). Cluster analysis. *Quality & Quantity, 14*(1), 75–100. doi:10.1007/BF00154794

Everitt, B. (Ed.). (1974). *Cluster analysis*. Heinmann Educational Books.

Fadli, C., & Thamrin, H. (2017). Room Arrangement of Vernacular Dwelling in Mandailing, Indonesia. *IACSIT International Journal of Engineering and Technology, 9*(4), 3427–3434. doi:10.21817/ijet/2017/v9i4/170904185

Compilation of References

Fan, Y.V., Jiang, P., Tan, R.R., Aviso, K.B., You, F., Zhao, X., Lee, C.T. & Klemes, J. J. (2022). Forecasting plastic waste generation and interventions for environmental hazard mitigation. *Journal of Hazardous Materials*, *424*(Part A), 1-14.

FAO, IFAD, UNICEF, WFP, & WHO. (2021). The State of Food Security and Nutrition in the World 2021. Transforming food systems for food security, improved nutrition and affordable healthy diets for all. FAO.

FAO. (2014). *Towards risk-based drought management in Europe and Central Asia*. https://www.fao.org/fileadmin/user_upload/Europe/documents/Events_2014/ECA2014/ECA_38_14_4_2_en.pdf

FAOSTAT. (2018). *Food and Agriculture Organization of the United Nations*. http://faostat3.fao.org/download/Q/QA/E

Farrell, M. J. (1957). The measurement of productivity efficiency. *Journal of the Royal Statistical Society. Series A (General)*, *120*(3), 253–281. doi:10.2307/2343100

Fathima, A. H., & Palanisamy, K. (2015). Optimization in microgrids with hybrid energy systems–A review. *Renewable & Sustainable Energy Reviews*, *45*, 431–446. doi:10.1016/j.rser.2015.01.059

Febrianti, F. D., Sugiyanto, S., & Fitria, J. R. (2020). Green Intellectual Capital Conservatism Earning Management, To Future Stock Return As Moderating Stock Return (Study Of Mining Companies In Indonesia Listed On Idx For The Period Of 2014-2019). *The Accounting Journal Of Binaniaga*, *5*(2), 141–154. doi:10.33062/ajb.v5i2.407

Feng, S., Krueger, A., & Oppenheimer, M. (2010). Linkages among climate change, crop yields and Mexico–US cross-border migration. *Proceedings of the National Academy of Sciences of the United States of America*, *107*(32), 14257–14262. doi:10.1073/pnas.1002632107 PMID:20660749

Fernandes, J., Mateus, R., Bragança, L., & Correia da Silva, J. J. (2015). Portuguese vernacular architecture: The contribution of vernacular materials and design approaches for sustainable construction. *Architectural Science Review*, *58*(4), 324–336. doi:10.1080/00038628.2014.974019

FEV. (2019). *El sector del vino y el papel de la FEV en los Objetivos de Desarrollo Sostenible*. Disponible en: http://www.fev.es/fev/sostenibilidad-y-responsabilidad/objetivos-de-desarrollo-sostenible_122_1_ap.html

Find calls for proposals and tenders. (n.d.). European Commission. Retrieved from https://ec.europa.eu/info/funding-tenders/opportunities/portal/screen/home

Fiore, M., Silvestri, R., Contò, F., & Pellegrini, G. (2017). Understanding the relationship between green approach and marketing innovations tools in the wine sector. *Journal of Cleaner Production*, *142*, 4085–4091. doi:10.1016/j.jclepro.2016.10.026

Ford, J. D., & Berrang-Ford, L. (2011). Introduction. In J. D. Ford & L. Berrang-Ford (Eds.), *Climate Change Adaptation in Developed Nations: From theory to practice*. Springer. doi:10.1007/978-94-007-0567-8_1

Fressoli, M., Arza, V., & Castillo, M. D. (2016). *Entina Argentina. The impact of citizen-generated data initiatives in Argentina*. Civicus Datashift.

Fridkin, K., Kenney, P. J., & Wintersieck, A. (2015). Liar, Liar, Pants on Fire: How Fact-Checking Influences Citizens' Reactions to Negative Advertising. *Political Communication*, *32*(1), 127–151. doi:10.1080/10584609.2014.914613

Fritz, S., See, L., Carlson, T., Haklay, M. M., Oliver, J. L., Fraisl, D., Mondarini, R., Brocklehurst, M., Shanley, L., Schade, S., When, U., Abrate, T., Anstee, J., Arnold, S., Billot, M., Campbell, J., Espey, J., Gold, M., Hager, G., & West, S. (2019). Citizen science and the United Nations Sustainable Development Goals. *Nature Sustainability*, *2*(10), 922–930. doi:10.103841893-019-0390-3

Galadima, A., & Lawal, A. M. (2017). Climate Change Situation in Zamfara State: Farmers' Awareness and Agricultural Implications. *Chemistry and Materials Research, 9*(8), 7–11. https://www.iiste.org/Journals/index.php/CMR/article/view/38210

Gamarra, C., & Guerrero, J. M. (2015). Computational optimization techniques applied to microgrids planning: A review. *Renewable & Sustainable Energy Reviews, 48*, 413–424. doi:10.1016/j.rser.2015.04.025

García Fernández, C. (2006). Cost-benefit analysis and the difficulty of its application to climate change [El análisis coste-beneficio y la dificultad de su aplicación al cambio climático]. *Estudios de Economía Aplicada, España, 24*(2), 751–762.

Gbenga, O., Opaluwa, H. I., Olabode, A., & Ayodele, O. J. (2020). Understanding the Effects of Climate Change on Crop and Livestock Productivity in Nigeria. *Asian Journal of Agricultural Extension. Economia e Sociologia, 38*(3), 83–92. doi:10.9734/ajaees/2020/v38i330327

GCF (Green Climate Fund). (2019). *Accredited Entity Directory.* https://www.greenclimate.fund/how-we-work/tools/entity-directory.

GCoD (Government of the Commonwealth of Dominica). (2018). *National Resilience Development Strategy Dominica 2030.* https://observatorioplanificacion.cepal.org/sites/default/files/plan/files/Dominica%202030The%20National%20Resilience%20Development%20Strategy.pdf

Ge, J., & Lin, B. (2021). Impact of public support and government's policy on climate change in China. *Journal of Environmental Management, 294*, 112983. doi:10.1016/j.jenvman.2021.112983 PMID:34119988

Georgakakos, A., Fleming, P., Dettinger, M., Peters-Lidard, C., Richmond, T. C., Reckhow, K., White, K., & Yates, D. (2014). Water resources. In J. M. Melillo, T. C., Richmond, & G. W. Yohe (Eds.), Climate Change Impacts in the United States: The Third National Climate Assessment (pp. 69-112). U.S. Global Change Research Program.

Get Involved. (n.d.). *The Australian Museum.* Retrieved February 8, 2022, from https://australian.museum/get-involved/

Ghavidel, S., Li, L., Aghaei, J., Yu, T., & Zhu, J. (2016, September). A review on the virtual power plant: Components and operation systems. In 2016 IEEE international conference on power system technology (POWERCON) (pp. 1-6). IEEE.

Giannakitsidou, O., Giannikos, I., & Chondrou, A. (2020). Ranking European Countries on the basis of their environmental and circular economy performance: A DEA application in MSW. *Waste Management (New York, N.Y.), 109*, 181–191. doi:10.1016/j.wasman.2020.04.055 PMID:32408101

Gilinsky, A. Jr, Newton, S., & Vega, R. (2016). Sustainability in the global wine industry: Concepts and cases. *Agriculture and Agricultural Science Procedia, 8*, 37–49. doi:10.1016/j.aaspro.2016.02.006

Giller, K. E., Hijbeek, R., Andersson, J. A., & Sumberg, J. (2021). Regenerative agriculture: An agronomic perspective. *Outlook on Agriculture, 50*(1), 13–25. doi:10.1177/00307270021998063 PMID:33867585

Glassie, H. (2000). *Vernacular architecture* (Vol. 2). Indiana University Press.

Godet, M. (2000). The Toolbox of Strategic Foresigh [La caja de herramientas de la prospectiva estratégica] (4th ed.). España: Laboratoire d'Investigation Prospectiva et Stratégique y Prospektiker – Instituto Europeo de Prospectiva y Estrategia.

Golkar, K. (2000). Sustainable urban designing in borders of deserts. *J. Fine Arts, 3*, 43–52.

Gomonov, K., Ratner, S., Lazanyuk, I., & Revinova, S. (2021). Clustering of EU countries by the level of circular economy: An object-oriented approach. *Sustainability, 13*(13), 1–20. doi:10.3390u13137158

Gong, J., Xie, D., Jiang, C., & Zhang, Y. (2011). Multiple objective compromised method for power management in virtual power plants. *Energies, 4*(4), 700–716. doi:10.3390/en4040700

Gosnell, H., Gill, N., & Voyer, M. (2019). Transformational adaptation on the farm: Processes of change and persistence in transitions to 'climate-smart' regenerative agriculture. *Global Environmental Change*, *59*, 101965. doi:10.1016/j.gloenvcha.2019.101965

Greenaway, D., & Milner, C. (2003). *A Grim REPA?* Leverhulme Centre for Research on Globalisation Economic Policy, University of Nottingham.

Griffith-Jones, S., Ocampo, J. A., & Spratt, S. (2012). *Financing renewable energy in developing countries: Mechanisms and responsibilities.* European Report on Development. Retrieved March 5, 2018, from https://policydialogue.org/files/publications/Financing_Renewable_Energy_in_Developing_Countries.pdf

Gronkiewicz-Waltz, H., Larsson, A., & Boni, A. (2020). *100 climate-neutral cities by 2030 - by and for the citizens: report of the mission board for climate-neutral and smart cities.* Publications Office. https://data.europa.eu/doi/10.2777/46063

Grose, V. L. (2018). *Five weaknesses of enterprise risk management.* Omega Systems Group Incorporated.

GTO (Geothermal Technologies Office). (2021). *Geothermal FAQs*. https://www.energy.gov/eere/geothermal/geothermal-faqs

Gu, T. J. (2021). *CATL breaks through the game by positioning itself as an upstream equipment vendor to LEAD.* Electronic Publishing House. https://kns.cnki.net/kcms/detail/detail.aspx?dbcode=CJFD&dbname=CJFDLAST2021&filename=YCYW202101014&uniplatform=NZKPT&v=iWDo6ZOTDFRkpBf0YWWr8P1mrVPhEcFfy_VKUBMk-B9tIrmynrXq8zWzxLFuYU3My

Guhl Corpas, A. (2008). Ethical aspects of global warming [Aspectos éticos del calentamiento climático global]. *Revista Latinoamericana de Bioética, Colombia*, *8*(2), 20–29.

Gui, E. M., & MacGill, I. (2018). Typology of future clean energy communities: An exploratory structure, opportunities, and challenges. *Energy Research & Social Science*, *35*, 94–107. doi:10.1016/j.erss.2017.10.019

Guillaud, H. (2014). Socio-cultural sustainability in vernacular architecture. In *Versus: heritage for tomorrow* (pp. 48–55). Firenze University Press. Retrieved from https://hal.archives-ouvertes.fr/hal-01159772

Gumilev, L. N. (2012a). *Ethnosphere: History of People and History of Nature.* Eksmo Publishing House.

Gumilev, L. N. (2012b). *Ethnogenesis and the Earth's Biosphere.* Eksmo Publishing House.

Gungor, V. C., Sahin, D., Kocak, T., Ergut, S., Buccella, C., Cecati, C., & Hancke, G. P. (2011). Smart grid technologies: Communication technologies and standards. *IEEE Transactions on Industrial Informatics*, *7*(4), 529–539. doi:10.1109/TII.2011.2166794

Guo, Y., Wu, S., He, Y.-B., Kang, F., Chen, L., Li, H., & Yang, Q.-H. (2022). Solid-state lithium batteries: Safety and prospects. *eScience*. doi:10.1016/j.esci.2022.02.008

Guo, J., Zhang, Y.-J., & Zhang, K.-B. (2018). The key sectors for energy conservation and carbon emissions reduction in China: Evidence from the input-output method. *Journal of Cleaner Production*, *179*, 180–190. doi:10.1016/j.jclepro.2018.01.080

Hachenberg, B., & Schiereck, D. (2018). Are Green Bonds Priced Differently from Conventional Bonds? *Journal of Asset Management*, *19*(6), 371–383. doi:10.105741260-018-0088-5

Haider, H. T., See, O. H., & Elmenreich, W. (2016). A review of residential demand response of smart grid. *Renewable & Sustainable Energy Reviews*, *59*, 166–178. doi:10.1016/j.rser.2016.01.016

Haile, G. G., Tang, Q., Hosseini-Moghari, S. M., Liu, X., Gebremicael, T. G., Leng, G., Kebede, A., Xu, X., & Yun, X. (2020). Projected impacts of climate change on drought patterns over East Africa. *Earth's Future*, *8*(7), e2020EF001502.

Hall, L. (2015, June 16). Aurorasaurus. *NASA*. Retrieved February 8, 2022, from https://www.nasa.gov/feature/aurorasaurus

Hamard, E., Cammas, C., Lemercier, B., Cazacliu, B., & Morel, J.-C. (2020). Micromorphological description of vernacular cob process and comparison with rammed earth. *Frontiers of Architectural Research*, *9*(1), 203–215. doi:10.1016/j.foar.2019.06.007

Hansen, Moore, & Reiter. (2022). *Decarbonizing the built environment: Takeaways from COP26*. McKinsey & Company.

Hantash, T. F. A. (2017). *Building A Zero Energy House For UAE : Traditional Architecture Revisited*. Presented at the 5th International Conference on Zero Energy Mass Customised Housing.

Hare, J., Shi, X., Gupta, S., & Bazzi, A. (2016). Fault diagnostics in smart micro-grids: A survey. *Renewable & Sustainable Energy Reviews*, *60*, 1114–1124. doi:10.1016/j.rser.2016.01.122

Hartmann, H., Moura, C. F., Anderegg, W. R. L., Ruehr, N. K., Salmon, Y., Allen, C. D., Arndt, S. K., Breshears, D. D., Davi, H., Galbraith, D., Ruthrof, K. X., Wunder, J., Adams, H. D., Bloemen, J., Cailleret, M., Cobb, R., Gessler, A., Grams, T. E. E., Jansen, S., ... O'Brien, M. (2018). Research frontiers for improving our understanding of drought-induced tree and forest mortality. *The New Phytologist*, *218*(1), 15–28. doi:10.1111/nph.15048 PMID:29488280

Hashemi, F. (2018). *Adapting vernacular strategies for the design of an energy efficient residential building in a hot and arid climate*. City of Yazd.

Hasni, T., Malek, R., & Zouioueche, N. (2021). *ALGERIA 100% RENEWABLE ENERGIES: Recommendations for a national energy transition strategy* [L'ALGÉRIE 100% ÉNERGIES RENOUVELABLES: Recommandations pour une stratégie nationale de transition énergétique]. Friedrich Ebert Stiftung.

Hegel, G. W. F. (1892). The science of logics. In *The Logic of Hegel* (2nd ed.). Oxford University Press. http://links.jstor.org/sici?sici=0022-3808(198610)94:5<1002:IRALG>2.0.CO;2-C

He, J., Li, J., Zhao, D., & Chen, X. (2022). Does oil price affect corporate innovation? Evidence from new energy vehicle enterprises in China. *Renewable & Sustainable Energy Reviews*, *156*, 111964. Advance online publication. doi:10.1016/j.rser.2021.111964

Henderson, S., Newman, S., Ward, D., Havens-Young, K., Alaback, P., & Meymaris, K. (2011). Project BudBurst: Continental-scale citizen science for all seasons. *AGU Fall Meeting Abstracts*.

Hepburn, C., Qi, Y., Stern, N., Ward, B., Xie, C., & Zenghelis, D. (2021). Towards carbon neutrality and China's 14th Five-Year Plan: Clean energy transition, sustainable urban development, and investment priorities. *Environmental Science and Ecotechnology*, *8*, 100130. Advance online publication. doi:10.1016/j.ese.2021.100130 PMID:36156997

Hernández, L., Baladron, C., Aguiar, J. M., Carro, B., Sanchez-Esguevillas, A., Lloret, J., Chinarro, D., Gomez-Sanz, J. J., & Cook, D. (2013). A multi-agent system architecture for smart grid management and forecasting of energy demand in virtual power plants. *IEEE Communications Magazine*, *51*(1), 106–113. doi:10.1109/MCOM.2013.6400446

Hertel, T. W., Burke, M. B., & Lobell, D. B. (2011). The poverty implications of climate-induced crop yield changes by 2030. *Global Environmental Change*, *20*(4), 577–585. doi:10.1016/j.gloenvcha.2010.07.001

Hirsch, A., Parag, Y., & Guerrero, J. (2018). Microgrids: A review of technologies, key drivers, and outstanding issues. *Renewable & Sustainable Energy Reviews*, *90*, 402–411. doi:10.1016/j.rser.2018.03.040

Hoffmann, R., Dimitrova, A., Muttarak, R., Cuaresma, J. C., & Peisker, J. (2020). A meta-analysis of country-level studies on environmental change and migration. *Nature Climate Change*, *10*(10), 904–912. doi:10.103841558-020-0898-6

Huang, C., & Kung, F. (2011). Environmental consciousness and intellectual capital management: Evidence from Taiwan's manufacturing industry. *Management Decision*, *49*(9), 1405–1425. doi:10.1108/00251741111173916

Hu, C.-F., Wang, H.-F., & Liu, T. (2022). Measuring efficiency of a recycling production system with imprecise data. *Numerical Algebra. Control and Optimization*, *12*(1), 79–91. doi:10.3934/naco.2021052

Hu, J.-W., Javaid, A., & Creutzig, F. (2021). Leverage points for accelerating adoption of shared electric cars: Perceived benefits and environmental impact of NEVs. *Energy Policy*, *155*, 112349. Advance online publication. doi:10.1016/j.enpol.2021.112349

Hulbert, J. M. (2016). Citizen science tools available for Ecological Research in South Africa. *South African Journal of Science*, *112*(5/6), 2. Advance online publication. doi:10.17159ajs.2016/a0152

Hulbert, J. M., Turner, S. C., & Scott, S. L. (2019). Challenges and solutions to establishing and sustaining citizen science projects in South Africa. *South African Journal of Science*, *115*(7/8). Advance online publication. doi:10.17159ajs.2019/5844

Hunt, V., Prince, S., Dixon-Fyle, S., & Dolan, K. (2020). *Diversity wins*. McKinsey.

Iacobucci, R., McLellan, B., & Tezuka, T. (2018). The synergies of shared autonomous electric vehicles with renewable energy in a virtual power plant and microgrid. *Energies*, *11*(8), 2016. doi:10.3390/en11082016

İçöz, C., & Er, F. (2020). A statistical analysis of municipal waste treatment types in European countries. *Environmental Research & Technology*, *3*(3), 113–118. doi:10.35208/ert.769634

Iglesias, A., Moneo, M., & Garrote, L. (2007). Defining the planning purpose, framework and concepts. In A. Iglesias, M. Moneo, & A. López-Francos, A. (Eds.), Drought management guidelines technical annex. CIHEAM/EC MEDA Water.

Ignatius, A. M. (2016). Rurality and climate change vulnerability in Nigeria: Assessment towards evidence based even rural development policy. *2016 Berlin Conference on Global Environmental Change*. 10.17169/REFUBIUM-21841

Impact, E. (2022). The ESG conundrum: how investors and companies can find common purpose in ESG. *Economist Impact*.

Inkwood Research. (2018). *Virtual Power Plant Market - Scope, Size, Share, Analysis by 2026*. Available at: https://www.inkwoodresearch.com/reports/global-virtual-power-plant-market/

International Council of Monuments and Sites. (2021). *Heritage and the sustainable development goals: policy guidance for heritage and development actors*. International Council on Monuments and Sites – ICOMOS.

IPCC. (2013). *Climate change 2013: The physical science basis*. In T. F. Stocker, D. Qin, G. K. Plattner, M. Tignor, S. K. Allen, J. Boschung, A. Nauels, Y. Xia, V. Bex, & P. M. Midgley (Eds.), Contribution of Working Group I to the Fifth Assessment Report of the Intergovernmental Panel on Climate Change. *Cambridge University Press*.

IPCC. (2014). *AR5 Climate Change 2014: Impacts, Adaptation, and Vulnerability*. https://www.ipcc.ch/report/ar5/wg2/

IPCC. (2014). Summary for policymakers. In C. B. Field, V. R. Barros, D. J. Dokken, K. J. Mach, M. D. Mastrandrea, T. E. Bilir, M. Chatterjee, K. L. Ebi, Y. O. Estrada, R. C. Genova, B. Girma, E. S. Kissel, A. N. Levy, S. MacCracken, P. R. Mastrandrea, & L. L. White (Eds.), Climate Change 2014: Impacts, Adaptation, and Vulnerability. Part A: Global and Sectoral Aspects (pp. 1-32). Contribution of Working Group II to the Fifth Assessment Report of the Intergovernmental Panel on Climate Change. Cambridge University Press.

IPCC. (2018). Summary for Policymakers. In V. Masson-Delmotte, P. Zhai, H. O. Pörtner, D. Roberts, J. Skea, P. R. Shukla, A. Pirani, W. Moufouma-Okia, C. Péan, R. Pidcock, S. Connors, J. B. R. Matthews, Y. Chen, X. Zhou, M. I. Gomis, E. Lonnoy, T. Maycock, M. Tignor, & T. Waterfield (Eds.), Global Warming of 1.5°C (pp. 1-32). An IPCC Special Report on the impacts of global warming of 1.5°C above pre-industrial levels and related global greenhouse gas emission pathways, in the context of strengthening the global response to the threat of climate change, sustainable development, and efforts to eradicate poverty. World Meteorological Organization.

IPCC. (2021). Summary for Policymakers. In V. Masson-Delmotte, P. Zhai, A. Pirani, S. L. Connors, C. Péan, S. Berger, N. Caud, Y. Chen, L. Goldfarb, M. I. Gomis, M. Huang, K. Leitzell, E. Lonnoy, J. B. R. Matthews, T. K. Maycock, T. Waterfield, O. Yelekçi, R. Yu, & B. Zhou (Eds.). Climate Change 2021: The Physical Science Basis. Contribution of Working Group I to the Sixth Assessment Report of the Intergovernmental Panel on Climate Change. Cambridge University Press.

IRENA & CPI. (2018). *Global Landscape of Renewable Energy Finance, 2018*. International Renewable Energy Agency. Retrieved March 5, 2021, from https://www.irena.org/publications/2018/jan/global-landscape-of-renewable-energy-finance

IRENA & CPI. (2020). *Global Landscape of Renewable Energy Finance, 2020*. International Renewable Energy Agency. Retrieved March 5, 2021, from https://www.irena.org/publications/2020/Nov/Global-Landscape-of-Renewable-Energy-Finance-2020

IRENA, Data, & Statistics. (2020). *International Renewable Energy Agency: Abu Dhabi, United Arab Emirates*. Available online: https://www.irena.org/Statistics

IRENA. (2016). *Unlocking renewable energy investment: The role of risk mitigation and structured finance*. IRENA. Retrieved January 25, 2018, from https://www.res4med.org/wp-content/uploads/2017/11/IRENA_Risk_Mitigation_and_Structured_Finance_2016.pdf

IRENA. (2020). *Renewable energy finance: Green Bonds*. International Renewable Energy Agency. Retrieved March 3, 2021, from https://www.irena.org/publications/2020/Jan/RE-finance-Green-bonds

IRNA. (2020). *The number of dams in the country reached 188*. Available at: https://www.irna.ir/news/84025654

Ishak, N. M., Bakar, A., & Yazid, A. (2014). Developing Sampling Frame for Case Study: Challenges and Conditions. *World Journal of Education*, 4(3), 29–35.

Jacques, J., & Preda, C. (2014). Model-Based clustering for multivariate functional data. *Computational Statistics & Data Analysis*, 71, 92–106. doi:10.1016/j.csda.2012.12.004

Jansen, L. (2003). The challenge of sustainable development. *Journal of Cleaner Production*, 11(3), 231–245. doi:10.1016/S0959-6526(02)00073-2

Ji, X. Y., Xu, W. H., Shen, P. Y., Ma, X. R., & Lu, R. (2020). *Analysis of the prospect of China's automobile industry under the adjustment of new energy subsidy policy - Taking BYD as an example*. Anhui University of Finance and Economics. https://kns.cnki.net/kcms/detail/detail.aspx?dbcode=CJFD&dbname=CJFDLAST2020&filename=YXJI202034009&uniplatform=NZKPT&v=7N3vuKiq-sacH2kA3cJtwPQMdOATxh1q0ch6TwSP-XYfyho7wOy0ChF-GwDpaNIV

Jiang, D., Li, W., Shen, Y., & Yu, S. (2022a). Does air pollution affect earnings management? Evidence from China. *Pacific-Basin Finance Journal*, 72, 101737. Advance online publication. doi:10.1016/j.pacfin.2022.101737

Jiang, T., Yu, Y., Jahanger, A., & Balsalobre-Lorente, D. (2022b). Structural emissions reduction of China's power and heating industry under the goal of "double carbon": A perspective from input-output analysis. *Sustainable Production and Consumption*, 31, 346–356. doi:10.1016/j.spc.2022.03.003

Jiang, W., & Chen, Y. (2022). The time-frequency connectedness among carbon, traditional/new energy and material markets of China in pre- and post-COVID-19 outbreak periods. *Energy*, *246*, 123320. Advance online publication. doi:10.1016/j.energy.2022.123320

Jiang, X., Chen, Y., Meng, X., Cao, W., Liu, C., Huang, Q., Naik, N., Murugadoss, V., Huang, M., & Guo, Z. (2022c). The impact of electrode with carbon materials on safety performance of lithium-ion batteries: A review. *Carbon*, *191*, 448–470. doi:10.1016/j.carbon.2022.02.011

Jin, B. (2021). Research on performance evaluation of green supply chain of automobile enterprises under the background of carbon peak and carbon neutralization. *Energy Reports*, *7*, 594–604. doi:10.1016/j.egyr.2021.10.002

Jing, L. (2018). *Does the Chinese Public Care About Climate Change?* https://www.chinadialogue.net/article/show/single/en/10831-Does-the-Chinese-public-care-about-climate-change-

Jirakraisiri, J., Badir, Y., & Frank, B. (2021). Translating green strategic intent into green process innovation performance: The role of green intellectual capital. *Journal of Intellectual Capital*, *22*(7), 43–67. doi:10.1108/JIC-08-2020-0277

Jirawuttinunt, S. (2018). The Relationship between Green Human Resource Management and Green Intellectual Capital of Certified ISO 14000 Businesses in Thailand. *St. Theresa. Journal of the Humanities and Social Sciences*, *4*(1), 20–37.

Jones, R. (2010). A risk management approach to climate change adaptation. In R. A. C. Nottage, D. S. Wratt, J. F. Bornman, & K. Jones (Eds.), *Climate Change Adaptation in New Zealand: Future Scenarios and Some Sectoral Perspectives* (pp. 10–25). New Zealand Climate Change Centre.

Jorqueta-Fontena, E., & Orrego.Verdugo, R. (2010). Impact of global warming on the phonology of a vine variety grown in southern Chile [Impacto del calentamiento global en la fonología de una variedad de vid cultivada en el sur de Chile]. *Agrociencia, México*, *44*(4), 427–436.

Ju, L., Zhao, R., Tan, Q., Lu, Y., Tan, Q., & Wang, W. (2019). A multi-objective robust scheduling model and solution algorithm for a novel virtual power plant connected with power-to-gas and gas storage tank considering uncertainty and demand response. *Applied Energy*, *250*, 1336–1355. doi:10.1016/j.apenergy.2019.05.027

Kahlen, M. T., Ketter, W., & van Dalen, J. (2018). Electric vehicle virtual power plant dilemma: Grid balancing versus customer mobility. *Production and Operations Management*, *27*(11), 2054–2070. doi:10.1111/poms.12876

Kant, I. (1781). *Critic of Pure Reason* [Critic der Reinen Vernunft]. Verlegts Johann Friedrich Hartknoch.

Karimi, V., Karami, E., Karami, S., & Keshavarz, M. (2020). Adaptation to climate change through agricultural paradigm shift. *Environment, Development and Sustainability*. Advance online publication. doi:10.100710668-020-00825-8

Karimi, V., Karami, E., & Keshavarz, M. (2018). Climate change and agriculture: Impacts and adaptive responses in Iran. *Journal of Integrative Agriculture*, *17*(1), 1–15. doi:10.1016/S2095-3119(17)61794-5

Kaur, A., Kaushal, J., & Basak, P. (2016). A review on microgrid central controller. *Renewable & Sustainable Energy Reviews*, *55*, 338–345. doi:10.1016/j.rser.2015.10.141

Kaygusuz, K. (2007). Energy for sustainable development: Key issues and challenges. *Energy Sources. Part B, Economics, Planning, and Policy*, *2*(1), 73–83.

Keshavarz, M. (2021). Investigating food security and food waste control of farm families under drought (A case of Kherameh County). *Quarterly Journal of Spatial Economy and Rural Development*, *34*, 83–106.

Keshavarz, M., & Karami, E. (2018). Drought and agricultural ecosystem services in developing countries. In S. Gaba, B. Smith, & E. Lichtfouse (Eds.), *Sustainable Agriculture Reviews 28: Ecology for Agriculture* (pp. 309–359). doi:10.1007/978-3-319-90309-5_9

Keshavarz, M., Karami, E., & Vanclay, F. (2013). Social experience of drought in rural Iran. *Land Use Policy*, *30*(1), 120–129. doi:10.1016/j.landusepol.2012.03.003

Khan, M. A., Riaz, H., Ahmed, M., & Saeed, A. (2021). Does green finance really deliver what is expected? An empirical perspective. *Borsa İstanbul Review*. doi:10.1016/j.bir.2021.07.006

Khan, A. A., Naeem, M., Iqbal, M., Qaisar, S., & Anpalagan, A. (2016). A compendium of optimization objectives, constraints, tools and algorithms for energy management in microgrids. *Renewable & Sustainable Energy Reviews*, *58*, 1664–1683. doi:10.1016/j.rser.2015.12.259

Klaassen, E., & van der Laan, M. (2019). *Energy and Flexibility Services for Citizens Energy Communities*. USEF.

Kneip, A., Park, B. U., & Simar, L. (1998). A note on the convergence of nonparametric DEA estimators for production efficiency scores. *Econometric Theory*, *14*(6), 783–793. doi:10.1017/S0266466698146042

Koehring, M., & Cornell, P. (2021, December). Delivering on a greener future: What to expect in 2022. *Economist Impact*. Retrieved from https://impact.economist.com/sustainability/net-zero-and-energy/delivering-on-a-greener-future-what-to-expect-in-2022?utm_medium=cpc.adword.pd&utm_source=google&ppccampaignID=17210591673&ppcadID=&utm_campaign=a.22brand_pmax&utm_content=conversion.direct-response.anonymous&gclid=CjwKCAjws--ZBhAX-EiwAv-RNL0YqvjUlcowUpazyaL4STGYz7JHtdVtHZHukHXbYx5g0DJnFqMqHKxoCg_8QAvD_BwE&gclsrc=aw.ds

Ko, J., & Yoon, Y. S. (2022). Lithium phosphorus oxynitride thin films for rechargeable lithium batteries: Applications from thin-film batteries as micro batteries to surface modification for large-scale batteries. *Ceramics International*, *48*(8), 10372–10390. doi:10.1016/j.ceramint.2022.02.173

Koona, R. K., Marshallb, S., Mornac, D., McCallumd, R., & Ashtinee, M. (2020). A Review of Caribbean Geothermal Energy Resource Potential. *West Indian Journal of Engineering*, *42*(2), 37–43.

Kuklinski, J. H., Quirk, P. J., Jerit, J., Schwieder, D., & Rich, R. F. (2000). Misinformation and the Currency of Democratic Citizenship. *The Journal of Politics*, *62*(3), 790–816. doi:10.1111/0022-3816.00033

Kummer, J. A., Bayne, E. M., & Machtans, C. S. (2016). Use of citizen science to identify factors affecting bird–window collision risk at houses. *The Condor*, *118*(3), 624–639. doi:10.1650/CONDOR-16-26.1

Lange, K. (1995). A gradient algorithm locally equivalent to the EM algorithm. *Journal of the Royal Statistical Society. Series B. Methodological*, *57*(2), 425–437. doi:10.1111/j.2517-6161.1995.tb02037.x

Laut, J., Porfiri, M., & Nov, O. (2014). Brooklyn Atlantis: a robotic platform for environmental monitoring with community participation. *Proceedings of the International Conference of Control, Dynamics Systems, and Robotics*.

Lawrence, A. (2010). Managing disputes with nonmarket stakeholders: Wage a fight, withdraw, wait, or work it out? *California Management Review*, *53*(1), 90–113. doi:10.1525/cmr.2010.53.1.90

Lee, C. C., & Lee, C. C. (2022). How does green finance affect green total factor productivity? Evidence from China. *Energy Economics*, *107*, 1–15.

Lehner, F., Coats, S., Stocker, T. F., Pendergrass, A. G., Sanderson, B. M., Raible, C. C., & Smerdon, J. E. (2017). Projected drought risk in 1.5°C and 2°C warmer climates. *Geophysical Research Letters*, *44*(14), 7419–7428. doi:10.1002/2017GL074117

Compilation of References

Lem, S. (1964). *Summa Technologiae*.

Levy, D. L., & Kolk, A. (2002). Strategic Responses to Global Climate Change: Conflucting Pressures on Multinatrionals in the Oil Industry. *Business and Politics*, *4*(3), 275–300. doi:10.2202/1469-3569.1042

Lewandowsky, S., Ecker, U. K. H., Seifert, C. M., Schwarz, N., & Cook, J. (2012). Misinformation and Its Correction: Continued Influence and Successful Debiasing. *Psychological Science in the Public Interest*, *13*(3), 106–131. doi:10.1177/1529100612451018 PMID:26173286

Li, W., Fan, J., & Zhao, J. (2022). Has green finance facilitated China's low-carbon economic transition? *Environmental Science and Pollution Research*. doi:10.1007/s11356-022-19891-8

Liang, H., & Zhuang, W. (2014). Stochastic modeling and optimization in a microgrid: A survey. *Energies*, *7*(4), 2027–2050. doi:10.3390/en7042027

Li, G., Luo, T., & Song, Y. (2021). Climate change mitigation efficiency of electric vehicle charging infrastructure in China: From the perspective of energy transition and circular economy. *Resources, Conservation and Recycling*. Advance online publication. doi:10.1016/j.resconrec.2021.106048

Li, J.-F., Ma, Z.-Y., Zhang, Y.-X., & Wen, Z.-C. (2018). Analysis on energy demand and CO2 emissions in China following the Energy Production and Consumption Revolution Strategy and China Dream target. *Advances in Climate Change Research*, *9*(1), 16–26. doi:10.1016/j.accre.2018.01.001

Li, J., & Sun, C. (2018). Towards a low carbon economy by removing fossil fuel subsidies? *China Economic Review*, *50*, 17–33. doi:10.1016/j.chieco.2018.03.006

Liu, C. (2010). Developing green intellectual capital in companies by AHP. In *2010 8th International Conference on Supply Chain Management and Information* (pp. 1-5). IEEE.

Liu, J., & Meng, Z. (2017). Innovation Model Analysis of New Energy Vehicles: Taking Toyota, Tesla and BYD as an Example. *Procedia Engineering*, *174*, 965–972. doi:10.1016/j.proeng.2017.01.248

Liu, W., Ma, Q., & Liu, X. (2022). Research on the dynamic evolution and its influencing factors of stock correlation network in the Chinese new energy market. *Finance Research Letters*, *45*, 102138. Advance online publication. doi:10.1016/j.frl.2021.102138

Liu, Y., Li, M., Lian, H., Tang, X., Liu, C., & Jiang, C. (2018). Optimal dispatch of virtual power plant using interval and deterministic combined optimization. *International Journal of Electrical Power & Energy Systems*, *102*, 235–244. doi:10.1016/j.ijepes.2018.04.011

Li, X., & Nam, K.-M. (2022). Environmental regulations as industrial policy: Vehicle emission standards and automotive industry performance. *Environmental Science & Policy*, *131*, 68–83. doi:10.1016/j.envsci.2022.01.015

Lu, S. T., & Liu, C. (2021). *BYD under the background of "One Belt One Road" Analysis of Competitive Advantages of New Energy Vehicle Exports*. Jilin International Studies University. https://kns.cnki.net/kcms/detail/detail.aspx?dbcode=CJFD&dbname=CJFDAUTO&filename=SCXH202202029&uniplatform=NZKPT&v=K0y7w0XLaFQ50_5cSpT1L5dyjnfS5EVUNqit6jJYWm1WUfwZf3NhVQbJeNCWw5dh

Luptacik, M. (Ed.). (2010). *Mathematical Optimization and Economic Analysis*. Springer. doi:10.1007/978-0-387-89552-9

Lu, W. (2020). Improved k-means clustering algorithm for big data mining under hadoop parallel framework. *Journal of Grid Computing*, *18*(3), 239–250. doi:10.100710723-019-09503-0

Lv, M., Lou, S., Liu, B., Fan, Z., & Wu, Z. (2017, October). Review on power generation and bidding optimization of virtual power plant. In *2017 International Conference on Electrical Engineering and Informatics (ICELTICs)* (pp. 66-71). IEEE. 10.1109/ICELTICS.2017.8253242

Macgregor, N. A., & Cowan, C. E. (2011). Government action to promote sustainable adaptation by the agriculture and land management sector in England. In J. D. Ford & L. Berrang-Ford (Eds.), *Climate Change Adaptation in Developed Nations: From theory to practice* (pp. 385–400). Springer. doi:10.1007/978-94-007-0567-8_28

Maitra, R., Melnykov, V., & Lahiri, S. N. (2012). Bootstrapping for significance of compact clusters in multidimensional datasets. *Journal of the American Statistical Association*, *107*(497), 378–392. doi:10.1080/01621459.2011.646935

Majumdar, J., Naraseeyappa, S., & Ankalaki, S. (2017). Analysis of agriculture data using data mining techniques: Application of big data. *Journal of Big Data*, *4*(1), 1–15. doi:10.118640537-017-0077-4

Makarichi, L., Jutidamrongphan, W., & Techato, K. (2018). The Evolution of Waste-To-Energy Incineration: A Review. *Renewable & Sustainable Energy Reviews*, *91*, 812–821. doi:10.1016/j.rser.2018.04.088

Malgwi, P. G., & Joshua, W. K. (2021). Assessment of the Perception and Awareness of Climate Change and the Influence of Information Amongst Tertiary Education Students in North-East Nigeria. *Library and Information Science Digest*, *14*, 14–24. https://lisdigest.org/index.php/lisd/article/view/154/139

Malik, S., Cao, Y., Mughal, Y., Kundi, G., Mughal, M., & Ramayah, T. (2020). Pathways towards sustainability in organizations: Empirical evidence on the role of green human resource management practices and green intellectual capital. *Sustainability*, *12*(8), 3228. doi:10.3390u12083228

Malone, T. W. (2018). *Superminds. The Surprising Power of People and Computers Thinking Together*. Little, Brown Spark.

Mancarella, P. (2012). Smart multi-energy grid: concepts, benefits and challenges. IEEE PES General Meeting, San Diego, CA, United States.

Mancarella, P. (2014). MES (multi-energy systems): An overview of concepts and evaluation models. *Energy*, *65*, 1–17. doi:10.1016/j.energy.2013.10.041

Mansoor, A., Jahan, S., & Riaz, M. (2021). Does green intellectual capital spur corporate environmental performance through green workforce? *Journal of Intellectual Capital*, *22*(5), 823–839. doi:10.1108/JIC-06-2020-0181

Marco-Lajara, B., Seva-Larrosa, P., Martínez-Falcó, J., & Sánchez-García, E. (2021b). How Has COVID-19 Affected The Spanish Wine Industry? An Exploratory Analysis. *NVEO-Natural Volatiles & Essential Oils Journal*.

Marco-Lajara, B., Zaragoza-Sáez, P. C., Martínez-Falcó, J., & Sánchez-García, E. (2022f). Does green intellectual capital affect green innovation performance? Evidence from the Spanish wine industry. British Food Journal, (ahead-of-print).

Marco-Lajara, B., Zaragoza-Saez, P., & Martínez-Falcó, J. (2022g). Green Innovation: Balancing Economic Efficiency With Environmental Protection. In Frameworks for Sustainable Development Goals to Manage Economic, Social, and Environmental Shocks and Disasters (pp. 239-254). IGI Global.

Marco-Lajara, B., Zaragoza-Saez, P., Falcó, J. M., & Millan-Tudela, L. A. (2022h). Analysing the Relationship Between Green Intellectual Capital and the Achievement of the Sustainable Development Goals. In Handbook of Research on Building Inclusive Global Knowledge Societies for Sustainable Development (pp. 111-129). IGI Global.

Marco-Lajara, B., Zaragoza-Sáez, P., Falcó, J. M., & Millan-Tudela, L. A. (2022i). Corporate Social Responsibility: A Narrative Literature Review. Frameworks for Sustainable Development Goals to Manage Economic, Social, and Environmental Shocks and Disasters, 16-34.

Compilation of References

Marco-Lajara, B., Zaragoza-Saez, P., Falcó, J. M., & Sánchez-García, E. (2022j). COVID-19 and Wine Tourism: A Story of Heartbreak. In Handbook of Research on SDGs for Economic Development, Social Development, and Environmental Protection (pp. 90-112). IGI Global.

Marco-Lajara, B., Zaragoza-Sáez, P., Martínez-Falcó, J., & Sánchez-García, E. (2022l). Green Intellectual Capital in the Spanish Wine Industry. In Innovative Economic, Social, and Environmental Practices for Progressing Future Sustainability (pp. 102-120). IGI Global.

Marco-Lajara, B., Falcó, J. M., Fernández, L. R., & Larrosa, P. S. (2022a). Evolución del pensamiento en la disciplina de dirección estratégica: la visión de la empresa basada en las capacidades dinámicas y en el conocimiento. [Evolution of thinking in the discipline of strategic management: the dynamic capabilities and knowledge-based view of the enterprise] In *Investigación y transferencia de las ciencias sociales frente a un mundo en crisis* [Social science research and transfer in the face of a world in crisis]. (pp. 1801–1826). Dykinson.

Marco-Lajara, B., Sáez, P. D. C. Z., Falcó, J. M., & García, E. S. (2022b). El capital intelectual verde como hoja de ruta para la sostenibilidad: El caso de bodegas Luzón [Green intellectual capital as a roadmap for sustainability: the case of bodegas Luzón]. *GeoGraphos*, *13*(147), 137–146.

Marco-Lajara, B., Sáez, P. D. C. Z., Falcó, J. M., & García, E. S. (2022c). Las rutas del vino de España: el impacto económico derivado de las visitas a bodegas y museos. [Spain's wine routes: the economic impact of visits to wineries and museums] In *Investigación y transferencia de las ciencias sociales frente a un mundo en crisis* [Social science research and transfer in the face of a world in crisis]. (pp. 1774–1800). Dykinson.

Marco-Lajara, B., Seva-Larrosa, P., Martínez-Falcó, J., & García-Lillo, F. (2022d). Wine clusters and Protected Designations of Origin (PDOs) in Spain: An exploratory analysis. *Journal of Wine Research*, *33*(3), 146–167.

Marco-Lajara, B., Seva-Larrosa, P., Ruiz-Fernández, L., & Martínez-Falcó, J. (2021a). The Effect of COVID-19 on the Spanish Wine Industry. In *Impact of Global Issues on International Trade* (pp. 211–232). IGI Global. doi:10.4018/978-1-7998-8314-2.ch012

Marco-Lajara, B., Zaragoza Sáez, P. D. C., & Martínez-Falcó, J. (2022e). Does Green Intellectual Capital Affect Green Performance? The Mediation of Green Innovation. *Telematique*, *21*(1), 4594–4602.

Marco-Lajara, B., Zaragoza-Sáez, P., Martínez-Falcó, J., & Ruiz-Fernández, L. (2022k). The Effect of Green Intellectual Capital on Green Performance in the Spanish Wine Industry: A Structural Equation Modeling Approach. *Complexity*, •••, 2022.

Martin, C., Elges, L., & Norwsworthy, B. (2014). *Protecting Climate Finance: An Anti-Corruption Assessment of the Global Environment Facility's Least Developed Countries Fund & Special Climate Change Fund*. https://www.academia.edu/30853719/PROTECTING_CLIMATE_FINANCE

Marx, K. (1995). *Capital: A New Abridgement*. Oxford University Press.

Mashhour, E., & Moghaddas-Tafreshi, S. M. (2009, January). A review on operation of micro grids and virtual power plants in the power markets. In *2009 2nd International Conference on Adaptive Science & Technology (ICAST)* (pp. 273-277). IEEE. 10.1109/ICASTECH.2009.5409714

Masiero, G., Ogasavara, M. H., Jussani, A. C., & Risso, M. L. (2016). Electric vehicles in China: BYD strategies and government subsidies. *RAI Revista de Administração e Inovação*, *13*(1), 3–11. doi:10.1016/j.rai.2016.01.001

Matemilola, S., Adedeji, O. H., Elegbede, I., & Kies, F. (2019). Mainstreaming Climate Change into the EIA Process in Nigeria: Perspectives from Projects in the Niger Delta Region. *Climate (Basel)*, *7*(2), 29. doi:10.3390/cli7020029

Matsumoto, K., Makridou, G., & Doumpos, M. (2020). Evaluating environmental performance using data envelopment analysis: The case of European Countries. *Journal of Cleaner Production, 272*, 1–13. doi:10.1016/j.jclepro.2020.122637

Ma, Y., Shi, T., Zhang, W., Hao, Y., Huang, J., & Lin, Y. (2019). Comprehensive policy evaluation of NEV development in China, Japan, the United States, and Germany based on the AHP-EW model. *Journal of Cleaner Production, 214*, 389–402. doi:10.1016/j.jclepro.2018.12.119

Mazraeh, H. M., & Pazhouhanfar, M. (2018). Effects of vernacular architecture structure on urban sustainability case study: Qeshm Island, Iran. *Frontiers of Architectural Research, 7*(1), 11–24. doi:10.1016/j.foar.2017.06.006

Mazzucato, M. (2021). *Debunking the Myth of Entrepreneurial State Public Sector-Private Sector Antagonism* [Girişimci Devlet Kamu Sektörü-Özel Sektör Karşıtlığı Masalının Çürütülmesi]. Koç Üniversitesi Yayınları.

McFarland, B. J. (2021). Blue Bonds and Seascape Bonds. In B. J. McFarlan (Ed.), *Conservation of Tropical Coral Reefs* (pp. 621–648). Palgrave Macmillan. doi:10.1007/978-3-030-57012-5_15

McGhie, H. A. (2019). *Museums and the Sustainable Development Goals: a How-to Guide for Museums, Galleries, the Cultural Sector and their Partners*. Curating Tomorrow.

McKinley, D. C., Miller-Rushing, A. J., Ballard, H. L., Bonney, R., Brown, H., Evans, D. M., French, R. A., Parrish, J., Phillips, T., Ryan, S., Shankley, L., Shirk, J., Stepenuck, K., Weltzin, J., Wiggins, A., Boyle, O., Briggs, R., Chapin, S., Hewitt, D., & Soukup, M. (2015). *Investing in citizen science can improve natural resources management and environmental protection. In Issues in Ecology*. Ecological Society of America.

McLean, S., & Charles, D. (2018). Caribbean Development Report: A Perusal of Public Debt in the Caribbean and its Impact on Economic Growth. *ECLAC – Studies and Perspectives Series, 70*, 1-39.

Mead, E., Roser-Renouf, C., Rimal, R. N., Flora, J. A., Maibach, E. W., & Leiserowitz, A. (2012). Information Seeking About Global Climate Change Among Adolescents: The Role of Risk Perceptions, Efficacy Beliefs, and Parental Influences. *Atlantic Journal of Communication, 20*(1), 31–52. doi:10.1080/15456870.2012.637027 PMID:22866024

Mekonnen, M. M., & Hoekstra, A. Y. (2016). Four billion people facing severe water scarcity. *Science Advances, 2*(2), e1500323. doi:10.1126ciadv.1500323 PMID:26933676

Memarian, G., & Brown, F. E. (2003). Climate, Culture, And Religion: Aspects Of The Traditional Courtyard House In Iran. *Journal of Architectural and Planning Research, 20*(3), 181–198.

MENR. (2022). *YEKA Modeli*. Retrieved January 11, 2022, from https://enerji.gov.tr/eigm-yenilenebilir-enerji-uretim-faaliyetleri-yeka-modeli

Mesjasz-Lech, A. (2014). Municipal waste management in context of sustainable urban development. *Procedia: Social and Behavioral Sciences, 151*, 244–245. doi:10.1016/j.sbspro.2014.10.023

Mirimanoff, D. (1917). The antinomies of Russel and Burali-Forti and the fundamental problem of set theory [Les antinomies de Russel et de Burali-Forti et le probleme fundamental de la théorie des ensembles]. *L'Enseygnement Mathematiques, 19*, 37–52.

Mission climate-neutral and Smart Cities: Info kit for cities now available. Research and innovation. (2021, October 29). Retrieved from https://research-and-innovation.ec.europa.eu/news/all-research-and-innovation-news/mission-climate-neutral-and-smart-cities-info-kit-cities-now-available-2021-10-29_en

Mitchell, T. (2015). *Debt Swaps for Climate Change Adaptation and Mitigation: A Commonwealth Proposal*. Commonwealth Secretariat.

Miyan, M. A. (2015). Droughts in Asian least developed countries: Vulnerability and sustainability. *Weather and Climate Extremes*, 7, 8–23. doi:10.1016/j.wace.2014.06.003

Mnatsakanyan, A., & Kennedy, S. W. (2014). A novel demand response model with an application for a virtual power plant. *IEEE Transactions on Smart Grid*, 6(1), 230–237. doi:10.1109/TSG.2014.2339213

Mohamed, T., & Hanane, A. (2021). The energy transition in algeria: How to prepare after oil on the horizon 2030? [La transition énergétique en Algérie: comment préparer l'après pétrole à l'horizon 2030?]. *Journal of Economic Sciences Institute*, 24(1), 1367–1382.

Mokyr, J. (2005). The intellectual origins of modern economic growth. *Journal of Economic History*, 65(2), 285 – 351.

Mokyr, J. (2009). *The Enlightened Economy*. Yale University Press.

Mongabay. (2021). *Dominica Forest Information and Data*. https://rainforests.mongabay.com/deforestation/2000/Dominica.htm

Morin, E. (2006). Organization and Complexity. *Annals of the New York Academy of Sciences*, 879(1), 115–121. doi:10.1111/j.1749-6632.1999.tb10410.x

Moser, S. C. (2011). Entering the period of consequences: The explosive us awakening to the need for adaptation. In J. D. Ford & L. Berrang-Ford (Eds.), *Climate Change Adaptation in Developed Nations: From theory to practice*. Springer. doi:10.1007/978-94-007-0567-8_3

Moskalenko, N. (2014). *Optimal Dynamic Energy Management System in Smart Homes*. Otto-von-Guericke-Universität Magdeburg.

Motealleh, P., Zolfaghari, M., & Parsaee, M. (2018). Investigating climate responsive solutions in vernacular architecture of Bushehr city. *HBRC Journal*, 14(2), 215–223. doi:10.1016/j.hbrcj.2016.08.001

Mouritsen, J. (2003). Intellectual capital and the capital market: The circulability of intellectual capital. *Accounting, Auditing & Accountability Journal*, 16(1), 18–30. doi:10.1108/09513570310464246

MT. (2021). *Yeşil Mutabakat Eylem Planı*. Retrieved October 1, 2021, from https://ticaret.gov.tr/data/60f1200013b876eb28421b23/MUTABAKAT%20YE%C5%9E%C4%B0L.pdf

MTF. (2021). *Republic of Turkey Sustainable Finance Framework*. Retrieved January 11, 2022, from https://ms.hmb.gov.tr/uploads/2021/11/Republic-of-Turkey-Sustainable-Finance-Framework.pdf

Murray, G. (1996). A synthesis of six exploratory, European case studies of successfully-exited, venture capital-financed, new technology-based firms. *Entrepreneurship Theory and Practice*, 20(4), 41–60. doi:10.1177/104225879602000404

Nabakov, P. (1999). Encyclopedia of Vernacular Architecture of the World. Traditional Dwellings and Settlements Review, 10(2), 69–75.

Naval, N., & Yusta, J. M. (2021). Virtual power plant models and electricity markets-A review. *Renewable & Sustainable Energy Reviews*, 149, 111393. doi:10.1016/j.rser.2021.111393

Nawrin, S., Rahman, M. R., & Akhter, S. (2017). Exploring k-means with internal validity indexes for data clustering in traffic management system. *International Journal of Advanced Computer Science and Applications*, 8(3), 264–272. doi:10.14569/IJACSA.2017.080337

Nazari, B., Liaghat, A., Akbari, M. R., & Keshavarz, M. (2018). Irrigation water management in Iran: Strategic planning for improving water use efficiency. *Agricultural Water Management*, 208, 7–18. doi:10.1016/j.agwat.2018.06.003

Nelson, G. C., Van der Mensbrugghe, D., Ahammad, H., Blanc, E., Calvin, K., Hasegawa, T., Havlik, P., Heyhoe, E., Kyle, P., Lotze-Campen, H., von Lampe, M., Mason d'Croz, D., van Meijl, H., Müller, C., Reilly, J., Robertson, R., Sands, R. D., Schmitz, C., Tabeau, A., ... Willenbockel, D. (2014). Agriculture and climate change in global scenarios: Why don't the models agree. *Agricultural Economics*, *45*(1), 85–101. doi:10.1111/agec.12091

Nguyen, A. T., Truong, N. S. H., Rockwood, D., & Tran Le, A. D. (2019). Studies on sustainable features of vernacular architecture in different regions across the world: A comprehensive synthesis and evaluation. *Frontiers of Architectural Research*, *8*(4), 535–548. doi:10.1016/j.foar.2019.07.006

Nikonowicz, Ł., & Milewski, J. (2012). Virtual power plants-general review: structure, application and optimization. *Journal of Power Technologies*, *92*(3).

Niroumand, H., Zain, M. F. M., & Jamil, M. (2013). A guideline for assessing of critical parameters on Earth architecture and Earth buildings as a sustainable architecture in various countries. *Renewable & Sustainable Energy Reviews*, *28*, 130–165. doi:10.1016/j.rser.2013.07.020

Nisar, Q. A., Haider, S., Ali, F., Jamshed, S., Ryu, K., & Gill, S. (2021). Green human resource management practices and environmental performance in Malaysian green hotels: The role of green intellectual capital and pro-environmental behavior. *Journal of Cleaner Production*, *311*, 127504. doi:10.1016/j.jclepro.2021.127504

Norouzzadeh, M. S., Nguyen, A., Kosmala, M., Swanson, A., Palmer, M. S., Packer, C., & Clune, J. (2018). Automatically identifying, counting, and describing wild animals in camera-trap images with Deep Learning. *Proceedings of the National Academy of Sciences of the United States of America*, *115*(25). Advance online publication. doi:10.1073/pnas.1719367115 PMID:29871948

Nosratabadi, S. M., Hooshmand, R. A., & Gholipour, E. (2017). A comprehensive review on microgrid and virtual power plant concepts employed for distributed energy resources scheduling in power systems. *Renewable & Sustainable Energy Reviews*, *67*, 341–363. doi:10.1016/j.rser.2016.09.025

O'Brien, J., & Stoch, H. (2021). ESG: Getting serious about decarbonisation: From Strategy to Execution. Deloitte Insights.

O'Neill; K. & Rudden, PJ. (2013). *Environmental Benchmarking & Best Practice Report RPS Group, Ireland*. European Green Capital Award 2012 & 2013.

OECD. (2021). *Government at a Glance 2021*. OECD Publishing. doi:10.1787/1c258f55-

OIV. (2004). *Resolution CST 1/2004-Develpment of Sustainable Vitivinivulture*. Pareis.

OIV. (2008). *Resolution CST/2008-OIV Guidelines for Sustainable Vitiviniculture: Production, Processing and Packaging of Products*. Verone/it.

OIV. (2016). *Resolution CST 518/2016-OIV General Principles of Sustainable Vitiviniculture – Environmental - Social - Economic and Cultural Aspects*. Brento Gonçalves.

Olukoya, O. A. P., & Atanda, J. O. (2020). Assessing the Social Sustainability Indicators in Vernacular Architecture—Application of a Green Building Assessment Approach. *Environments*, *7*(9), 67. Advance online publication. doi:10.3390/environments7090067

Öner, Y., & Bulut, H. (2021). A robust EM clustering approach: ROBEM. *Communications in Statistics. Theory and Methods*, *50*(19), 4587–4605. doi:10.1080/03610926.2020.1722840

Orakpo, E. (2017, January 4). *Earthquake in Nigeria: Measures to avert devastating impacts— Experts*. Vanguard News. https://www.vanguardngr.com/2017/01/earthquake-nigeria-measures-avert-devastating-impacts-experts/

Ortega Díaz, A., & Casamadrid Gutiérrez, E. (2018). Competing actors in the climate change arena in Mexico: A network analysis. *Journal of Environmental Management*, *215*, 239–247. doi:10.1016/j.jenvman.2018.03.056 PMID:29573674

Overmyer, T. C. (2015). *Urban resilience: Reframing climate change for action and advocacy* [Unpublished MSc. Thesis]. Purdue University, West Lafayette, IN, United States.

Oxfam. (2018). *Climate Finance Shadow Report 2018*. https://d1tn3vj7xz9fdh.cloudfront.net/s3fs-public/file_attachments/bp-climate-finance-shadow-report-030518-en.pdf

PACT (Protected Areas Conservation Trust). (2011). *Strategic Plan 2011 - 2016*. https://issuu.com/pact.belize/docs/strategicplan1116

PACT (Protected Areas Conservation Trust). (2017). *Home*. https://www.pactbelize.org/

Painter, J., & Gavin, N. T. (2015). Climate Skepticism in British Newspapers, 2007–2011. *Environmental Communication*, *10*(4), 432–452. doi:10.1080/17524032.2014.995193

Palizban, O., Kauhaniemi, K., & Guerrero, J. M. (2014). Microgrids in active network management—Part I: Hierarchical control, energy storage, virtual power plants, and market participation. *Renewable & Sustainable Energy Reviews*, *36*, 428–439. doi:10.1016/j.rser.2014.01.016

Pal, N. R., & Biswas, J. (1997). Cluster validation using graph theoretic concepts. *Pattern Recognition*, *30*(4), 847–857. doi:10.1016/S0031-3203(96)00127-6

Pardo, L. E., Bombaci, S. P., Huebner, S., Somers, M. J., Fritz, H., Downs, C., Guthmann, A., Hetem, R. S., Keith, M., le Roux, A., Mgqatsa, N., Packer, C., Palmer, M. S., Parker, D. M., Peel, M., Slotow, R., Strauss, W. M., Swanepoel, L., Tambling, C., ... Venter, J. A. (2021). Snapshot safari: A large-scale collaborative to monitor Africa's remarkable biodiversity. *South African Journal of Science*, *117*(1/2). Advance online publication. doi:10.17159ajs.2021/8134

Patton, M. (2002). *Qualitative Research and Evaluation Methods* (3rd ed.). Sage.

Pedrasa, M. A. A., Spooner, T. D., & MacGill, I. F. (2011). A novel energy service model and optimal scheduling algorithm for residential distributed energy resources. *Electric Power Systems Research*, *81*(12), 2155–2163. doi:10.1016/j.epsr.2011.06.013

Pennington-Gray, L. (2017). Reflections to move forward: Where destination crisis management research needs to go. *Tourism Management Perspectives*, *25*, 136–139. doi:10.1016/j.tmp.2017.11.013

Pérez Salazar, B. (2007). Review of "An Inconvenient Truth. The Planetary Crisis of Global Warming and How to Deal with It" by Al Gore [Reseña de "Una verdad incómoda. La crisis planetaria del calentamiento global y cómo afrontarla" de Al Gore]. *Revista de Economía Institucional, Colombia*, *9*(17), 385–395.

Peri, M. (2019). The Green Advantage: Exploring the Convenience of Issuing Green Bonds. *Journal of Cleaner Production*, *219*, 127–135. doi:10.1016/j.jclepro.2019.02.022

Pew Research Center. (2012, July 16). *YouTube & News*. https://www.pewresearch.org/journalism/2012/07/16/youtube-news/

Pew Research Center. (2015, December 2). *Women more than men see climate change as personal threat*. https://www.pewresearch.org/fact-tank/2015/12/02/women-more-than-men-say-climate-change-will-harm-them-personally/

Philokyprou, M., Michael, A., Savvides, A., & Malaktou, E. (2015). Evaluation of bioclimatic design features of vernacular architecture in Cyprus. *Case studies from rural settlements in different climatic regions*.

Piaget, J. V. F. (2008). *The Psychology of Intelligence*. Direct-Media.

Pichler, M., Krenmayr, N., Schneider, E., & Brand, U. (2021). EU industrial policy: Between modernization and transformation of the automotive industry. *Environmental Innovation and Societal Transitions*, *38*, 140–152. doi:10.1016/j.eist.2020.12.002

Pilipenko, A. I. (2020). Education and theory of psychological and cognitive barriers: human capital as driver of stable economic growth. In Social, Economic, and Environmental Impacts Between Sustainable Financial Systems and Financial Markets. IGI Global. doi:10.4018/978-1-7998-1033-9.ch011

Pilipenko, A. I., Pilipenko, Z. A., & Pilipenko, O. I. (2022a). Patterns of Self-Sufficient Companies' Network Interactions Reorganization due to COVID-19: Dialectics of Organizational Structures Optimization. In Challenges and Emerging Strategies for Global Networking Post COVID-19. IGI Global.

Pilipenko, A. I., Pilipenko, Z. A., & Pilipenko, O. I. (2022b). Dialectics of Self-Movement of Resilient Companies in the Economy and Society post COVID-19: Patterns of Organizational Transformations of Networking Interactions. In Challenges and Emerging Strategies for Global Networking Post COVID-19. IGI Global.

Pilipenko, A., Pilipenko, O., & Pilipenko, Z. (2019). Education and inclusive development: puzzle of low-learning equilibrium. In Modeling Economic Growth in Contemporary Russia. Bradford, UK: Emerald Publishing Limited. doi:10.1108/978-1-78973-265-820191006

Pilipenko, O. I., Pilipenko, Z. A., & Pilipenko, A. I. (2021). Theory of Shocks, COVID-19, and Normative Fundamentals for Policy Responses. IGI Global.

Pilipenko, O. I., Pilipenko, Z. A., & Pilipenko, A. I. (2022). COVID-19 Shock and Subsequent Crisis: How It Was. In M. Khosrow-Pour (Ed.), *Research Anthology on Business Continuity and Navigating Times of Crisis* (Vols. 1–4). Information Resources Management Association. doi:10.4018/978-1-6684-4503-7.ch073

Pineda-Escobar, M. (2019). Moving the 2030 agenda forward: SDG implementation in Colombia. *Corporate Governance: The International Journal of Business in Society*, *19*(1), 176-188.

Pinsky, A.A. (1978). Methodology as a science. *Soviet Pedagogy*, 12.

Plan, A. (2011). *Communication from the Commission to the European Parliament, the Council, the European Economic and Social Committee and the Committee of the Regions*. European Commission.

Plastic spotter: Spot plastic in the canals of Leiden. (2019, November 27). Leiden University. Retrieved from https://www.universiteitleiden.nl/en/news/2019/11/plastic-spotters-launch

Podder, A. K., Islam, S., Kumar, N. M., Chand, A. A., Rao, P. N., Prasad, K. A., Logeswaran, T., & Mamun, K. A. (2020). Systematic categorization of optimization strategies for virtual power plants. *Energies*, *13*(23), 6251. doi:10.3390/en13236251

Porter, J. R., Xie, L., Challinor, A. J., Cochrane, K., Howden, S. M., Iqbal, M. M., Lobell, D. B., & Travasso, M. I. (2014). Food security and food production systems. In C. B. Field, V. R. Barros, D. J. Dokken, K. J. Mach, M. D. Mastrandrea, T. E. Bilir, M. Chatterjee, K. L. Ebi, Y. O. Estrada, R. C. Genova, B. Girma, E. S. Kissel, A. N. Levy, S. MacCracken, P. R. Mastrandrea, & L. L. White (Eds.), Climate Change 2014: Impacts, Adaptation, and Vulnerability. Part A: Global and Sectoral Aspects (pp. 485-535). Contribution of Working Group II to the Fifth Assessment Report of the Intergovernmental Panel on Climate Change. Cambridge University Press.

Potvin, A. (1997). The arcade environment. *Architectural Research Quarterly*, *2*(4), 64–79. doi:10.1017/S1359135500001603

Prayag, G. (2018). Symbiotic relationship or not? Understanding resilience and crisis management in tourism. *Tourism Management Perspectives*, *25*, 133–135. doi:10.1016/j.tmp.2017.11.012

Prigogine, I., & Stengers, I. (2018). *Order out of chaos: Man's new dialogue with nature*. Verso Books.

Proverbs, D. (2010). The Impact of Climate Change on Local Scale Flood Risk for Indivual developments. School of Engineering and the Built Enviroment, University of Wolverhampton, United Kingdom. *Systematic Reviews*, 89.

Pudjianto, D., Ramsay, C., & Strbac, G. (2007). Virtual power plant and system integration of distributed energy resources. *IET Renewable Power Generation*, *1*(1), 10–16. doi:10.1049/iet-rpg:20060023

Pu, G., Zhu, X., Dai, J., & Chen, X. (2022). Understand technological innovation investment performance: Evolution of industry-university-research cooperation for technological innovation of lithium-ion storage battery in China. *Journal of Energy Storage*, *46*, 103607. Advance online publication. doi:10.1016/j.est.2021.103607

Qiu, J., Meng, K., Zheng, Y., & Dong, Z. Y. (2017). Optimal scheduling of distributed energy resources as a virtual power plant in a transactive energy framework. *IET Generation, Transmission & Distribution*, *11*(13), 3417–3427. doi:10.1049/iet-gtd.2017.0268

Raab, A. F., Ferdowsi, M., Karfopoulos, E., Unda, I. G., Skarvelis-Kazakos, S., Papadopoulos, P., . . . Strunz, K. (2011, September). Virtual power plant control concepts with electric vehicles. In *2011 16th International Conference on Intelligent System Applications to Power Systems* (pp. 1-6). IEEE. 10.1109/ISAP.2011.6082214

Race to Zero Campaign. (n.d.). Retrieved from https://unfccc.int/climate-action/race-to-zero-campaign

Ramasamy, S., & Nirmala, K. (2020). Disease prediction in data mining using association rule mining and keyword-based clustering algorithms. *International Journal of Computers and Applications*, *42*(1), 1–8. doi:10.1080/1206212X.2017.1396415

Rasmussen, Z. A. (2015). Coyotes on the Web: Understanding human-coyote interaction and online education using citizen science. *Dissertations and Theses, Paper 2643*. . doi:10.15760/etd.2639

Rasoulinezhad, E., & Taghizadeh-Hesary. (2022). Role of green finance in improving energy efficiency and renewable energy development. *Energy Efficiency*, *15*(14), 1–12.

Recioui, A., & Bentarzi, H. (Eds.). (2020). *Optimizing and Measuring Smart Grid Operation and Control*. IGI Global.

Recovery and resilience facility. (2022, July 29). Retrieved from https://ec.europa.eu/info/business-economy-euro/recovery-coronavirus/recovery-and-resilience-facility_en

Reisinger, A., Kitching, R. L., Chiew, F., Hughes, L., Newton, P. C. D., Schuster, S. S., Tait, A., & Whetton, P. (2014). Australia. In V. R. Barros, C. B. Field, D. J. Dokken, M. D. Mastrandrea, K. J. Mach, T. E. Bilir, M. Chatterjee, K. L. Ebi, Y. O. Estrada, R. C. Genova, B. Girma, E. S. Kissel, A. N. Levy, S. MacCracken, P. R. Mastrandrea, & L. L. White (Eds.), Climate Change 2014: Impacts, Adaptation, and Vulnerability (pp. 1371-1438). Part B: Regional Aspects. Contribution of Working Group II to the Fifth Assessment Report of the Intergovernmental Panel on Climate Change. Cambridge University Press.

REN-21. (2016). *Renewables – Global Status Report: 2016*. Retrieved June 4, 2017, from http://www.ren21.net/wp-content/uploads/2016/05/GSR_2016_Full_Report_lowres.pdf

Richter, A. (2018). *World Bank Has Approved Funding of $27 Million for the Development of a 7 MW Geothermal Power Plant in the Commonwealth of Dominica in the Caribbean*. https://www.thinkgeoenergy.com/world-bank-approves-27m-for-7-mw-dominica-geothermal-project-caribbean/

Richter, A., Moskalenko, N., Hauer, I., Schröter, T., & Wolter, M. (2017, June). Technical integration of virtual power plants into German system operation. In *2017 14th International Conference on the European Energy Market (EEM)* (pp. 1-6). IEEE. 10.1109/EEM.2017.7981876

Rios, A. M., & Picazo-Tadeo, A. J. (2021). Measuring environmental performance in the treatment of municipal solid waste: The case of the European Union-28. *Ecological Indicators*, *123*, 1–10. doi:10.1016/j.ecolind.2020.107328

Rogers, P. P., Jalal, K. F., & Boyd, J. A. (Eds.). (2008). An introduction to sustainable development. Earthscan (Glen Educational Foundation).

Romer, P. M. (1990). Endogenous technological change. *Journal of Political Economy*, *98*(5), S71-S102.

Romer, P.M. (1988). Increasing returns and long-run growth. *The Journal of Political Economy*, *94*(5), 1002-1037.

Romo, R., & Micheloud, O. (2015). Power quality of actual grids with plug-in electric vehicles in presence of renewables and microgrids. *Renewable & Sustainable Energy Reviews*, *46*, 189–200. doi:10.1016/j.rser.2015.02.014

Rosati, F., & Faria, L. (2019). Addressing the SDGs in sustainability reports: The relationship with institutional factors. *Journal of Cleaner Production*, *215*, 1312–1326. doi:10.1016/j.jclepro.2018.12.107

Rosenthal, I., Byrnes, J., Cavanaugh, K., Bell, T., Harder, B., Haupt, A., Rassweiler, A., Perez-Matus, A., Assis, J., Swanson, A., Boyer, A., McMaster, A., Trouille, L. (2018). *Floating forests: Quantitative validation of citizen science data generated from consensus classifications*. Academic Press.

Rowley, J., Callaghan, C., Cutajar, T., Portway, C., Potter, K., Mahony, S., Trembath, D., Flemons, P., & Woods, A. (2019). FrogID: Citizen scientists provide validated biodiversity data on frogs of Australia. *Herpetological Conservation and Biology*, *14*(1), 155–170.

Roy, J., Tschakert, P., Waisman, H., Abdul Halim, S., Antwi-Agyei, P., Dasgupta, P., Hayward, B., Kanninen, M., Liverman, D., Okereke, C., Pinho, P. F., Riahi, K., & Suarez Rodriguez, A. G. (2018). Sustainable Development, Poverty Eradication and Reducing Inequalities. In V. Masson-Delmotte, P. Zhai, H. O. Pörtner, D. Roberts, J. Skea, P. R. Shukla, A. Pirani, W. Moufouma-Okia, C. Péan, R. Pidcock, S. Connors, J. B. R. Matthews, Y. Chen, X. Zhou, M. I. Gomis, E. Lonnoy, T. Maycock, M. Tignor, & T. Waterfield (Eds.), Global Warming of 1.5°C. An IPCC Special Report on the impacts of global warming of 1.5°C above pre-industrial levels and related global greenhouse gas emission pathways, in the context of strengthening the global response to the threat of climate change, sustainable development, and efforts to eradicate poverty. World Meteorological Organization.

Rozenblum, Y., Dalyot, K., Lachman, E., & Baram-Tsabari, A. (2018). Gendered engagement with posts authored by female scientists on Facebook. *Proceedings of the 15th Chais Conference for the Study of Innovation and Learning Technologies: Learning in the Digital Era*.

Sachs, J. D. (Ed.). (2015). *The age of sustainable development*. Columbia University Press. doi:10.7312ach17314

Sahebzadeh, S., Heidari, A., Kamelnia, H., & Baghbani, A.-N. (2017). Sustainability Features of Iran's Vernacular Architecture: A Comparative Study between the Architecture of Hot–Arid and Hot–Arid–Windy Regions. *Sustainability*, *9*(5), 749. doi:10.3390u9050749

Salmani, M. A., Tafreshi, S. M., & Salmani, H. (2009). Operation optimization for a virtual power plant. *Proceedings of the 2009 IEEE PES/IAS Conference on Sustainable Alternative Energy (SAE)*, 1–6. 10.1109/SAE.2009.5534848

Salman, M. (2018). Sustainability and vernacular architecture: Rethinking what identity is. In *Urban and Architectural Heritage Conservation within Sustainability*. IntechOpen.

Saoud, A., & Recioui, A. (2017). A review on Data communication in smart grids. *Algerian Journal of Signals and Systems*, *2*(3), 162–179. doi:10.51485/ajss.v2i3.42

Schäufele, I., & Hamm, U. (2017). Consumers' perceptions, preferences and willingness-to-pay for wine with sustainability characteristics: A review. *Journal of Cleaner Production*, *147*, 379–394. doi:10.1016/j.jclepro.2017.01.118

Compilation of References

Schipper, E. L. F. (2006). Conceptual History of Adaptation in the UNFCCC Process. *Review of European Community & International Environmental Law*, *15*(1), 82–92. doi:10.1111/j.1467-9388.2006.00501.x

Schleicher, K., & Schmidt, C. (2020). Citizen science in Germany as research and Sustainability Education: Analysis of the main forms and foci and its relation to the Sustainable Development Goals. *Sustainability*, *12*(15), 6044. doi:10.3390u12156044

Schmid-Petri, H. (2017). Do Conservative Media Provide a Forum for Skeptical Voices? The Link Between Ideology and the Coverage of Climate Change in British, German, and Swiss Newspapers. *Environmental Communication*, *11*(4), 554–567. doi:10.1080/17524032.2017.1280518

Schwab, K., & Malleret, T. (2020, July 14). *COVID-19's legacy: This is how to get the Great Reset right.* World Economic Forum. Retrieved from https://www.weforum.org/agenda/2020/07/covid19-this-is-how-to-get-the-great-reset-right

Scott, K., & Schulz, L. (2017). Lookit (part 1): A new online platform for Developmental Research. *Open Mind: Discoveries in Cognitive Science*, *1*(1), 4–14. doi:10.1162/OPMI_a_00002

Scott-Parker, B., Nunn, P. D., Mulgrew, K., Hine, D., Marks, A., Mahar, D., & Tiko, L. (2016). Pacific Islanders' understanding of climate change: Where do they source information and to what extent do they trust it? *Regional Environmental Change*, *17*(4), 1005–1015. doi:10.100710113-016-1001-8

Secretaría de Economía. (2009, June 30). Agreement establishing the stratification of micro, small and medium-sized enterprises [Acuerdo por el que se establece la estratificación de las micro, pequeñas y medianas empresas]. *Diario Oficial de la Federación*.

Secretaría de Economía. (2012). *Mexican Business Information System* [Sistema de Información Empresarial Mexicano]. Recovered from http://www.siem.gob.mx/siem/

Seo, Y.-C., Alam, T., & Yang, W.-S. (2017). Gasification of Municipal Solid Waste. In Y. Yun (Ed.), *Gasification for Low-grade Feedstock* (pp. 115–136). Intechopen.

Sharafi, L., Zarafshani, K., Keshavarz, M., Azadi, H., & Van Passel, S. (2020). Drought risk assessment: Towards drought early warning system and sustainable environment in western Iran. *Ecological Indicators*, *114*, 106276. doi:10.1016/j.ecolind.2020.106276

Sharafi, L., Zarafshani, K., Keshavarz, M., Azadi, H., & Van Passel, S. (2021). Farmers' decision to use drought early warning system in developing countries. *The Science of the Total Environment*, *758*, 142761. doi:10.1016/j.scitotenv.2020.142761 PMID:33183818

Shchedrovitsky, G. P. (1997). *Philosophy. Science. Methodology* [Filosofija, nauka, metodologija]. Shkoly kulturnoi politiki.

Shen, J., Chang, S. I., Lee, E. S., Deng, Y., & Brown, S. J. (2005). Determination of cluster number in clustering microarray data. *Applied Mathematics and Computation*, *169*(2), 1172–1185. doi:10.1016/j.amc.2004.10.076

Sheptulin, A. P. (1975). Dialectics of the Special and of the General. Krasnoyarsk.

Sheptulin, A. P. (1978). *Marxist-Leninist Philosophy*. Progress Publishers.

Shi, H., Chai, J., Lu, Q., Zheng, J., & Wang, S. (2022). The impact of China's low-carbon transition on economy, society and energy in 2030 based on CO2 emissions drivers. *Energy*, *239*, 122336. Advance online publication. doi:10.1016/j.energy.2021.122336

SHURA Enerji Dönüşümü Merkezi. (2019). *Türkiye'de Enerji Dönüşümünün Finansmanı*. Retrieved July 25, 2020, from https://www.shura.org.tr/wp-content/uploads/2019/10/Turkiyede_Enerji_Donusumunun_Finansmani.pdf

SHURA Enerji Dönüşümü Merkezi. (2020a). *Enerji verimliliği çözümü: Finansman mekanizmaları*. Retrieved June 20, 2021, from https://www.shura.org.tr/wp-content/uploads/2020/10/SHURA_FinansmanMekanizmalari.pdf

SHURA Enerji Dönüşümü Merkezi. (2020b). *Türkiye enerji dönüşümünü hızlandırmak için 2020 yılı sonrası düzenleyici politika mekanizması seçenekleri: Şebeke ölçeğinde ve dağıtık güneş ve rüzgar enerjisi kapasite kurulumları*. Retrieved January 10, 2022, from https://www.shura.org.tr/wp-content/uploads/2021/01/2020_yili_sonrasi_duzenleyici_politika.pdf?_ga=2.202753873.1740824367.1645079503-919185812.1644914363

SHURA Enerji Dönüşümü Merkezi. (2021). *Türkiye'de Yenilenebilir Enerji Tedariki ve Belgelenmesi*. Retrieved January 10, 2022, from https://shura.org.tr/turkiyede-yenilenebilir-enerji-tedariki-ve-belgelenmesi/

Siddaiah, R., & Saini, R. P. (2016). A review on planning, configurations, modeling and optimization techniques of hybrid renewable energy systems for off-grid applications. *Renewable & Sustainable Energy Reviews, 58*, 376–396. doi:10.1016/j.rser.2015.12.281

Siddiqi, T. (2011). China and India: More Cooperation than Competition in Energy and Climate Change. *Journal of International Affairs, 64*(2), 73–90.

Simpson, B., & Burpee, G. (2014). *Adaptation under the new normal of climate change: The future of agricultural extension and advisory services*. USAID.

Singh, K., Kelly, S. O., & Sastry, M. K. S. (2009). Municipal Solid Waste to Energy: An Economic and Environmental Assessment for Application in Trinidad and Tobago. *The Journal of the Association of Professional Engineers of Trinidad and Tobago, 38*(1), 42–49.

Singh, M. K., Mahapatra, S., & Atreya, S. K. (2010). Thermal performance study and evaluation of comfort temperatures in vernacular buildings of North-East India. *1st International Symposium on Sustainable Healthy Buildings, 45*(2), 320–329. 10.1016/j.buildenv.2009.06.009

Skarvelis-Kazakos, S., Rikos, E., Kolentini, E., Cipcigan, L. M., & Jenkins, N. (2013). Implementing agent-based emissions trading for controlling Virtual Power Plant emissions. *Electric Power Systems Research, 102*, 1–7. doi:10.1016/j.epsr.2013.04.004

Slobodchicov, V. I., & Isaev, E. I. (1995). Human Psychology. Academic Press.

Slobodchicov, V. I., & Isaev, E. I. (2000). Psychology of Human Development. Academic Press.

Slobodchicov, V. I., & Isaev, E. I. (2013). *Psychology of Human Education. Formation of Subjectivity in Educational Processes*. Academic Press.

Smart Cities Community. (n.d.). Institute of Electrical and Electronics Engineers (IEEE). Retrieved from https://www.ieee.org/membership-catalog/productdetail/showProductDetailPage.html?product=CMYSC764

Smart sustainable cities: An analysis of definitions. (2014). ITU-T Focus Group on Smart Sustainable Cities.

Soflaei, F., Shokouhian, M., & Soflaei, A. (2017). Traditional courtyard houses as a model for sustainable design: A case study on BWhs mesoclimate of Iran. *Frontiers of Architectural Research, 6*(3), 329–345. doi:10.1016/j.foar.2017.04.004

Söküt Açar, T., & Ayman Öz, N. (2020). The determination of optimal cluster number by silhouette index at clustering of the European Union Member Countries and candidate Turkey by waste indicators. *Pamukkale University Journal of Engineering Sciences, 26*(3), 481–487. doi:10.5505/pajes.2019.49932

Strasser, B. J., Baudry, J., Mahr, D., Sanchez, G., & Tancoigne, E. (2018). "Citizen science"? Rethinking science and public participation. *Science & Technology Studies*, 52–76. doi:10.23987ts.60425

Compilation of References

Sun, S., Jin, C., He, W., Li, G., Zhu, H., & Huang, J. (2021). Management status of waste lithium-ion batteries in China and a complete closed-circuit recycling process. *The Science of the Total Environment*, *776*, 145913. doi:10.1016/j.scitotenv.2021.145913 PMID:33639457

Sussman, F. G., & Freed, J. R. (2008). *Adapting to climate change: A business approach*. Pew Center on Global Climate Change. Recovered from https://www.c2es.org/business-adaptation

Sutcliffe, H. (2020). COVID-19: The 4 building blocks of the Great Reset. *Sustainable Development Impact Summit*.

Su, W., & Wang, J. (2012). Energy management systems in microgrid operations. *The Electricity Journal*, *25*(8), 45–60. doi:10.1016/j.tej.2012.09.010

Szolnoki, G. (2013). A cross-national comparison of sustainability in the wine industry. *Journal of Cleaner Production*, *53*, 243–251. doi:10.1016/j.jclepro.2013.03.045

T.C. EPDK. (2005). *Law on the Use of Renewable Energy Sources for Electrical Energy Production* [Yenilenebilir Enerji Kaynaklarının Elektrik Enerjisi Üretimi Amaçlı Kullanımına İlişkin Kanun]. Kanun No: 5346. Retrieved June 20, 2020, from https://www.epdk.gov.tr/Detay/Icerik/3-0-0-122/yenilenebilir-enerji-kaynaklari-destekleme-mekanizmasi-yekdem

T.C. SBB. (2019). *Eleventh Development Plan (2019-2023)* [On Birinci Kalkınma Planı (2019-2023)]. Retrieved June 20, 2020, from https://www.sbb.gov.tr/wp-content/uploads/2021/12/On_Birinci_Kalkinma_Plani-2019-2023.pdf

Taber, C. S., & Lodge, M. (2006). Motivated Skepticism in the Evaluation of Political Beliefs. *American Journal of Political Science*, *50*(3), 755–769. doi:10.1111/j.1540-5907.2006.00214.x

Taleb, N. N. (2012). *Antifragile: Things That Gain from Disorder*. Random House Trade Paperbacks.

Tan, R., Tang, D., & Lin, B. (2018). Policy impact of new energy vehicles promotion on air quality in Chinese cities. *Energy Policy*, *118*, 33–40. doi:10.1016/j.enpol.2018.03.018

Tapia, A., Lalone, N., MacDonald, E., Hall, M., Case, N., & Heavner, M. (2014). *AURORASAURUS:* Citizen science, early warning systems and space weather. *Proceedings of the AAAI Conference on Human Computation and Crowdsourcing*.

Tascikaraoglu, A., Erdinc, O., Uzunoglu, M., & Karakas, A. (2014). An adaptive load dispatching and forecasting strategy for a virtual power plant including renewable energy conversion units. *Applied Energy*, *119*, 445–453. doi:10.1016/j.apenergy.2014.01.020

Tasseron, P., Zinsmeister, H., Rambonnet, L., Hiemstra, A.-F., Siepman, D., & van Emmerik, T. (2020). Plastic hotspot mapping in Urban Water Systems. *Geosciences*, *10*(9), 342. doi:10.3390/geosciences10090342

Tehran Times. (2021). *Rouhani inaugurates 2nd, 3rd phases of Persian Gulf water transfer project*. Https://www.tehrantimes.com/news/459096

Thanassoulis, E. (Ed.). (2001). *Introduction to the theory and application of data envelopment analysis: a foundation text with integrated software*. Springer. doi:10.1007/978-1-4615-1407-7

The business case for inclusive growth. (2018) *Deloitte Global Inclusive Growth Survey*.

The Economist Intelligence Unit. (2021). *Global Food Security Index 2020*. Author.

The People's Government of Hainan Province. (2021). *Hainan Province New Energy Vehicles for Electricity Application Pilot Implementation Plan*. Retrieved from https://www.hainan.gov.cn/hainan/tingju/202103/2c7e2327668348e2ac8b5126c471ed93.shtml

The People's Government of Hainan Province. (2022). *More than 20,000 new charging posts in Hainan province by 2021*. Retrieved from https://www.hainan.gov.cn/hainan/tingju/202201/0f3a34589322481ca2cfbddc3fc90ecb.shtml

The People's Government of Hainan Province. (2022). *Several Measures to Encourage the Use of New Energy Vehicles in Hainan Province in 2022*. Retrieved from https://www.hainan.gov.cn/hainan/tingju/202203/7119c88fcac64961ace0 2720f3229b6f.shtml

The urban agenda for the EU. (n.d.). Retrieved from https://ec.europa.eu/regional_policy/en/policy/themes/urban-development/agenda/#:~:text=The%20Urban%20Agenda%20for%20the%20EU%20is%20an%20integrated%20and,of%20 life%20in%20urban%20areas

The World Bank. (2021). World Development Indicators: *Trends in greenhouse gas emissions. The World Bank*. Retrieved from http://wdi.worldbank.org/table/3.9

Thiagarajan, A., & Sekkizhar, J. (2017). The Impact of Green Intellectual Capital on Integrated Sustainability Performance in the Indian Auto-component Industry. *Journal of Contemporary Research in Management*, *12*(4), 21–78.

Thompson, A. A., Jr., Strickland, A. J., III, & Gamble, J. E. (2008). Strategic Management [Administración Estratégica]. In Teoría y casos (15th ed.). McGraw Hill.

Thravalou, S., Philokyprou, M., Michael, A., & Savvides, A. (2015). *The role of semi-open spaces as thermal environment modifiers in vernacular rural architecture of Cyprus*. Biocultural.

Tian, X., Zhang, Q., Chi, Y., & Cheng, Y. (2021). Purchase willingness of new energy vehicles: A case study in Jinan City of China. *Regional Sustainability*, *2*(1), 12–22. doi:10.1016/j.regsus.2020.12.003

Tone, K. (2001). On returns to scale under weight restrictions in data envelopment analysis. *Journal of Productivity Analysis*, *16*(1), 31–47. doi:10.1023/A:1011147118637

Tone, K., Toloo, M., & Izadikhah, M. (2020). A modified slacks-based measure of efficiency in data envelopment analysis. *European Journal of Operational Research*, *287*(2), 560–571. doi:10.1016/j.ejor.2020.04.019

Topfer, K. (2009). *Energy efficiency in buildings: transforming the market*. WBCSD.

Towards A Green Economy in Europe. (2013). EU Environmental Policy Targets and Objectives 2010-2050. European Environment Agency, Report 8.

TSKB. (2020). *Energy Outlook 2020* [Enerji Görünümü 2020]. Retrieved July 10, 2021, from https://www.tskb.com.tr/i/assets/document/pdf/enerji-sektor-gorunumu-2020.pdf

TSKB. (2021). *Energy Outlook 2021* [Enerji Görünümü 2021]. Retrieved January 20, 2022, from https://www.tskb.com.tr/i/assets/document/pdf/enerji-sektor-gorunumu-2021.pdf

Tzanakakis, V. A., Paranychianakis, N. V., & Angelakis, A. N. (2020). Water Supply and Water Scarcity. *Water (Basel)*, *12*(9), 2347. doi:10.3390/w12092347

Uitto, J. I., & Batra, G. (2022). Transformational Change for People and the Planet: Evaluating. *Environmental Development*.

Ullah, H., Wang, Z., Bashir, S., Khan, A., Riaz, M., & Syed, N. (2021). Nexus between IT capability and green intellectual capital on sustainable businesses: Evidence from emerging economies. *Environmental Science and Pollution Research International*, *28*(22), 27825–27843. doi:10.100711356-020-12245-2 PMID:33515153

Umegbolu, E. (2020). Awarness and Knowledge of Health Implications of Climate Change in Oji River Lga of Enugu State, Southeast Nigeria. *European Journal Pharmaceutical and Medical Research*, 7(7), 204–210. https://www.researchgate.net/profile/Emmanuel-Umegbolu-2/publication/342611614_AWARENESS_AND_KNOWLEDGE_OF_HEALTH_IMPLICATIONS_OF_CLIMATE_CHANGE_IN_OJI_RIVER_LGA_OF_ENUGU_STATE_SOUTHEAST_NIGERIA/links/5efcea56a6fdcc4ca4411954/AWARENESS-AND-KNOWLEDGE-OF-HEALTH-IMPLICATIONS-OF-CLIMATE-CHANGE-IN-OJI-RIVER-LGA-OF-ENUGU-STATE-SOUTHEAST-NIGERIA.pdf

UNEP. (2016). *Definitions and concepts*. Background note. Inquiry working paper 16/13. https://www.unep.org/resources/report/definitions-and-concepts-background-note-inquiry-working-paper-1613

UNESCO. (2016a). *Contemporary learning crisis: education' failure in "creating sustainable future for all"*. In The Global Education Monitoring Report (2nd ed.). UNESCO Publishing.

UNESCO. (2016b). *Education for People and Planet: Creating Sustainable Futures for All*. Global Education Monitoring Report 2016, Paris: UNESCO Publishing. Retrieved from: https://unesdoc.unesco.org/images/0024/002457/245752e.pdf

UNFCCC (United Nations Framework Convention on Climate Change). (2015a). *Background on the UNFCCC: The International Response to Climate Change*. Accessed December 13, 2015. https://unfccc.int/essential_background/items/6031.php

UNFCCC (United Nations Framework Convention on Climate Change). (2015b). *Intended Nationally Determined Contribution (INDC) of the Commonwealth of Dominica*. https://www4.unfccc.int/sites/ndcstaging/PublishedDocuments/Dominica%20First/Commonwealth%20of%20Dominica-%20Intended%20Nationally%20Determined%20Contributions%20(INDC).pdf

United Cities and Local Governments. (2018). *Culture in the sustainable development goals: a guide for local action*. UCLG Committee on Culture.

United Nations Educational Scientific and Cultural Organization. (2014). *Gender equality. Heritage and creativity*. UNESCO.

United Nations Educational Scientific and Cultural Organization. (2016). *Cultural urban future: Global report on culture for sustainable urban development*. UNESCO.

United Nations Framework Convention on Climate Change. (2021). *Nationally determined contributions under the Paris Agreement: Synthesis report by the secretariat*. https://unfccc.int/sites/default/files/resource/cma2021_02_adv_0.pdf

United Nations. (2012). *Back to Our Common Future Sustainable Development in the 21st Century (SD21) Project*. UN.

United Nations. (n.d.). *Global indicator framework for the Sustainable Development Goals and targets of the 2030 Agenda for Sustainable Development*. Retrieved February 8, 2022, from https://unstats.un.org/sdgs/indicators/Global%20Indicator %20Framework%20after%20refinement_Eng.pdf

United Nations. (n.d.a). https://www.un.org/en/observances/world-population-day

United Nations. (n.d.a). *The 17 goals | sustainable development*. United Nations. Retrieved from https://sdgs.un.org/goals

United Nations. (n.d.b). https://sdgs.un.org/goals

United Nations. (n.d.b). *Transforming our world: the 2030 Agenda for Sustainable Development*. Retrieved from https://sdgs.un.org/2030agenda

United Nations. (n.d.c). https://www.un.org/development/desa/disabilities/envision2030-goal12.html

Uscinski, J. E., & Parent, J. M. (2014). *American conspiracy theories*. Oxford University Press., doi:10.1093/acprof:oso/9780199351800.001.0001

USDOE (United States Department of Energy). (2014). Costs Associated with Compressed Natural Gas Vehicle Fueling Infrastructure. United States Department of Energy.

USDOE. (2018). *Global LNG Fundamentals*. United States Department of Energy.

USDS (United States Department of State). (2010). *Background Note: Dominica*. https://www.state.gov/r/pa/ei/bgn/2295.htm

Ushinsky, K. D. (2005). Selected Works (vol. 4). Academic Press.

Van Aelst, P., Strömbäck, J., Aalberg, T., Esser, F., de Vreese, C., Matthes, J., Hopmann, D., Salgado, S., Hubé, N., Stępińska, A., Papathanassopoulos, S., Berganza, R., Legnante, G., Reinemann, C., Sheafer, T., & Stanyer, J. (2017). Political communication in a high-choice media environment: A challenge for democracy? *Annals of the International Communication Association*, *41*(1), 3–27. doi:10.1080/23808985.2017.1288551

Van den Hove, S., Le Menestrel, M., & de Bettignies, H.-C. (2002). The oil industry and climate change: Strategies and ethical dilemmas. *Climate Policy*, *2*(1), 3–18. doi:10.3763/cpol.2002.0202

Van der Waal, J., & Thijssens, T. (2020). Corporate involvement in sustainable development goals: Exploring the territory. *Journal of Cleaner Production*, *252*, 119625. doi:10.1016/j.jclepro.2019.119625

Vandoorn, T. L., De Kooning, J. D. M., Meersman, B., & Vandevelde, L. (2013). Review of primary control strategies for islanded microgrids with power-electronic interfaces. *Renewable & Sustainable Energy Reviews*, *19*, 613–628. doi:10.1016/j.rser.2012.11.062

VanGrasstek, C. (2013). *The History and Future of the World Trade Organization*. World Trade Organization. doi:10.30875/14b6987e-en

Verkade, N., & Höffken, J. (2019). Collective energy practices: A practice-based approach to civic energy communities and the energy system. *Sustainability*, *11*(11), 3230. doi:10.3390u11113230

Verma, T., Kamal, M. A., & Brar, T. (2022). An Appraisal of Vernacular Architecture of Bikaner. *Climatic Responsiveness and Thermal Comfort of Havelis*, *9*, 41–60.

Vernadsky, V. I. (1960). Biosphere. Selected works. Academic Press.

Vernadsky, V. I. (2018). *Philosophy of Science. Selected Works*. Youwright Publisher.

Vohland, K., Land-Zandstra, A., Ceccaroni, L., Lemmens, R., Perello, J., Ponti, M., Samson, R., & Wagenknecht, K. (2021). *The science of citizen science*. Springer International Publishing. doi:10.1007/978-3-030-58278-4

Von der Leyen, U. (n.d.). *Beautiful, sustainable, together*. New European Bauhaus. Retrieved from https://new-european-bauhaus.europa.eu/index_en

von Mises, L. H. E. (1998). Human Action: A Treatise on Economics. Yale University Press.

Von Schelling, F. W. J. (1993). *System of transcendental idealism (1800)*. University of Virginia Press.

Vrijhoef, R., & Koskela, L. (2000). The four roles of supply chain management in construction. *European Journal of Purchasing & Supply Management*, *6*(3), 169–178. doi:10.1016/S0969-7012(00)00013-7

Vygotsky, L. S. (1960). *Development of Higher Mental Functions*. Nauka Publisher.

Compilation of References

Wang, F., Cai, W., & Elahi, E. (2021). Do green finance and environmental regulation play a crucial role in the reduction of Emissions? An empirical analysis of 126 Chinese Cities. *Sustainability*, *13*(23), 1–20.

Wang, J., & Li, L. (2016). Sustainable energy development scenario forecasting and energy saving policy analysis of China. *Renewable & Sustainable Energy Reviews*, *58*, 718–724. doi:10.1016/j.rser.2015.12.340

Wang, M., & Feng, C. (2020). The impacts of technological gap and scale economy on the low-carbon development of China's industries: An extended decomposition analysis. *Technological Forecasting and Social Change*, *157*, 120050. Advance online publication. doi:10.1016/j.techfore.2020.120050

Wang, T., & Liu, Y. P. (2013). Ecological Design Strategies of Vernacular Architecture in Shanxi, China. *Applied Mechanics and Materials*, *275–277*, 2773–2776. . doi:10.4028/www.scientific.net/AMM.275-277.2773

Wang, X., Huang, L., Daim, T., Li, X., & Li, Z. (2021). Evaluation of China's new energy vehicle policy texts with quantitative and qualitative analysis. *Technology in Society*, *67*, 101770. Advance online publication. doi:10.1016/j.techsoc.2021.101770

Wang, Y.-J., Chen, Y., Hewitt, C., Ding, W.-H., Song, L.-C., Ai, W.-X., Han, Z.-Y., Li, X.-C., & Huang, Z.-L. (2021). Climate services for addressing climate change: Indication of a climate livable city in China. *Advances in Climate Change Research*, *12*(5), 744–751. doi:10.1016/j.accre.2021.07.006

Weber, M. (2009). *The theory of social and economic organization*. Simon and Schuster.

WEF. (2018). *The Inclusive Growth and Development Report*. The World Economic Forum.

WEF. (2020). *The Future of Nature and Business*. New Nature Economy Report II. World Economic Forum.

Weitzman, M. L. (2009). On modeling and interpreting the economics of catastrophic climate change. *The Review of Economics and Statistics*, *91*(1), 1–19. doi:10.1162/rest.91.1.1

Welcome to the new European Bauhaus prizes 2022. (2022). Retrieved from http://prizes.new-european-bauhaus.eu/home

West, S., & Paterman, R. (2017). *How could citizen science support the sustainable development goals?* Stockholm Environment Institute.

Wilhite, D. (2016). *Drought- management policies and preparedness plans: Changing the paradigm from crisis to risk management. In Land Restoration*. Elsevier Inc. doi:10.1016/B978-0-12-801231-4.00007-0

Willi, M., Pitman, R. T., Cardoso, A. W., Locke, C., Swanson, A., Boyer, A., Veldthuis, M., & Fortson, L. (2018). Identifying animal species in camera trap images using Deep Learning and Citizen Science. *Methods in Ecology and Evolution*, *10*(1), 80–91. doi:10.1111/2041-210X.13099

Wiryadinata, S., Morejohn, J., & Kornbluth, K. (2019). Pathways to carbon neutral energy systems at the University of California, Davis. *Renewable Energy*, *130*, 853–866. doi:10.1016/j.renene.2018.06.100

World Bank. (2018). *World Development Report 2018: Learning to Realize Education's Promise*. World Bank.

World Data. (2021a). *Energy Consumption in Dominica*. https://www.worlddata.info/america/dominica/energy-consumption.php

World Data. (2021b). *Tourism in Dominica*. https://www.worlddata.info/america/dominica/tourism.php

Worldometer. (n.d.). https://www.worldometers.info/world-population/

Wreford, A., Moran, D., & Adger, N. (2010). *Climate Change and Agriculture: Impacts, Adaptation and Mitigation*. OECD Publications. doi:10.1787/9789264086876-en

WTS (Waste Today Staff). (2015). *A Look at CNG Station Project Costs*. https://www.wastetodaymagazine.com/article/rew0315-cng-fueling-costs/

Wu, Z., Shao, Q., Su, Y., & Zhang, D. (2021). A socio-technical transition path for new energy vehicles in China: A multi-level perspective. *Technological Forecasting and Social Change, 172*, 121007. Advance online publication. doi:10.1016/j.techfore.2021.121007

Wyns, A., & Beagley, J. (2021). COP26 and beyond: Long-term climate strategies are key to safeguard health and equity. *The Lancet. Planetary Health, 5*(11), e752–e754. doi:10.1016/S2542-5196(21)00294-1 PMID:34774113

Xin-gang, Z., & Wei, W. (2020). Driving force for China's photovoltaic industry output growth: Factor-driven or technological innovation-driven? *Journal of Cleaner Production, 274*, 122848. Advance online publication. doi:10.1016/j.jclepro.2020.122848

Xu, L., & Jordan, M. I. (1996). On convergence properties of the EM algorithm for gaussian mixtures. *Neural Computation, 8*(1), 129–151. doi:10.1162/neco.1996.8.1.129

Xu, W. W., Park, J. Y., & Park, H. W. (2015). The networked cultural diffusion of Korean wave. *Online Information Review, 39*(1), 43–60. doi:10.1108/OIR-07-2014-0160

Yadiati, W., Nissa, N., Paulus, S., Suharman, H., & Meiryani, M. (2019). The role of green intellectual capital and organizational reputation in influencing environmental performance. *International Journal of Energy Economics and Policy, 9*(3), 261–267. doi:10.32479/ijeep.7752

Yang, M.-S., Chang-Chien, S.-J., & Hung, W.-L. (2017). Learning-based EM clustering for data on the unit hypersphere with application to exoplanet data. *Applied Soft Computing, 60*, 101–114. doi:10.1016/j.asoc.2017.06.037

Yang, M.-S., Lai, C.-Y., & Lin, C.-Y. (2012). A robust EM clustering algorithm for gaussian mixture models. *Pattern Recognition, 45*(11), 3950–3961. doi:10.1016/j.patcog.2012.04.031

Yang, X., Zhao, C., Xu, H., Liu, K., & Zha, J. (2021). Changing the industrial structure of tourism to achieve a low-carbon economy in China: An industrial linkage perspective. *Journal of Hospitality and Tourism Management, 48*, 374–389. doi:10.1016/j.jhtm.2021.07.006

Yang, Z., Li, K., & Foley, A. (2015). Computational scheduling methods for integrating plug-in electric vehicles with power systems: A review. *Renewable & Sustainable Energy Reviews, 51*, 396–416. doi:10.1016/j.rser.2015.06.007

Yanine, F. F., Caballero, F. I., Sauma, E. E., & Córdova, F. M. (2014). Building sustainable energy systems: Homeostatic control of grid-connected microgrids, as a means to reconcile power supply and energy demand response management. *Renewable & Sustainable Energy Reviews, 40*, 1168–1191. doi:10.1016/j.rser.2014.08.017

Yavuz, L., Önen, A., Muyeen, S. M., & Kamwa, I. (2019). Transformation of microgrid to virtual power plant–a comprehensive review. *IET Generation, Transmission & Distribution, 13*(11), 1994–2005. doi:10.1049/iet-gtd.2018.5649

Yin, R. (2012). Case study methods. In H. Cooper, P. M. Camic, D. L. Long, A. T. Panter, D. Rindskopf, & K. J. Sher (Eds.), APA handbook of research methods in psychology, Vol. 2. Research designs: Quantitative, qualitative, neuropsychological, and biological (pp. 141–155). American Psychological Association. doi:10.1037/13620-009

Yong, J., Yusliza, M., Ramayah, T., & Fawehinmi, O. (2019). Nexus between green intellectual capital and green human resource management. *Journal of Cleaner Production, 215*, 364–374. doi:10.1016/j.jclepro.2018.12.306

You, S., Træholt, C., & Poulsen, B. (2009, June). A market-based virtual power plant. In *2009 International Conference on Clean Electrical Power* (pp. 460-465). IEEE.

Compilation of References

Yousaf, W., Asghar, E., Meng, H., Songyuan, Y., & Fang, F. (2017, October). Intelligent control method of distributed generation for power sharing in virtual power plant. In *2017 IEEE International Conference on Unmanned Systems (ICUS)* (pp. 576-581). IEEE. 10.1109/ICUS.2017.8278411

Yu, P., Weng, Y., & Ahuja, A. (2022). Carbon Financing and the Sustainable Development Mechanism. In Handbook of Research on Energy and Environmental Finance 4.0 (pp. 301-332). Academic Press.

Yu, P., Jiao, A., & Sampat, M. (2022). The Effect of Chinese Green Transformation on Competitiveness and the Environment. In *Handbook of Research on Green* (pp. 257–279). Circular, and Digital Economies as Tools for Recovery and Sustainability. doi:10.4018/978-1-7998-9664-7.ch014

Yusliza, M., Yong, J., Tanveer, M., Ramayah, T., Faezah, J., & Muhammad, Z. (2020). A structural model of the impact of green intellectual capital on sustainable performance. *Journal of Cleaner Production*, *249*, 119334. doi:10.1016/j.jclepro.2019.119334

Yusoff, Y., Nejati, M., Kee, D., & Amran, A. (2020). Linking green human resource management practices to environmental performance in hotel industry. *Global Business Review*, *21*(3), 663–680. doi:10.1177/0972150918779294

Zamani, A. G., Zakariazadeh, A., & Jadid, S. (2016). Day-ahead resource scheduling of a renewable energy based virtual power plant. *Applied Energy*, *169*, 324–340. doi:10.1016/j.apenergy.2016.02.011

Zameer, H., Wang, Y., & Yasmeen, H. (2020). Reinforcing green competitive advantage through green production, creativity and green brand image: Implications for cleaner production in China. *Journal of Cleaner Production*, *247*, 119119. doi:10.1016/j.jclepro.2019.119119

Zarate L. (2015). Right to the city for all: a manifesto for social justice in an urban century. *The Nature of Cities*.

Zarei, Z., Karami, E., & Keshavarz, M. (2020). Co-production of knowledge and adaptation to water scarcity in developing countries. *Journal of Environmental Management*, *262*, 110283. doi:10.1016/j.jenvman.2020.110283 PMID:32090886

Zeng, Z., Yang, H., Zhao, R., & Cheng, C. (2013). Topologies and control strategies of multi-functional grid-connected inverters for power quality enhancement: A comprehensive review. *Renewable & Sustainable Energy Reviews*, *24*, 223–270. doi:10.1016/j.rser.2013.03.033

Zhai, Z., & Previtali, J. M. (2010). Ancient vernacular architecture: Characteristics categorization and energy performance evaluation. *Energy and Building*, *42*(3), 357–365. doi:10.1016/j.enbuild.2009.10.002

Zhang, G., Jiang, C., & Wang, X. (2019). Comprehensive review on structure and operation of virtual power plant in electrical system. *IET Generation, Transmission & Distribution*, *13*(2), 145–156. doi:10.1049/iet-gtd.2018.5880

Zhang, S., Bai, X., Zhao, C., Tan, Q., Luo, G., Wu, L., Xi, H., Li, C., Chen, F., Ran, C., Liu, M., Gong, S., & Song, F. (2022). China's carbon budget inventory from 1997 to 2017 and its challenges to achieving carbon neutral strategies. *Journal of Cleaner Production*, *347*, 130966. Advance online publication. doi:10.1016/j.jclepro.2022.130966

Zhang, Y., Sun, J., Yang, Z., & Wang, Y. (2020). Critical success factors of green innovation: Technology, organization and environment readiness. *Journal of Cleaner Production*, *264*, 121701. doi:10.1016/j.jclepro.2020.121701

Zhao, P., & Zhang, M. (2018). The impact of urbanisation on energy consumption: A 30-year review in China. *Urban Climate*, *24*, 940–953. doi:10.1016/j.uclim.2017.11.005

Zheng, G. W., Siddik, A. B., Masukujjaman, M., Fatema, N., & Alam, S. S. (2021). Green Finance Development in Bangladesh: The role of private commerical banks (PCBs). *Sustainability*, *13*(2), 1–17.

Zhou, H., Yang, Y., Chen, Y., & Zhu, J. (2018). data envelopment analysis application in sustainability: the origins, development and future directions. *European Journal of Operational Research*, *264*, 1-16.

Zhou, S., & Xu, Z. (2018). A novel internal validity index based on the cluster centre and the nearest neighbour cluster. *Applied Soft Computing*, *71*, 78–88. doi:10.1016/j.asoc.2018.06.033

Zhou, X., Tang, X., & Zhang, R. (2020). Impact of green finance on economic development and environmental quality: A study based on provincial panel data from China. *Environmental Science and Pollution Research International*, *27*, 19915–19932.

About the Contributors

Adebowale Adetayo is an academic staff of Adeleke University. His research interest is Library Science, Social media, Knowledge Management, and Business Information Management. He has published many articles in reputable journals and currently working on projects relating to pandemics, vaccines, and virtual learning. He has a Master's Degree in Information Resources Management from Babcock University, PhD in Library and Information Science from Adeleke University.

Antonio Alcázar Blanco is a Substitute Professor in the Department of Financial Economics and Accounting at the University of Extremadura (UEX). PhD in Financial Economics and Accounting from the University of Extremadura. Diploma in Business Sciences, Degree in Actuarial and Financial Sciences, Master's Degree in Research in Social and Legal Sciences from the University of Extremadura, University Master's Degree in Teacher Training for Compulsory Secondary Education and Baccalaureate, Vocational Training and Language Teaching from the Antonio de Nebrija University (Universidad Antonio de Nebrija).

Fatma Gül Altin is an assistant professor at Mehmet Akif Ersoy University, Turkey. She earned her Ph.D. degree from Suleyman Demirel University, Turkey. Her research interests include production management, supply chain management and logistics. She has published several papers, book chapters and conference papers.

Margarita Angelidou is an architect, urban planner and expert in smart and sustainable cities and regions (MBA, MSc, PhD). Margarita has been involved as a researcher, principal investigator and project manager in the smart and sustainable cities domain since 2007, counting 16 international and national projects. She has been awarded over 10 personal awards and scholarships for scientific excellence, including three individual post-doctoral research grants from the Greek State. She has published over 50 articles in international journals, books and conferences on topics related to urban and regional innovation: smart city policies and governance, sustainable and resilient cities, social innovation in cities, Do-It-Yourself urbanism, public space, sustainable transport in cities circular city development, etc. Next to the previous, since 2018 she has been serving as an expert monitor and evaluator for the European Commission and national organizations across Europe.

Konstantinos Asikis is a PhD-cand. Researcher on Urban Planning & Regional Development, Architecture sch. NTUA, Civil Engineer-Transport MEng AUTH, MSc Spatial Planning UTH-Clermont-Ferrand Univ. EIT H2020 Professional Certification on Innovative Financing for Climate Action, UTH

DEVOPS Certification Smart Cities Planner. WHO-UN Habitat- Healthy Cities Network Working Groups. Head of Strategic/Operational Planning & ICT dpt. Municipality of Farkadona, Project Manager, EU Projects Partner Coordinator.

Abata-Ebire Blessing Damilola is academic staff of Federal Polytechnic Ede, Osun State. Her research interest is Library Science, Social media, Knowledge Management, and Health Information System. She has published many articles in reputable journals and currently working on projects relating to virtual learning, turnover intention and Emerging Technologies,. She is a graduate of Taisolarin University of Education.

Don Charles has a PhD in Economics (2017), an MSc in Economics (2009), and a BSc in Economics (2006), all of which were earned from the University of the West Indies (UWI) St. Augustine Campus. His research interests are mainly in the areas of energy economics, econometrics, international trade and value chains, climate change policy, and portfolio finance. By the end of 2021 he had 19 years of work experience which covers economic research, lecturing, industrial relations, project management, procurement, event management, and administration. His work experience covers employment in the public service, the private sector, UN organizations, and in academia.

Ziling Cheng is an independent researcher. His research interests include Carbon Finance, Chinese Economy, and Sustainability.

Seyda Emekci, MSc, PhD, received her bachelor's, master and doctoral degrees from Middle East Technical University (METU). Worked as an associate at the Development Agency and the General Director of Nature Conservation and National Parks, in Turkey from 2014 to 2017. Postdoctoral researcher at East London University, in the UK in 2019. Currently teaches in the Architecture Department at Ankara Yıldırım Beyazıt University (AYBU). Has been involved in various change projects at the different architectural levels in public and private sectors and with civil associations in Turkey since 2011. Specialization areas include sustainability, smart homes, energy efficiency, housing economics.

Eleni Karachaliou holds a MEng in Urban/Spatial Planning & Development Engineering (5 year degree) from the Engineering Faculty of Aristotle University of Thessaloniki (AUTH) and an additional Master's degree in Geoinformatics (emphasis on Management of Photogrammetric production and Remote Sensing with GIS), also from AUTH, as a fellow of the Onassis Foundation (2014-2015). Over the last 8 years, she has been participating in national and international research, development, and innovation projects in several sectors (i.e. Transport, ICT, Environment, Energy, Health, etc.). as a research associate, consultant, and project manager. Her research and professional interests focus on the broader field of collection, distribution, storage, analysis, processing, presentation of geographic data or geographic information using a variety of methods and tools such as GIS, remote sensing, laser scanning, photogrammetry, UAV platforms, and others.

Ezatollah Karami is a professor of agricultural development and extension at the College of Agriculture, Shiraz University, Iran. He is a specialist in the human dimensions of resource conservation, social impacts of climate change, water scarcity, and social impact assessment. He has published widely on human dimensions of resource conservation, sustainability, drought, climate change impact, and related

About the Contributors

issues in international journals. He has received national and international awards for his contributions in the field including the ECO international award (Economic Cooperation Organization Award in Agriculture and Environment) and the Robert H. Whittaker Award, from the Ecological Society of America.

Bartolomé Marco-Lajara (PhD) is Professor at the University of Alicante. His research interests are strategic management and tourism management. He is the author of several books, book chapters and international articles related to teaching methodology and the areas above mentioned. He is a member of the Tourism Research Institute at the University of Alicante since its foundation and the main researcher of a European project 'Next Tourism Generation Alliance' and of the public competitive project for the creation of the Tourist Observatory of the Valencian Community (Spain). He has taken part in others public and private projects, such as the development of the strategic plan of the Alicante province for the period 2010-2020. He is the Assistant Dean of the Economics Faculty at the University of Alicante.

Javier Martínez Falcó is Assistant Professor in the Department of Management at the University of Alicante. In the field of research, he focuses his interest on issues related to the Strategic Management of the Company, specifically in the areas of Knowledge Management and Corporate Sustainability of wine companies, on which he has written several publications in the form of articles, book chapters and contributions to conferences. He has also participated in several national research projects and teaches Strategic Management on the ADE, TADE and DADE degree courses.

Yetunde Omodele Oladipo, a Graduate Student of Salford University, Manchester, UK. As an erudite scholar, her research interests are Library and Information Sciences, Catalogue Systems, Knowledge Management and Archival Management. She has published articles in peer-reviewed journals and currently working on projects relating to managing innovative management system, virtual learning and emerging technologies. She is a graduate of Bayero University, Kano, Nigeria.

Iria Paz-Gil is a Professor at the Rey Juan Carlos University in Madrid (Spain), in the Department of Business Economics, within the area of Marketing. in Business Economics. Master´s degree in Marketing. Participation in prestigious projects developed by the Royal Academy of Economic and Financial Sciences, or by the Spanish Agency for International Cooperation, among others. Author of several articles published in prestigious journals, such as Perspectivas, JMBE, such as Springer. Member of the organising committee of national and international conferences of the AEDEM academy. She has taught on specialisation and master's courses, such as a specialist course in Sports Management or Master in Foreign Trade.

Olga I. Pilipenko graduated from Moscow State Institute for Foreign Relations (MGIMO University) in 1975. In 1981 she got her degree of Candidate of Economic Sciences at the Institute of Latin America of the USSR Academy of Sciences. In 1994 she got a degree of Doctor of Economic Sciences at Lomonosov Moscow State University. She worked as full Professor for many Russian Universities: All-Russian Correspondence Financial and Economic Institute (VSFEI), People's Friendship University of Russia (RUDN), Lomonosov Moscow State University, Russian Presidential Academy of National Economy and Public Administration (RANERA) (last work). In the late 1990s, she was invited as a researcher to the Institute of Technology of Hearning and to Aarhus University (Denmark) under the interstate exchange program. She has more than 130 publications (in Russian and in English), including ones indexed by RISC (RF), Scopus, Web of Science. Her scientific interests lie in the field of bifurcation effects' modeling in global financial markets in connection with shocks; interaction of factors of global financial stability, monetary circulation, and public finance; of dialectic lows of self-movement of human-created systems in economy, society, technological structures; of economic and social policies interaction for the coronavirus pandemic; of self-organization of economic and financial systems in the context of the theory of shocks.

Zoya A. Pilipenko received the following degrees at Lomonosov Moscow State University: Doctor of Economic Sciences (Finance and World Economy) (2013), Candidate of Economic Sciences (Finance and World Economy) (2004), Master in Economics (2003) and Bachelor in Economics (2001). She started her professional career in Finance at Insurance Company "Gefest" (Russia, Moscow, in 2003) and at the Joint Stock Company "Sberbank" (Russia, Moscow, in 2004), where she estimated the required rate of return on offered banking services and actual profitability of credit services in particular. For the last 15 years she has been working for the Central Bank of Russia. In her current role as Advisor to the Head of Platforms Operators and Information Services Supervision Division at the Financial Market Infrastructure Department she is deeply involved in the process of CRA's supervision enhancement and standardization, methodologies validation requirements' formation, as well as mapping framework development and market structure analysis. She has teaching and research experience: more than 60 publications in Russian and in English (indexed by RISC (RF), Scopus, etc.). She has got certificates in Banking and Finance from practical seminars in Great Britain, Luxembourg, Austria, Italy, and France. She has got certificates in Banking and Finance from practical seminars in Great Britain, Luxembourg, Austria, Italy and France. Her research interests lie in the field of shocks theory and self-organization laws of systemic integrity in economy and finance, as well as monetary and fiscal policies' formation on financial markets at national and global levels due to the coronavirus pandemic.

Alberto Prado Román is a Professor at the Rey Juan Carlos University in Madrid (Spain), in the Department of Business Economics, within the area of Marketing. in Business Economics. Master's degree in Marketing. Participation in prestigious projects developed by the Royal Academy of Economic and Financial Sciences, or by the Spanish Agency for International Cooperation, among others. Author of several articles published in prestigious journals, such as Revista de Administração de Empresas or Revista Europea de Dirección y Economía de la Empresa, and publishers, such as Springer. Member of the organising committee of national and international conferences of the AEDEM academy. He has taught on specialisation and master's courses, such as a specialisation course in Sports Management, a Master's Degree in Senior Management or a Master's Degree in Foreign Trade.

Miguel Prado Román is a Professor at the Rey Juan Carlos University in Madrid (Spain), in the Department of Business Economics, within the area of Accounting and Finance. PhD. in Business Economics. Master's degree in Auditing and Higher Accounting. Participation in prestigious projects developed by the Royal Academy of Economic and Financial Sciences, or by the Spanish Agency for International Cooperation, among others. Author of several articles published in prestigious journals, such as Revista de Administração de Empresas or Revista Europea de Dirección y Economía de la Empresa, and publishers, such as Springer. Member of the organising committee of national and international conferences of the AEDEM academy. He has taught on specialisation and master's courses, such as a

About the Contributors

specialist course in Sports Management, University Master's Degree in Senior Management or University Master's Degree in Auditing and Higher Accounting.

Abdelmadjid Recioui is a Professor at the Institute of Electrical Engineering and Electronics University of Boumerdes, Algeria. He obtained a PhD degree in electrical and electronic engineering option telecommunications from the Institute of Electrical Engineering and Electronics, University of Boumerdes in 2011. He holds also Master (Magister) Electronic System Engineering degree which has been achieved at the Institute of Electrical Engineering and Electronics, University of Boumerdes in 2006. In June 2002, he finished his engineering studies at the institute of Electrical Engineering and Electronics, University of Boumerdes. He is a research Assistant at the laboratory signals and systems from January 2008 to Present in Laboratory: signals and systems, Inst. of Electrical Engineering and electronics, University of Boumerdes. His research interests include Antennas, Wireless Communication Systems, antenna array synthesis and design, capacity enhancement, system optimization, smart antennas, power system protection, power system optimization, power system communications.

Lorena Ruiz Fernández is Assistant Professor in the Department of Business Organization at the University of Alicante (UA). PhD in Business, Economics and Society from the University of Alicante, she has focused her line of research in the area of Strategic Management, and specifically in the Dynamic Capabilities View focused on the tourism sector. She is interested in the use of cutting-edge quantitative methodology. In her research career, she has published articles in scientific journals such as International Journal of Emerging Markets, Sustainability, Business Ethics: A European Review, Journal of Fashion Marketing and Management, indexed in the WoS database, as well as different book chapters in the field of Social Sciences. She has also participated as a speaker in several national and international congresses, conferences and/or seminars. Among the latter, it can highlight SKM Symposium, EIBA or ISEOR. Currently, she is part of the DECI-GLOBAL research group and participates in the European Project THE NEXT TOURISM GENERATION ALLIANCE (NTG).

Michael Sampat is an independent researcher working in several interdisciplinary capacities. Current projects include books on business geography and mass media psychological conditioning.

Alexandros Skondras is a PhD candidate at the School of Spatial Planning and Development, Aristotle University of Thessaloniki, Greece. He holds a BSc in Architectural Engineering from Politecnico di Torino and a MSc in Sustainable Design and Development from Politecnico di Torino where he worked as a teaching fellow from 2014 to 2016 in the fields of photogrammetry, BIM and urban design. His research interests focus on the use of UAV for collection, visualization, processing and presentation of spatial information as tools for strategic planning in the fields of urban design and sustainable development. He is also an architectural photographer with focus on the visualization and transformations of the built environment, holding a degree from the University of Arts in London, and over the last 4 years he has been published by a multitude of architecture and design journals.

Alexia Spyridonidou is Spatial/Urban Planning & Development MEng AUTH, MSc Planning, Organization & Management of Transport Systems AUTH, MSc Techniques for Spatial Analysis, Planning & Management AUTH, MSc City/Urban, Community & Regional Planning INSA Strasburg. CIVINET Greece-Cyprus network co-founder (national branch of CIVITAS initiative of the EC). Sustainable

Urban Mobility Plan (SUMP) approach Ambassador for Greece (as defined in Eltis - Urban Mobility Observatory). Business Development Manager at EPLO European Public Law Organization - Institute of Circular Economy and Climate Change. Jury member for the Innovation in Politics Awards 2021. Member of the Technical Secretariat of the global initiative for the protection of cultural and natural heritage from climate change (Greek government, UNESCO, WMO, UNFCCC) and organising team of COP26 EU side event. Chair of the Smart Cities pillar of Circle the Med Forum. Member of Net Zero Cities panel for the EC Mission 100 climate neutral and smart cities.

Efstratios Stylianidis is a Professor at the School of Spatial Planning and Development of the Aristotle University, Greece, where he leads the Laboratory of Geoinformatics (LabGeo). This period (2019-2023) he is acting as Vice Rector for Research and Lifelong Learning. His research focuses on the intersection of Geospatial Sciences and ICT for the benefit of natural and built environment. He is the author of 2 theses, 2 books, 84 scientific publications in peer reviewed journals and conferences proceedings, 10 invited chapters in books as well as the editor of 3 books. He has participated in more than 51 national and mainly international research projects (H2020, FP7, Erasmus+, etc.), 9 of which as the scientific and project coordinator. He has participated in 59 national and international conferences, while in 29 he has a member of the organizing and/or scientific committee. He is a reviewer in 18 scientific journals, while he participates in the editorial board of 4 scientific journals. In 2017, he received the European Satellite Navigation Competition 2017 award (Madrid Challenge) for the H2020 EU co-funded LARA project, while he is holding 1 patent and 1 more under filing. He was a visiting Professor at Columbia University (2018), while as an invited speaker he offered lectures in many universities, including Columbia University, Princeton University, and University of California Los Angeles (UCLA).

Ioannis Tavantzis is an Urban/Spatial Planning & Development engineer. He is a PhD candidate in the Laboratory of Geoinformatics (LabGeo) at the School of Spatial Planning and Development, Faculty of Engineering, Aristotle University of Thessaloniki, Greece specializing on Geoinformatics and Smart Cities. He is an active member of the research team of the LabGeo with participation in the execution of a Horizon 2020 project, two Erasmus+ projects, one "Research-Innovate-Create - ESPA 2014-2020" (NSRF) project and one "Investment Innovation Plans - ESPA 2014-2020" project.

Nikolaos Tokas is a Production and Management Engineer from Democritus University of Thrace. He contributes as a research assistant at the Laboratory of Geoinformatics in the department of Spatial Planning and Development Engineering of the Aristotle University of Thessaloniki. He focuses on the development and software programming of algorithms, specifically designed to tackle engineering problems and contribute in further research. He has developed algorithms using numerical methods to solve linear systems with sparse matrices for fluid and solid simulations, as well as algorithms of computational intelligence, computer vision and other fields of engineering.

José G. Vargas-Hernández, M.B.A., Ph.D., Member of the National System of Researchers of Mexico and a research professor at Tecnológico Mario Molina Unidad Zapopan formerly at University Center for Economic and Managerial Sciences, University of Guadalajara. Professor Vargas-Hernández has a Ph. D. in Public Administration and a Ph.D. in Organizational Economics. He has undertaken studies in Organisational Behaviour and has a Master of Business Administration, published four books and more than 200 papers in international journals and reviews (some translated to English, French,

About the Contributors

German, Portuguese, Farsi, Chinese, etc.) and more than 300 essays in national journals and reviews. He has obtained several international Awards and recognition.

Shucai Xu is an associate researcher and PhD supervisor of Tsinghua University. He is currently the Editorial Director of Journal of Automobile Safety and Energy Conservation of Tsinghua University, Vice President of Suzhou Automobile Research Institute of Tsinghua University, Deputy Secretary General of Automobile Safety Technology Branch of China Society of Automotive Engineering, Deputy Director of Structure and Simulation Laboratory of National Automobile Quality Supervision and Inspection Center (Beijing). Prof. Xu is mainly engaged in the design of intelligent safe driving assistance and intelligent occupant restraint system, vehicle lightweight and structural crashworthiness design and optimization. Prof. Xu has presided over more than 30 local & international scientific research projects. Prof. Xu has published more than 50 SCI / EI papers and 4 research books, and has more than 20 national invention patents.

Poshan (Sam) Yu is a Lecturer in Accounting and Finance in the International Cooperative Education Program of Soochow University (China). He is also an External Professor of FinTech and Finance at SKEMA Business School (China), a Visiting Professor at Krirk University (Thailand) and a Visiting Researcher at the Australian Studies Centre of Shanghai University (China). Sam leads FasterCapital (Dubai, UAE) as a Regional Partner (China) and serves as a Startup Mentor for AIC RAISE (Coimbatore, India). His research interests include financial technology, regulatory technology, public-private partnerships, mergers and acquisitions, private equity, venture capital, start-ups, intellectual property, art finance, and China's "One Belt One Road" policy.

Patrocinio Zaragoza-Sáez (PhD, University of Alicante, Spain) is a Professor at the Department of Management at the University of Alicante, Spain. Her primary research interests are knowledge management and intellectual capital, strategic management and tourism management. She has published research papers in international journals including Journal of Knowledge Management, Journal of Intellectual Capital, Journal of Business Research, Regional Studies, Knowledge Management Research and Practice, International Journal of Knowledge Management Studies, Business Strategy & Environment, Intangible Capital, Current Issues in Tourism, Cornell Hospitality Quarterly, Tourism Economics and Tourism Management. She is the Coordinator of the University Master's Degree in Tourism Management and Planning.

Index

A

Algeria 89-90, 97-106
Attitude 73-75, 78-83, 88, 118, 188, 202

B

Biodiversity 61, 135, 144-145, 147, 149-150, 186, 190, 192, 200, 211, 232, 253
Biodiversity Conservation 144
Business Planning 152, 155, 158, 162

C

Carbon Neutral 97, 109-110, 114-115, 118-120, 131
Carbon Peak 109-110, 113-115, 118-120, 129
Carbon Reduction 109
Caribbean 22, 24-26, 37-43, 52
Case Study 19-20, 118, 127, 130, 132-147, 163, 165, 167-168, 174, 176, 179, 227-228
Challenges 19, 48, 56, 61, 66, 79, 90, 92-93, 102, 105-106, 120, 126, 131, 146, 148, 164-165, 170, 176, 182, 209, 212-213, 218, 232, 239-240, 243-245, 248-250, 252-253, 257
Circular Economy 3-4, 14, 17, 19, 129, 132, 171, 173-174, 181, 244
Cities 1-2, 85, 130, 141, 150, 164, 173, 201, 218, 223, 231, 242-245, 247-248, 250, 252, 254-257
Citizen Science 132-151
Clean Energy 89, 105, 119-121, 128, 138, 158, 174, 201, 237
Climate Change 21-26, 36-43, 46-53, 55-63, 65-71, 73-88, 100, 103, 109-110, 112, 114-115, 120, 122, 124, 126-127, 129-130, 132, 134, 137, 141-144, 151-162, 164-165, 171, 174-175, 185, 188, 200-201, 209, 214-218, 229-230, 232, 237, 239, 243-245, 247, 255
Climate Change Mitigation 79, 81, 84, 129, 132, 143, 151

Climate Finance 21, 37-40, 42, 45, 230
Clustering 1-5, 7, 10, 15-19
Conservation Finance 21, 36, 40
Crisis Management 46, 56-63, 65, 68, 70-71

D

DEA 1-4, 8, 10, 13, 15, 17
Dialectics of Human Education and Socialization 183
Dominica 21-22, 24-29, 31-44
Donor Financing 21, 36, 38, 40

E

Economic Impact Analysis 27
EM 1-2, 4-6, 10, 12, 16-19
Energy Efficiency 90, 94, 97-98, 100, 114, 122, 138-139, 158-159, 169, 215, 217-218, 225-226, 228, 231-232, 241
Energy Policy 128, 130
Energy Sector 21, 27, 32-33, 35, 39, 89, 100-102
Energy Transition 41, 89, 104-106, 128-129, 244
Environmental Health 132, 253
EU 1-4, 10, 12-18, 25, 129, 226, 229, 237, 239-240, 244, 247-248, 252, 256-257
European Green Deal 229-230, 232, 237, 239-240, 244, 248, 257

F

Fact-Checking 73, 75, 78, 80-81, 83-85, 88

G

Global Warming 46-48, 67, 69, 71, 85, 110, 126-127, 151-155, 157-161, 215, 229
Green Bonds 36, 40, 42, 158, 229, 231, 235, 237-240, 254
Green Deal Action Plan 229-230, 235, 239

Index

Green Economy 18, 29, 44, 229, 232, 239
Green Finance 229-231, 235-237, 239-242
Green Human Capital 163, 165, 180-181
Green Intellectual Capital 163-164, 170, 175-181
Green Investment 229, 232-237, 239
Green Relational Capital 163, 165, 180, 182
Green Structural Capital 163, 165, 180-181
Greenhouse Gas Emissions 55, 89, 114-115, 130, 153, 159, 215-216, 224, 234

H

Healthy 68, 75, 135, 227, 245, 248, 252, 254
High-Level Political Forum of Sustainable Development (HLPF) 134, 151

I

Information Sources 73, 75, 77, 79-80, 87

L

Landfill 2, 4, 10, 20, 29, 32, 34
Low-Carbon Economy 109-110, 112, 115, 127, 131, 239

M

Mexican Companies 152, 160
Microgrids 89-95, 102-108
Micro-Level Management 46, 61
Municipal Waste 1-2, 4, 10, 12-17, 19-20, 141
Municipalities 2, 20, 243-249, 251, 253, 255-256

N

National Science Foundation (NSF) 139-140, 151
Nationally Determined Contributions 21-24, 39, 41, 214
NDC Implementation 21, 24, 36
New Energy Vehicle 109-110, 114-115, 117-120, 122, 124-126, 128-130
New Normal Management 46, 57-58, 61, 63-64
Nigeria 73-76, 79-87

O

Optimization 17, 89, 91-95, 104-107, 112, 126, 205, 212, 218

P

Participatory Research 146

Q

Quality of Life 136, 152-153, 158-160, 162, 218, 248

R

Recovery of Waste 20
Recycling of Waste 20
Renewable Energies 89, 98, 104-105
Renewable Energy 33, 38, 41, 89-90, 93-95, 97-98, 102-105, 107-108, 114-115, 131, 138, 159, 217, 230, 232-236, 240-241
Research Funding Organizations (RFO) 147, 151
Research Performing Organizations (RPO) 147, 151
Risk Management 46, 56-57, 68-69, 71, 94, 136, 141

S

Self-Organization and Self-Development of a Human Being 183
Smart 89-90, 92-93, 96-97, 103-107, 243-245, 248, 252, 254-255, 257
Social Sustainability 175, 216, 222, 227
Specialisation 243
Strategic Planning 70, 152, 155, 157, 160, 244
Sustainability 8, 17, 19, 63, 65, 70, 87, 89, 94, 104, 108, 121, 130-131, 134-135, 138, 143, 147-149, 154, 164, 166, 169, 171, 173-182, 190, 211-212, 215-218, 222-227, 232, 236, 242-245, 253, 256
Sustainable Cities 173, 201, 248, 257
Sustainable Development 1-2, 14-18, 20, 60, 69, 71, 100, 109-110, 112, 115, 122, 124, 126, 131-134, 136, 143-151, 163-166, 171, 175, 177, 179-180, 182, 186, 200, 209-211, 213-214, 217, 223, 230-231, 237, 240, 243-244, 248, 257
Sustainable Development Goals (SDG) 151

T

Transformation 19, 59, 89, 108, 112, 115, 120-121, 124, 126, 129, 131, 194, 197, 201-202, 222, 225, 229, 236, 243-244, 248, 250-251, 253
Transition 41, 47, 66, 89, 102, 104-106, 112, 115, 118, 121-122, 128-131, 134, 173, 187-188, 194, 196, 207-208, 231-232, 241, 243-245, 248-250

U

Urban 4, 17, 70, 85, 119, 121, 124, 128, 133, 135-136, 139, 141-142, 145, 150, 159, 221, 223, 226-228, 243-245, 248-250, 254, 257

V

Vernacular Architecture 215-220, 222-228
Virtual Power Plants 89-92, 103-107

W

Waste Disposal 20

Waste Treatment 4, 10, 17, 19-20
Wine 163-164, 167-182
Wine industry 163-164, 167-168, 170-171, 174-178, 180
Winery 165, 167-174, 180-182

Recommended Reference Books

IGI Global's reference books are available in three unique pricing formats:
Print Only, E-Book Only, or Print + E-Book.
Shipping fees may apply.

www.igi-global.com

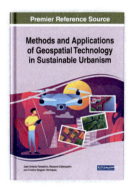

Methods and Applications of Geospatial Technology in Sustainable Urbanism

ISBN: 9781799822493
EISBN: 9781799822516
© 2021; 684 pp.
List Price: US$ 195

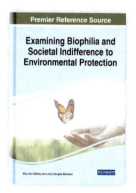

Examining Biophilia and Societal Indifference to Environmental Protection

ISBN: 9781799844082
EISBN: 9781799844099
© 2021; 279 pp.
List Price: US$ 195

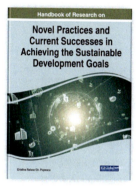

Novel Practices and Current Successes in Achieving the Sustainable Development Goals

ISBN: 9781799884262
EISBN: 9781799884286
© 2021; 461 pp.
List Price: US$ 295

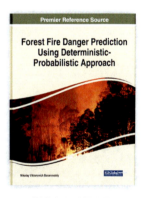

Forest Fire Danger Prediction Using Deterministic-Probabilistic Approach

ISBN: 9781799872504
EISBN: 9781799872528
© 2021; 297 pp.
List Price: US$ 195

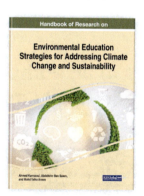

Environmental Education Strategies for Addressing Climate Change and Sustainability

ISBN: 9781799875123
EISBN: 9781799875192
© 2021; 416 pp.
List Price: US$ 295

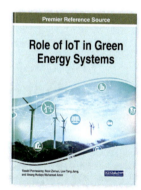

Role of IoT in Green Energy Systems

ISBN: 9781799867098
EISBN: 9781799867111
© 2021; 405 pp.
List Price: US$ 195

Do you want to stay current on the latest research trends, product announcements, news, and special offers?
Join IGI Global's mailing list to receive customized recommendations, exclusive discounts, and more.
Sign up at: **www.igi-global.com/newsletters**.

Publisher of Timely, Peer-Reviewed Inclusive Research Since 1988

www.igi-global.com | Sign up at www.igi-global.com/newsletters | facebook.com/igiglobal | twitter.com/igiglobal | linkedin.com/igiglobal

Ensure Quality Research is Introduced to the Academic Community

Become an Evaluator for IGI Global Authored Book Projects

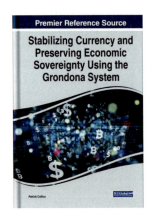 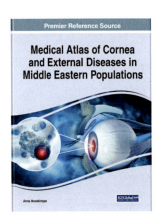 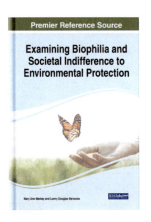 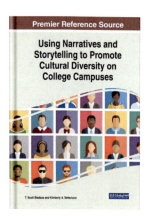

The overall success of an authored book project is dependent on quality and timely manuscript evaluations.

Applications and Inquiries may be sent to:
development@igi-global.com

Applicants must have a doctorate (or equivalent degree) as well as publishing, research, and reviewing experience. Authored Book Evaluators are appointed for one-year terms and are expected to complete at least three evaluations per term. Upon successful completion of this term, evaluators can be considered for an additional term.

If you have a colleague that may be interested in this opportunity, we encourage you to share this information with them.

Recommended Reference Books

IGI Global's reference books are available in three unique pricing formats:
Print Only, E-Book Only, or Print + E-Book.

Shipping fees may apply.

www.igi-global.com

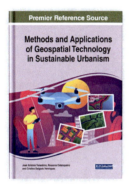

ISBN: 9781799822493
EISBN: 9781799822516
© 2021; 684 pp.
List Price: US$ 195

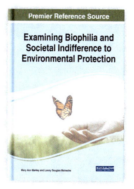

ISBN: 9781799844082
EISBN: 9781799844099
© 2021; 279 pp.
List Price: US$ 195

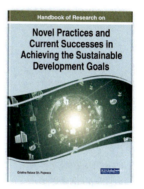

ISBN: 9781799884262
EISBN: 9781799884286
© 2021; 461 pp.
List Price: US$ 295

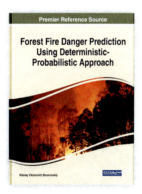

ISBN: 9781799872504
EISBN: 9781799872528
© 2021; 297 pp.
List Price: US$ 195

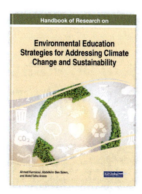

ISBN: 9781799875123
EISBN: 9781799875192
© 2021; 416 pp.
List Price: US$ 295

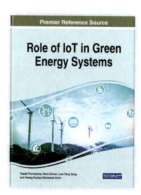

ISBN: 9781799867098
EISBN: 9781799867111
© 2021; 405 pp.
List Price: US$ 195

Do you want to stay current on the latest research trends, product announcements, news, and special offers?
Join IGI Global's mailing list to receive customized recommendations, exclusive discounts, and more.
Sign up at: **www.igi-global.com/newsletters**.

Publisher of Timely, Peer-Reviewed Inclusive Research Since 1988

Ensure Quality Research is Introduced to the Academic Community

Become an Evaluator for IGI Global Authored Book Projects

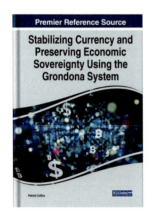 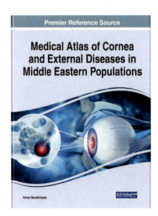 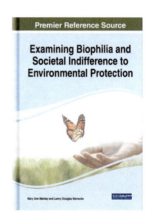 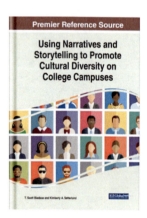

The overall success of an authored book project is dependent on quality and timely manuscript evaluations.

Applications and Inquiries may be sent to:
development@igi-global.com

Applicants must have a doctorate (or equivalent degree) as well as publishing, research, and reviewing experience. Authored Book Evaluators are appointed for one-year terms and are expected to complete at least three evaluations per term. Upon successful completion of this term, evaluators can be considered for an additional term.

If you have a colleague that may be interested in this opportunity, we encourage you to share this information with them.

Easily Identify, Acquire, and Utilize Published Peer-Reviewed Findings in Support of Your Current Research

IGI Global OnDemand

Purchase Individual IGI Global OnDemand Book Chapters and Journal Articles

For More Information:
www.igi-global.com/e-resources/ondemand/

Browse through 150,000+ Articles and Chapters!

Find specific research related to your current studies and projects that have been contributed by international researchers from prestigious institutions, including:

- Accurate and Advanced Search
- Affordably Acquire Research
- Instantly Access Your Content
- Benefit from the InfoSci Platform Features

" It really provides **an excellent entry into the research literature of the field**. It presents a manageable number of **highly relevant sources** on topics of interest to a wide range of researchers. The sources are **scholarly, but also accessible** to 'practitioners'. "

- Ms. Lisa Stimatz, MLS, University of North Carolina at Chapel Hill, USA

Interested in Additional Savings?

Subscribe to
IGI Global OnDemand Plus

Learn More

Acquire content from over 128,000+ research-focused book chapters and 33,000+ scholarly journal articles for as low as US$ 5 per article/chapter (original retail price for an article/chapter: US$ 37.50).

6,600+ E-BOOKS.
ADVANCED RESEARCH.
INCLUSIVE & ACCESSIBLE.

IGI Global e-Book Collection

- **Flexible Purchasing Options** (Perpetual, Subscription, EBA, etc.)
- Multi-Year Agreements with **No Price Increases** Guaranteed
- **No Additional Charge** for Multi-User Licensing
- No Maintenance, Hosting, or Archiving Fees
- Transformative **Open Access Options** Available

Request More Information, or Recommend the IGI Global e-Book Collection to Your Institution's Librarian

Among Titles Included in the IGI Global e-Book Collection

Research Anthology on Racial Equity, Identity, and Privilege (3 Vols.)
EISBN: 9781668445082
Price: US$ 895

Handbook of Research on Remote Work and Worker Well-Being in the Post-COVID-19 Era
EISBN: 9781799867562
Price: US$ 265

Research Anthology on Big Data Analytics, Architectures, and Applications (4 Vols.)
EISBN: 9781668436639
Price: US$ 1,950

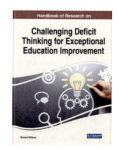

Handbook of Research on Challenging Deficit Thinking for Exceptional Education Improvement
EISBN: 9781799888628
Price: US$ 265

Acquire & Open

When your library acquires an IGI Global e-Book and/or e-Journal Collection, your faculty's published work will be considered for immediate conversion to Open Access *(CC BY License)*, at no additional cost to the library or its faculty *(cost only applies to the e-Collection content being acquired)*, through our popular **Transformative Open Access (Read & Publish) Initiative**.

For More Information or to Request a Free Trial, Contact IGI Global's e-Collections Team: eresources@igi-global.com | 1-866-342-6657 ext. 100 | 717-533-8845 ext. 100

Have Your Work Published and Freely Accessible
Open Access Publishing

With the industry shifting from the more traditional publication models to an open access (OA) publication model, publishers are finding that OA publishing has many benefits that are awarded to authors and editors of published work.

 Freely Share Your Research

 Higher Discoverability & Citation Impact

 Rigorous & Expedited Publishing Process

 Increased Advancement & Collaboration

Acquire & Open

 When your library acquires an IGI Global e-Book and/or e-Journal Collection, your faculty's published work will be considered for immediate conversion to Open Access *(CC BY License)*, at no additional cost to the library or its faculty *(cost only applies to the e-Collection content being acquired)*, through our popular **Transformative Open Access (Read & Publish) Initiative**.

- Provide Up To **100%** OA APC or CPC Funding
- Funding to Convert or Start a Journal to **Platinum OA**
- Support for Funding an **OA Reference Book**

IGI Global publications are found in a number of prestigious indices, including Web of Science™, Scopus®, Compendex, and PsycINFO®. The selection criteria is very strict and to ensure that journals and books are accepted into the major indexes, IGI Global closely monitors publications against the criteria that the indexes provide to publishers.

WEB OF SCIENCE™  **Scopus®**

PsycINFO®

Learn More Here:

For Questions, Contact IGI Global's Open Access Team at openaccessadmin@igi-global.com

Printed in the United States
by Baker & Taylor Publisher Services